Tucholsky Wagner Zola Scott Sydow Freud Schlegel
Turgenev Wallace Fonatne

Twain Walther von der Vogelweide Fouqué Friedrich II. von Preußen
Weber Freiligrath Frey
Kant Ernst
Fechner Fichte Weiße Rose von Fallersleben Richthofen Frommel
Hölderlin
Engels Fielding Eichendorff Tacitus Dumas
Fehrs Faber Flaubert
Eliasberg Ebner Eschenbach
Feuerbach Maximilian I. von Habsburg Fock Eliot Zweig
Ewald Vergil
Goethe Elisabeth von Österreich London
Mendelssohn Balzac Shakespeare
Lichtenberg Rathenau Dostojewski Ganghofer
Trackl Stevenson Doyle Gjellerup
Tolstoi Hambruch
Mommsen Thoma Lenz Hanrieder Droste-Hülshoff
von Arnim Hägele Hauff Humboldt
Dach Verne Reuter
Karrillon Garschin Rousseau Hagen Hauptmann Gautier
Defoe Baudelaire
Damaschke Descartes Hebbel
Hegel Kussmaul Herder
Wolfram von Eschenbach Dickens Schopenhauer
Bronner Darwin Melville Grimm Jerome Rilke George
Bebel
Campe Horváth Aristoteles Proust
Bismarck Vigny Barlach Voltaire Federer Herodot
Gengenbach Heine
Storm Casanova Tersteegen Gilm Grillparzer Georgy
Chamberlain Lessing Langbein Gryphius
Brentano Lafontaine
Strachwitz Claudius Schiller Kralik Iffland Sokrates
Katharina II. von Rußland Bellamy Schilling
Gerstäcker Raabe Gibbon Tschechow
Löns Hesse Hoffmann Gogol Wilde Gleim Vulpius
Luther Heym Hofmannsthal
Klee Hölty Morgenstern
Roth Heyse Klopstock Goedicke
Luxemburg Puschkin Homer Kleist
La Roche Horaz Mörike Musil
Machiavelli
Navarra Aurel Musset Kierkegaard Kraft Kraus
Nestroy Marie de France Lamprecht Kind Kirchhoff Hugo Moltke
Laotse Ipsen Liebknecht
Nietzsche Nansen Ringelnatz
Marx Lassalle Gorki Klett
von Ossietzky May Leibniz Irving
vom Stein Lawrence
Petalozzi Knigge
Platon
Sachs Poe Pückler Michelangelo Kock Kafka
Liebermann Korolenko
de Sade Praetorius Mistral Zetkin

Mary at the Farm and Book of Recipes Compiled during Her Visit among the "Pennsylvania Germans"

Edith Thomas

Imprint

This book is part of the TREDITION CLASSICS series.

Author: Edith Thomas
Cover design: toepferschumann, Berlin (Germany)

Publisher: tredition GmbH, Hamburg (Germany)
ISBN: 978-3-8495-1404-4

www.tredition.com
www.tredition.de

We love our Pennsylvania, grand old Keystone State;
Land of far famed rivers, and rock-ribbed mountains great.
With her wealth of "Dusky Diamonds" and historic valleys fair,
Proud to claim her as our birthplace; land of varied treasures rare.

PREFACE

The incidents narrated in this book are based on fact, and, while not absolutely true in every particular, the characters are all drawn from real life. The photographs are true likenesses of the people they are supposed to represent, and while in some instances the correct names are not given (for reasons which the reader will readily understand), the various scenes, relics, etc., are true historically and geographically. The places described can be easily recognized by any one who has ever visited the section of Pennsylvania in which the plot (if it can really be called a plot) of the story is laid. Many of the recipes given Mary by Pennsylvania German housewives, noted for the excellence of their cooking, have never appeared in print.

THE AUTHOR.

THIS BOOK IS DEDICATED TO MY FRIENDS WITH GRATITUDE FOR THEIR MANY HELPFUL KINDNESSES.

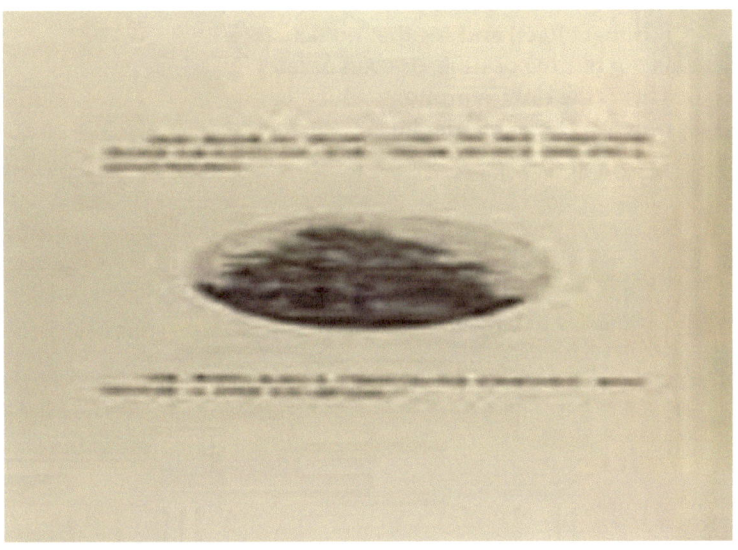

"HE WHO HAS A THOUSAND FRIENDS, HAS NEVER A ONE TO SPARE."

THE HOUSEKEEPER'S SYMPHONY

"To do the best that I can, from morn till night.
And pray for added strength with coming light;
To make the family income reach alway,
With some left over for a rainy day;
To do distasteful things with happy face,
To try and keep the odds and ends in place.
To smile instead of frown at Fate,
Which placed me in a family always late
For meals; to do the sewing, mending and
The thousand small things always near at hand,
And do them always with a cheerful heart,
Because in life they seem to be my part;
To know the place of everything and keep

It there, to think, to plan, to cook, to sweep,
To brew, to bake, to answer questions,
To be the mainspring of the family clock.
(Or that effect) and see that no tick, tock
Is out of time or tune, or soon or late,
This is the only symphony which I
Can ever hope to operate."

MARION WILEY.

CONTENTS

PREFACE

CONTENTS

ILLUSTRATIONS

CHAPTER I. Mary's Letter Received at Clear Spring Farm

CHAPTER II. Mary's Arrival at the Farm

CHAPTER III. Schuggenhaus Township

CHAPTER IV. John Landis

CHAPTER V. The Old Farm-House and Garden

CHAPTER VI. Mary Confides in "Aunt Sarah" and Gives Her Views on Suffrage for Women

CHAPTER VII Professor Schmidt

CHAPTER VIII. Uses of An Old-Fashioned Wardrobe

CHAPTER IX. Poetry and Pie

CHAPTER X. Sibylla Linsabigler

CHAPTER XI. New Colonial Rag Rugs

CHAPTER XII. Mary Imitates Navajo Blankets

CHAPTER XIII. "The Girls' Camp Fire" Organized by Mary

CHAPTER XIV. Mary Makes "Violet and Rose Leaf" Beads

CHAPTER XV. Mary and Elizabeth Visit Sadie Singmaster

CHAPTER XVI. The Old Parlor Made Beautiful (Modernized)

CHAPTER XVII. An Old Song Evening

CHAPTER XVIII. A Visit to the "Pennsylvania Palisades"

CHAPTER XIX. Mary Is Taught to Make Pastry, Patties and Rosenkuchen

CHAPTER XX. Old Potteries and Decorated Dishes

CHAPTER XXI. The Value of Wholesome, Nutritious Food

CHAPTER XXII. A Variety of Cakes Evolved From One Recipe

CHAPTER XXIII. The Old "Taufschien"

CHAPTER XXIV. The Old Store on the Ridge Road

CHAPTER XXV. An Elbadritchel Hunt

CHAPTER XXVI. The Old Shanghai Rooster

CHAPTER XXVII. A "Potato Pretzel"

CHAPTER XXVIII. Faithful Service

CHAPTER XXIX. Mary, Ralph, Jake and Sibylla Visit the Allentown

CHAPTER XXX. Fritz Schmidt Explores Durham Cave

CHAPTER XXXI. Mary's Marriage

MARY'S COLLECTION OF RECIPES

INDEX TO RECIPES

ILLUSTRATIONS

Mary
Aunt Sarah
The Old Spring House
The Old Mill Wheel
The Old Mill
Old Corn Crib
The New Red Barn
The Old Farm-House
Ralph Jackson
Rocky Valley
Professor Schmidt
Frau Schmidt
Old Time Patch-Work Quilts
Old Time Patch-Work
Home-Made Rag Carpet
A Hit-and-Miss Rug
A Brown and Tan Rug
A Circular Rug
Imitation of Navajo Blankets
Rug With Design
Rug With Swastika in Centre
Home Manufactured Silk Prayer Rug
Elizabeth Schmidt — "Laughing Water"
Articles in the Old Parlor Before It Was Modernized
Other Articles in the Old Parlor Before It Was Modernized
Palisades, or Narrows of Nockamixon
The Canal at the Narrows
The Narrows, or Pennsylvania Palisades
Top Rock
Ringing Rocks of Bucks County, Pennsylvania
High Falls
Big Rock at Rocky Dale
The Old Towpath at the Narrows
Old Earthenware Dish
Sgraffito Plate
Old Plates Found in Aunt Sarah's Corner Cupboard

Old Style Lamps
Old Taufschien
The Old Store on Ridge Road
Catching Elbadritchels
Old Egg Basket at the Farm
A Potato Pretzel
Loaf of Rye Bread
A "Brod Corvel," or Bread Basket
Church Which Sheltered Liberty Bell in 1777-78
Liberty Bell Tablet
Durham Cave
The Woodland Stream
Polly Schmidt
An Old-Fashioned Bucks County Bake-Oven

MARY

CHAPTER I

MARY'S LETTERS RECEIVED AT CLEAR SPRING FARM.

One morning in early spring, John Landis, a Pennsylvania German farmer living in Schuggenhaus Township, Bucks County, on opening his mail box, fastened to a tree at the crossroads (for the convenience of rural mail carriers) found one letter for his wife Sarah, the envelope addressed in the well-known handwriting of her favorite niece, Mary Midleton, of Philadelphia.

A letter being quite an event at "Clear Spring" farm, he hastened with it to the house, finding "Aunt Sarah," as she was called by every one (Great Aunt to Mary), in the cheery farm house kitchen busily engaged kneading sponge for a loaf of rye bread, which she carefully deposited on a well-floured linen cloth, in a large bowl for the final raising.

Carefully adjusting her glasses more securely over the bridge of her nose, she turned at the sound of her husband's footsteps. Seeing the letter in his hand she inquired: "What news, John?" Quickly opening the letter handed her, she, after a hasty perusal, gave one of the whimsical smiles peculiar to her and remarked decisively, with a characteristic nod of her head: "John, Mary Midleton intends to marry, else why, pray tell me, would she write of giving up teaching her kindergarten class in the city, to spend the summer with us on the farm learning, she writes, to keep house, cook, economize and to learn how to get the most joy and profit from life?"

"Well, well! Mary is a dear girl, why should she not think of marrying?" replied her husband; "she is nineteen. Quite time, I think, she should learn housekeeping—something every young girl should know. We should hear of fewer divorces and a less number of failures of men in business, had their wives been trained before marriage to be good, thrifty, economical housekeepers and, still more important, good homemakers. To be a helpmate in every sense of the word is every woman's duty, I think, when her husband works early and late to procure the means to provide for her comforts and luxuries and a competency for old age. Write Mary to come at once, and under your teaching she may, in time, become as capable a housekeeper and as good a cook as her Aunt Sarah; and, to my way of thinking, there is none better, my dear."

Praise from her usually reticent husband never failed to deepen the tint of pink on Aunt Sarah's still smooth, unwrinkled, youthful looking face, made more charming by being framed in waves of silvery gray hair, on which the "Hand of Time," in passing, had sprinkled some of the dust from the road of life.

In size, Sarah Landis was a little below medium height, rather stout, or should I say comfortable, and matronly looking; very erect for a woman of her age. Her bright, expressive, gray eyes twinkled humorously when she talked. She had developed a fine character by her years of unselfish devotion to family and friends. Her splendid sense of humor helped her to overcome difficulties, and her ability to rise above her environment, however discouraging their conditions, prevented her from being unhappy or depressed by the small annoyances met daily. She never failed to find joy and pleasure in the faithful performance of daily tasks, however small or insignificant. Aunt Sarah attributed her remarkably fine, clear complexion, seldom equalled in a woman of her years, to good digestion and excellent health; her love of fresh air, fruit and clear spring water. She usually drank from four to five tumblerfuls of water a day. She never ate to excess, and frequently remarked: "I think more people suffer from over-eating than from insufficient food." An advocate of deep breathing, she spent as much of her time as she could spare from household duties in the open air.

AUNT SARAH

Sarah Landis was not what one would call beautiful, but good and whole-souled looking. To quote her husband: "To me Sarah never looks so sweet and homelike when all 'fussed up' in her best black dress on special occasions, as she does when engaged in daily household tasks around home, in her plain, neat, gray calico dress."

This dress was always covered with a large, spotlessly clean, blue gingham apron of small broken check, and she was very particular about having a certain-sized check. The apron had a patch pocket, which usually contained small twists or little wads of cord, which, like "The Old Ladies in Cranford," she picked up and saved for a possible emergency.

One of Aunt Sarah's special economies was the saving of twine and paper bags. The latter were always neatly folded, when emptied, and placed in a cretonne bag made for that purpose, hanging in a convenient corner of the kitchen.

Aunt Sarah's gingham apron was replaced afternoons by one made from fine, Lonsdale cambric, of ample proportions, and on special occasions she donned a hemstitched linen apron, inset at upper edge of hem with crocheted lace insertion, the work of her own deft fingers. Aunt Sarah's aprons, cut straight, on generous lines, were a part of her individuality.

Sarah Landis declared: "Happiness consists in giving and in serving others," and she lived up to the principles she advocated. She frequently quoted from the "Sons of Martha," by Kipling:

> "Lift ye the stone or cleave the wood, to make a path more fair or flat,
> Not as a ladder from earth to heaven, not as an altar to any creed,
> But simple service, simply given, to his own kind in their human need."

"I think this so fine," said Aunt Sarah, "and so true a sentiment that I am almost compelled to forgive Kipling for saying 'The female of the species is more deadly than the male.'"

Aunt Sarah's goodness was reflected in her face and in the tones of her voice, which were soft and low, yet very decided. She possessed a clear, sweet tone, unlike the slow, peculiar drawl often aiding with the rising inflection peculiar to many country folk among the "Pennsylvania Germans."

The secret of Aunt Sarah's charm lay in her goodness. Being always surrounded by a cheery atmosphere, she benefited all with whom she came in contact. She took delight in simple pleasures. She had the power of extracting happiness from the common, little every-day tasks and frequently remarked, "Don't strive to live without work, but to find more joy in your work." Her opinions were highly respected by every one in the neighborhood, and, being possessed of an unselfish disposition, she thought and saw good in every one; brought out the best in one, and made one long to do better, just to gain her approval, if for no higher reward. Sarah Landis was a loyal friend and one would think the following, by Mrs. Craik, applied to her:

"Oh, the comfort, the inexpressible comfort, of feeling safe with a person—having neither to weigh thoughts nor measure words, but pouring them all right out, just as they are—chaff and grain together, certain that a faithful hand will take and sift them, keep what is worth keeping and then with the breath of kindness blow the rest away."

She was never so happy as when doing an act of kindness for some poor unfortunate, and often said. "If 'twere not for God and good people, what would become of the unfortunate?" and thought like George McDonald, "If I can put one touch of rosy sunset into the life of any man or woman (I should add child) I shall feel that I have worked with God."

Aunt Sarah's sweet, lovable face was the first beheld by many a little, new-born infant; her voice, the first to hush its wailing cries as she cuddled it up to her motherly breast, and oft, with loving hands, softly closed the lids over eyes no longer able to see; whom the Gracious Master had taken into His keeping.

One day I overheard Aunt Sarah quote to a sorrowing friend these fine, true lines from Longfellow's "Resignation": "Let us be patient, these severe afflictions not from the ground arise, but celestial benedictions assume the dark disguise."

THE OLD SPRING HOUSE

CHAPTER II.

MARY'S ARRIVAL AT THE FARM.

The day preceding that of Mary's arrival at the farm was a busy one for Aunt Sarah, who, since early morning, had been preparing the dishes she knew Mary enjoyed. Pans of the whitest, flakiest rolls, a large loaf of sweetest nut-brown, freshly-baked "graham bread," of which Mary was especially fond; an array of crumb-cakes and pies of every description covered the well-scrubbed table in the summer kitchen, situated a short distance from the house. A large, yellow earthenware bowl on the table contained a roll of rich, creamy "smier kase" just as it had been turned from the muslin bag, from which the "whey" had dripped over night; ready to be mixed with cream for the supper table. Pats of sweet, freshly-churned butter, buried in clover blossoms, were cooling in the old spring-house near by.

The farm house was guiltless of dust from cellar to attic. Aunt Sarah was a model housekeeper; she accomplished wonders, yet never appeared tired or flurried as less systematic housekeepers often do, who, with greater expenditure of energy, often accomplish less work. She took no unnecessary steps; made each one count, yet never appeared in haste to finish her work.

Said Aunt Sarah, "The lack of system in housework is what makes it drudgery. If young housekeepers would sit down and plan their work, then do it, they would save time and labor. When using the fire in the range for ironing or other purposes, use the oven for preparing dishes of food which require long, slow cooking, like baked beans, for instance. Bake a cake or a pudding, or a pan of quickly-made corn pone to serve with baked beans, for a hearty meal on a cold winter day. A dish of rice pudding placed in the oven requires very little attention, and when baked may be placed on ice until served. If this rule be followed, the young housewife will be surprised to find how much easier will be the task of preparing a meal later in the day, especially in hot weather."

The day following, John Landis drove to the railroad station, several miles distant, to meet his niece. As Mary stepped from the train into the outstretched arms of her waiting Uncle, many admiring glances followed the fair, young girl. Her tan-gold naturally wavy, masses of hair rivaled ripened grain. The sheen of it resembled corn silk before it has been browned and crinkled by the sun. Her eyes matched in color the exquisite, violet-blue blossoms of the chicory weed. She possessed a rather large mouth, with upturned corners, which seemed made for smiles, and when once you had been charmed with them, she had made an easy conquest of you forever. There was a sweet, winning personality about Mary which was as impossible to describe as to resist. One wondered how so much adorable sweetness could be embodied in one small maid. But Mary's sweetness of expression and charming manner covered a strong will and tenacity of purpose one would scarcely have believed possible, did they not have an intimate knowledge of the young girl's disposition. Her laugh, infectious, full of the joy of living, the vitality of youth and perfect health and happiness, reminded one of the lines: "A laugh is just like music for making living sweet."

Seated beside her Uncle in the carriage, Mary was borne swiftly through the town out into the country. It was one of those preternaturally quiet, sultry days when the whole universe appears lifeless and inert, free from loud noise, or sound of any description, days which we occasionally have in early Spring or Summer, when the stillness is oppressive.

Frequently at such times there is borne to the nostrils the faint, stifling scent of burning brush, indicating that land is being cleared by the forehanded, thrifty farmer for early planting. Often at such times, before a shower, may be distinctly heard the faintest twitter and "peep, peep" of young sparrows, the harsh "caw, caw" of the crow, and the song of the bobolink, poised on the swaying branch of a tall tree, the happiest bird of Spring; the dozy, drowsy hum of bees; the answering call of lusty young chanticleers, and the satisfied cackle of laying hens and motherly old biddies, surrounded by broods of downy, greedy little newly-hatched chicks. The shrill whistle of a distant locomotive startles one with its clear, resonant intonation, which on a less quiet day would pass unnoticed. Mary, with the zest of youth, enjoyed to the full the change from the past months of confinement in a city school, and missed nothing of the beauty of the country and the smell of the good brown earth, as her Uncle drove swiftly homeward.

"Uncle John," said Mary, "'tis easy to believe God made the country."

"Yes," rejoined her Uncle, "the country is good enough for me."

"With the exception of the one day in the month, when you attend the 'Shriners' meeting' in the city," mischievously supplemented Mary, who knew her Uncle's liking for the Masonic Lodge of which he was a member, "and," she continued, "I brought you a picture for your birthday, which we shall celebrate tomorrow. The picture will please you, I know. It is entitled, 'I Love to Love a Mason, 'Cause a Mason Never Tells.'"

They passed cultivated farms. Inside many of the rail fences, inclosing fields of grain or clover, were planted numberless sour cherry trees, snowy with bloom, the ground underneath white with fallen petals. The air was sweet with the perfume of the half-opened buds on the apple trees in the near-by orchards and rose-like pink blossoms of the "flowering" crab-apple, in the door yards. Swiftly they drove through cool, green, leafy woods, crossing a wooden bridge spanning a small stream, so shallow that the stones at the bottom were plainly to be seen. A loud splash, as the sound of carriage wheels broke the uninterrupted silence, and a commotion in the water gave evidence of the sudden disappearance of several green-backed frogs, sunning themselves on a large, moss-grown rock, projecting above the water's edge; from shady nooks and crevices peeped clusters of early white violets; graceful maidenhair ferns, and hardier members of the fern family, called "Brake," uncurled their graceful, sturdy fronds from the carpet of green moss and lichen at the base of tree trunks, growing along the water's edge. Partly hidden by rocks along the bank of the stream, nestled a few belated cup-shaped anemones or "Wind Flowers," from which most of the petals had blown, they being one of the earliest messengers of Spring. Through the undergrowth in the woods, in passing, could be seen the small buds of the azalea or wild honeysuckle, "Sheep's Laurel," the deep pink buds on the American Judas tree, trailing vines of "Tea Berry," and beneath dead leaves one caught an occasional glimpse of fragrant, pink arbutus. In marshy places beside the creek, swaying in the wind from slender stems, grew straw-colored, bell-shaped blossoms of "Adder's Tongue" or "Dog Tooth Violet," with their mottled green, spike-shaped leaves. In the shadow of a large rock grew dwarf huckleberry bushes, wild strawberry vines, and among grasses of many varieties grew patches of white and pink-tinted Alsatian clover.

Leaving behind the spicy, fragrant, "woodsy" smell of wintergreen, birch and sassafras, and the faint, sweet scent of the creamy, wax-like blossoms of "Mandrake" or May apple, peeping from beneath large, umbrella-like, green leaves they emerged at last from the dim, cool shadows of the woods into the warm, bright sunlight again.

Almost before Mary realized it, the farm house could be seen in the distance, and her Uncle called her attention to his new, red barn, which had been built since her last visit to the farm, and which, in her Uncle's estimation, was of much greater importance than the house.

Mary greeted with pleasure the old landmarks so familiar to her on former visits. They passed the small, stone school house at the crossroads, and in a short time the horses turned obediently into the lane leading to the barn a country lane in very truth, a tangle of blackberry vines, wild rose bushes, by farmers called "Pasture Roses," interwoven with bushes of sumach, wild carrots and golden rod.

Mary insisted that her Uncle drive directly to the barn, as was his usual custom, while she was warmly welcomed at the farm house gate by her Aunt. As her Uncle led away the horses, he said, "I will soon join you, Mary, 'to break of our bread and eat of our salt,' as they say in the 'Shrine.'"

On their way to the house, Mary remarked: "I am so glad we reached here before dusk. The country is simply beautiful! Have you ever noticed, Aunt Sarah, what a symphony in green is the yard? Look at the buds on the maples and lilacs—a faint yellow green—and the blue-green pine tree near by; the leaves of the German iris are another shade; the grass, dotted with yellow dandelions, and blue violets; the straight, grim, reddish-brown stalks of the peonies before the leaves have unfolded, all roofed over with the blossom-covered branches of pear, apple and 'German Prune' trees. Truly, this must resemble Paradise!"

"Yes," assented her Aunt, "I never knew blossoms to remain on the pear trees so long a time. We have had no 'blossom shower' as yet to scatter them, but there will be showers tonight, I think, or I am no prophet. I feel rain in the atmosphere, and Sibylla said a few moments ago she heard a 'rain bird' in the mulberry tree."

"Aunt Sarah," inquired Mary, "is the rhubarb large enough to use?"

"Yes, indeed, we have baked rhubarb pies and have had a surfeit of dandelion salad or 'Salat,' as our neighbors designate it. Your Uncle calls 'dandelion greens' the farmers' spring tonic; that and 'celadine,' that plant you see growing by the side of the house. Later in the season it bears small, yellow flowers not unlike a very small buttercup blossom, and it is said to be an excellent remedy for chills and fevers, and it tastes almost as bitter as quinine. There are bushels of dandelion blossoms, some of which we shall pick tomorrow, and from them make dandelion wine."

"And what use will my thrifty Aunt make of the blue violets?" mischievously inquired Mary.

"The violets," replied her Aunt, "I shall dig up carefully with some earth adhering to their roots and place them in a glass bowl for a centrepiece on the table for my artistic and beauty-loving niece; and if kept moist, you will be surprised at the length of time they will remain 'a thing of beauty' if not 'a joy forever.' And later, Mary, from them I'll teach you to make violet beads."

"Aunt Sarah, notice that large robin endeavoring to pull a worm from the ground. Do you suppose the same birds return here from the South every Summer?"

"Certainty, I do."

"That old mulberry tree, from the berries of which you made such delicious pies and marmalade last Summer, is it dead?"

"No; only late about getting its Spring outfit of leaves."

CHAPTER III.

SCHUGGENHAUS TOWNSHIP.

"Schuggenhaus," said Sarah Landis, speaking to her niece, Mary Midleton, "is one of the largest and most populous townships in Bucks County, probably so named by the early German settlers, some of whom, I think, were my father's ancestors, as they came originally from Zweibrucken, Germany, and settled in Schuggenhaus Township. Schuggenhaus is one of the most fertile townships in Bucks County and one of the best cultivated; farming is our principal occupation, and the population of the township today is composed principally of the descendants of well-to-do Germans, frequently called 'Pennsylvania Dutch.'"

"I have often heard them called by that name," said Mary. "Have you forgotten, Aunt Sarah, you promised to tell me something interesting about the first red clover introduced in Bucks County?"

"Red clover," replied her Aunt, "that having bright, crimson-pink heads, is the most plentiful and the most common variety of clover; but knowing how abundantly it grows in different parts of the country at the present time, one would scarcely have believed, in olden times, that it would ever be so widely distributed as it now is.

"One reason clover does so well in this country is that the fertilization of the clover is produced by pollenation by the busy little bumble-bee, who carries the pollen from blossom to blossom, and clover is dependent upon these small insects for fertilization, as without them clover would soon die out."

"I admire the feathery, fuzzy, pink-tipped, rabbit-foot clover," said Mary; "it is quite fragrant, and usually covered with butterflies. It makes such very pretty bouquets when you gather huge bunches of it."

THE OLD MILL WHEEL

"No, Mary, I think you are thinking of Alsatian clover, which is similar to white clover. The small, round heads are cream color, tinged with pink; it is very fragrant and sweet and grows along the roadside and, like the common white clover, is a favorite with bees. The yellow hop clover we also find along the roadside. As the heads of clover mature, they turn yellowish brown and resemble dried hops; sometimes yellow, brown and tan blossoms are seen on one branch. The cultivation of red clover was introduced here a century ago, and when in bloom the fields attracted great attention. Being the first ever grown in this part of Bucks County, people came for miles to look at it, the fence around the fields some days being lined with spectators, I have been told by my grandfather. I remember when a child nothing appeared to me more beautiful than my father's fields of flax; a mass of bright blue flowers. I also remember the fields of broom-corn. Just think! We made our own brooms, wove linen from the flax raised on our farm and made our own tallow candles. Mary, from what a thrifty and hard-working lot of ancestors you are descended! You inherit from your mother your love of work and from your father your love of books. Your father's uncle was a noted Shakespearean scholar."

Many old-time industries are passing away. Yet Sarah Landis, was a housewife of the old school and still cooked apple butter, or "Lodt Varrik," as the Germans call it; made sauerkraut and hard soap, and naked old-fashioned "German" rye bread on the hearth, which owed its excellence not only to the fact of its being hearth baked but to the rye flour being ground in an old mill in a near-by town, prepared by the old process of grinding between mill-stones instead of the more modern roller process. This picture of the old mill, taken by Fritz Schmidt, shows it is not artistic, but, like most articles of German manufacture, the mill was built more for its usefulness than to please the eye.

THE OLD MILL

"Aunt Sarah, what is pumpernickel?" inquired Mary, "is it like rye bread?"

"No, my dear, not exactly, it is a dark-colored bread, used in some parts of Germany. Professor Schmidt tells me the bread is usually composed of a mixture of barley flour and rye flour. Some I have eaten looks very much like our own brown bread. Pumpernickel is considered a very wholesome bread by the Germans—and I presume one might learn to relish it, but I should prefer good, sweet, home-made rye bread. I was told by an old gentleman who came to this country from Germany when a boy, that pumpernickel was used in the German army years ago, and was somewhat similar to 'hard tack,' furnished our soldiers in the Civil War. But I cannot vouch for the truth of this assertion."

"Aunt Sarah," said Mary later, "Frau Schmidt tells me the Professor sends his rye to the mill and requests that every part of it be ground without separating—making what he calls 'whole rye flour,' and from this Frau Schmidt bakes wholesome, nutritious bread which they call 'pumpernickel,' She tells me she uses about one-third of this 'whole rye flour' to two-thirds white bread flour when baking bread, and she considers bread made from this whole grain more wholesome and nutritious than the bread made from our fine rye flour."

CHAPTER IV.

JOHN LANDIS.

The Bucks County farmer, John Landis, rather more scholarly in appearance than men ordinarily found in agricultural districts, was possessed of an adust complexion, caused by constant exposure to wind and weather; tall and spare, without an ounce of superfluous fat; energetic, and possessed of remarkable powers of endurance. He had a kindly, benevolent expression; his otherwise plain face was redeemed by fine, expressive brown eyes. Usually silent and preoccupied, and almost taciturn, yet he possessed a fund of dry humor. An old-fashioned Democrat, his wife was a Republican. He usually accompanied Aunt Sarah to her church, the Methodist, although he was a member of the German Reformed, and declared he had changed his religion to please her, but change his politics, never. A member of the Masonic Lodge, his only diversion was an occasional trip to the city with a party of the "boys" to attend a meeting of the "Shriners."

Aunt Sarah protested. "The idea, John, at your age, being out so late at night and returning from the city on the early milk train the following morning, and then being still several miles from home. It's scandalous!"

He only chuckled to himself; and what the entertainment had been, which was provided at Lulu Temple, and which he had so thoroughly enjoyed, was left to her imagination. His only remark when questioned was: "Sarah, you're not in it. You are not a 'Shriner.'" And as John had in every other particular fulfilled her ideal of what constitutes a good husband, Sarah, like the wise woman she was, allowed the subject to drop.

A good, practical, progressive farmer, John Landis constantly read, studied and pondered over the problem of how to produce the largest results at least cost of time and labor. His crops were skillfully planted in rich soil, carefully cultivated and usually harvested earlier than those of his neighbors. One summer he raised potatoes so large that many of them weighed one pound each, and new potatoes and green peas, fresh from the garden, invariably appeared on Aunt Sarah's table the first of July, and sometimes earlier. I have known him to raise cornstalks which reached a height of thirteen feet, which were almost equaled by his wife's sunflower stalks, which usually averaged nine feet in height.

Aunt Sarah, speaking one day to Mary, said: "Your Uncle John is an unusually silent man. I have heard him remark that when people talk continuously they are either *very* intelligent or tell untruths." He, happening to overhear her remark, quickly retorted:

"The man who speaks a dozen tongues,
When all is said and done,
Don't hold a match to him who knows
How to keep still in one."

When annoyed at his wife's talkativeness, her one fault in her husband's eyes, if he thought she had a fault, he had a way of saying, "Alright, Sarah, Alright," as much as to say "that is final; you have said enough," in his peculiar, quick manner of speaking, which Aunt Sarah never resented, he being invariably kind and considerate in other respects.

John Landis was a successful farmer because he loved his work, and found joy in it. While not unmindful of the advantages possessed by the educated farmer of the present day, he said, "'Tis not college lore our boys need so much as practical education to develop their efficiency. While much that we eat and wear comes out of the ground, we should have more farmers, the only way to lower the present high cost of living, which is such a perplexing problem to the housewife. There is almost no limit to what might be accomplished by some of our bright boys should they make agriculture a study. Luther Burbank says, 'To add but one kernel of corn to each ear grown in this country in a single year would increase the supply five million bushels.'"

CHAPTER V.

THE OLD FARM HOUSE AND GARDEN.

Old Corn Crib

The New Barn

The old un-paint ed farm house, built of logs a century ago, had changed in the passing years to a grayish tint. An addition had been built to the house several years before Aunt Sarah's occupancy, The sober hue of the house harmonized with the great, gnarled old trunk of the meadow willow near-by. Planted when the house was built, it spread its great branches protectingly over it. A wild clematis growing at the foot of the tree twined its tendrils around the massive trunk until in late summer they had become an inseparable part of it, almost covering it with feathery blossoms.

Near by stood an antique arbor, covered with thickly-clustering vines, in season bending with the weight of "wild-scented" grapes, their fragrance mingling with the odor of "Creek Mint" growing near by a small streamlet and filling the air with a delicious fragrance. The mint had been used in earlier years by Aunt Sarah's grandfather as a beverage which he preferred to any other.

From a vine clambering up the grape arbor trellies, in the fall of the year, hung numerous orange-colored balsam apples, which opened, when ripe, disclosing bright crimson interior and seeds. These apples, Aunt Sarah claimed, if placed in alcohol and applied externally, possessed great medicinal value as a specific for rheumatism.

THE OLD FARM HOUSE

A short distance from the house stood the newly-built red barn, facing the pasture lot. On every side stretched fields which, in summer, waved with wheat, oats, rye and buckwheat, and the corn crib stood close by, ready for the harvest to fill it to overflowing. Beside the farm house door stood a tall, white oleander, planted in a large, green-painted wooden tub. Near by, in a glazed earthenware pot, grew the old-fashioned lantana plant, covered with clusters of tiny blossoms, of various shades of orange, red and pink.

In flower beds outlined by clam shells which had been freshly whitewashed blossomed fuchsias, bleeding hearts, verbenas, dusty millers, sweet clove-scented pinks, old-fashioned, dignified, purple digitalis or foxglove, stately pink Princess Feather, various brilliant-hued zinnias, or more commonly called "Youth and Old Age," and as gayly colored, if more humble and lowly, portulacas; the fragrant white, star-like blossoms of the nicotiana, or "Flowering Tobacco," which, like the yellow primrose, are particularly fragrant at sunset. Geraniums of every hue, silver-leaved and rose-scented; yellow marigolds and those with brown, velvety petals; near by the pale green and white-mottled leaves of the plant called "Snow on the Mountain" and in the centre of one of the large, round flower beds, grew sturdy "Castor Oil Beans," their large, copper-bronze leaves almost covering the tiny blue forget-me-nots growing beneath. Near the flower bed grew a thrifty bush of pink-flowering almonds; not far distant grew a spreading "shrub" bush, covered with fragrant brown buds, and beside it a small tree of pearly-white snowdrops.

Sarah Landis loved the wholesome, earthy odors of growing plants and delighted in her flowers, particularly the perennials, which were planted promiscuously all over the yard. I have frequently heard her quote: "One is nearer God's heart in a garden than any place else on earth." And she would say, "I love the out-of-door life, in touch with the earth; the natural life of man or woman." Inside the fence of the kitchen garden were planted straight rows of both red and yellow currants, and several gooseberry bushes. In one corner of the garden, near the summer kitchen, stood a large bush of black currants, from the yellow, sweet-scented blossoms of which Aunt Sarah's bees, those "Heaven instructed mathematicians," sucked honey. Think of Aunt Sarah's buckwheat cakes, eaten with honey made from currant, clover, buckwheat and dandelion blossoms!

Her garden was second to none in Bucks County. She planted tomato seeds in boxes and placed them in a sunny window, raising her plants early; hence she had ripe tomatoes before any one else in the neighborhood. Her peas were earlier also, and her beets and potatoes were the largest; her corn the sweetest; and, as her asparagus bed was always well salted, her asparagus was the finest to be had.

Through the centre of the garden patch, on either side the walk, were large flower beds, a blaze of brilliant color from early Spring, when the daffodils blossomed, until frost killed the dahlias, asters, scarlet sage, sweet Williams, Canterbury bells, pink and white snapdragon, spikes of perennial, fragrant, white heliotrope; blue larkspur, four o'clocks, bachelor buttons and many other dear, old-fashioned flowers. The dainty pink, funnel-shaped blossoms of the hardy swamp "Rose Mallow'" bloomed the entire Summer, the last flowers to be touched by frost, vying in beauty with the pink monthly roses planted near by.

Children who visited Aunt Sarah delighted in the small Jerusalem cherry tree, usually covered with bright, scarlet berries, which was planted near the veranda, and they never tired pinching the tiny leaves of the sensitive plant to see them quickly droop, as if dead, then slowly unfold and straighten as if a thing of life.

Visitors to the farm greatly admired the large, creamy-white lily-like blossoms of the datura. Farthest from the house were the useful herb beds, filled with parsley, hoarhound, sweet marjoram, lavender, saffron, sage, sweet basil, summer savory and silver-striped rosemary or "old man," as it was commonly called by country folk.

Tall clusters of phlox, a riot of color in midsummer, crimson-eyed, white and rose-colored blossoms topping the tall steins, and clusters of brilliant-red bergamot near by had been growing, from time immemorial, a cluster of green and white-striped grass, without which no door yard in this section of Bucks County was considered complete in olden times. Near by, silvery plumes of pampas grass gently swayed on their reed-like stems. Even the garden was not without splashes of color, where, between rows of vegetables, grew pale, pink-petaled poppies, seeming to have scarcely a foothold in the rich soil. But the daintiest, sweetest bed of all, and the one that Mary enjoyed most, was where the lilies of the valley grew in the shade near a large, white lilac bush. Here, on a rustic bench beneath an old apple tree, stitching on her embroidery, she dreamed happy dreams of her absent lover, and planned for the life they were to live together some day, in the home he was striving to earn for her by his own manly exertions; and she assiduously studied and pondered over Aunt Sarah's teaching and counsel, knowing them to be wise and good.

A short distance from the farm house, where the old orchard sloped down to the edge of the brook, grew tall meadow rue, with feathery clusters of green and white flowers; and the green, gold-lined, bowl-shaped blossoms of the "Cow Lily," homely stepsisters of the fragrant, white pond lily, surrounded by thick, waxy, green leaves, lazily floated on the surface of the water from long stems in the bed of the creek, and on the bank a carpet was formed by golden-yellow, creeping buttercups.

In the side yard grew two great clumps of iris, or, as it is more commonly called, "Blue Flag." Its blossoms, dainty as rare orchids, with lily-like, violet-veined petals of palest-tinted mauve and purple.

On the sunny side of the old farm house, facing the East, where at early morn the sun shone bright and warm, grew Aunt Sarah's pansies, with velvety, red-brown petals, golden-yellow and dark purple. They were truly "Heart's Ease," gathered with a lavish hand, and sent as gifts to friends who were ill. The more she picked the faster they multiplied, and came to many a sick bed "sweet messengers of Spring."

If Aunt Sarah had a preference for one particular flower, 'twas the rose, and they well repaid the time and care she lavished on them. She had pale-tinted blush roses, with hearts of deepest pink; rockland and prairie and hundred-leaf roses, pink and crimson ramblers, but the most highly-prized roses of her collection were an exquisite, deep salmon-colored "Marquis De Sinety" and an old-fashioned pink moss rose, which grew beside a large bush of mock-orange, the creamy blossoms of the latter almost as fragrant as real orange blossoms of the sunny Southland. Not far distant, planted in a small bed by themselves, grew old-fashioned, sweet-scented, double petunias, ragged, ripple, ruffled corollas of white, with splotches of brilliant crimson and purple, their slender stems scarcely strong enough to support the heavy blossoms.

In one of the sunniest spots in the old garden grew Aunt Sarah's latest acquisition. "The Butterfly Bush," probably so named on account of its graceful stems, covered with spikes of tiny, lilac-colored blossoms, over which continually hovered large, gorgeously-hued butterflies, vying with the flowers in brilliancy of color, from early June until late Summer.

Aunt Sarah's sunflowers, or "Sonnen Blume," as she liked to call them, planted along the garden fence to feed chickens and birds alike, were a sight worth seeing. The birds generally confiscated the larger portion of seeds. A pretty sight it was to see a flock of wild canaries, almost covering the tops of the largest sunflowers, busily engaged picking out the rich, oily seeds.

Aunt Sarah loved the golden flowers, which always appeared to be nodding to the sun, and her sunflowers were particularly fine, some being as much as fifty inches in circumference.

A bouquet of the smaller ones was usually to be seen in a quaint, old, blue-flowered, gray jar on the farm house veranda in Summertime. Earlier in the season blossoms of the humble artichoke, which greatly resemble small sunflowers, or large yellow daisies, filled the jar. Failing either of these, she gathered large bouquets of golden-rod or wild carrot blossoms, both of which grew in profusion along the country lanes and roadside near the farm. But the old gray jar never held a bouquet more beautiful than the one of bright, blue "fringed gentians," gathered by Aunt Sarah in the Fall of the year, several miles distant from the farm.

CHAPTER VI.

MARY CONFIDES IN AUNT SARAH AND GIVES HER VIEWS ON SUFFRAGE FOR WOMEN.

"There's no deny'n women are foolish,
God A'mighty made them to match the men."

A short time after her arrival at the farm Mary poured into the sympathetic ear of Aunt Sarah her hopes and plans. Her lover, Ralph Jackson, to whom she had become engaged the past Winter, held a position with the Philadelphia Electric Company, and was studying hard outside working hours. His ambition was to become an electrical engineer. He was getting fair wages, and wished Mary to marry him at once. She confessed she loved Ralph too well to marry him, ignorant as she was of economical housekeeping and cooking.

Mary, early left an orphan, had studied diligently to fit herself for a kindergarten teacher, so she would be capable of earning her own living on leaving school, which accounted for her lack of knowledge of housework, cooking, etc.

Aunt Sarah, loving Mary devotedly, and knowing the young man of her choice to be clean, honest and worthy, promised to do all in her power to make their dream of happiness come true. Learning from Mary that Ralph was thin and pale from close confinement, hard work and study, and of his intention of taking a short vacation, she determined he should spend it on the farm, where she would be able to "mother him."

"You acted sensibly, Mary," said her Aunt, "in refusing to marry Ralph at the present time, realizing your lack of knowledge of housework and inability to manage a home. Neither would you know how to spend the money provided by him economically and wisely, and, in this age of individual efficiency, a business knowledge of housekeeping is almost as important in making a happy home as is love. I think it quite as necessary that a woman who marries should understand housekeeping in all its varied branches as that the man who marries should understand his trade or profession; for, without the knowledge of means to gain a livelihood (however great his love for a woman), how is the man to hold that woman's love and affection unless he is able by his own exertions to provide her with necessities, comforts, and, perhaps, in later years, luxuries? And in return, the wife should consider it her duty and pleasure to know how to do her work systematically; learn the value of different foods and apply the knowledge gained daily in preparing them; study to keep her husband in the best of health, physically and mentally. Then will his efficiency be greater and he will be enabled to do his 'splendid best' in whatever position in life he is placed, be he statesman or hod-carrier. What difference, if an honest heart beat beneath a laborer's hickory shirt, or one of fine linen? 'One hand, if it's true, is as good as another, no matter how brawny or rough.' Mary, do not think the trivial affairs of the home beneath your notice, and do not imagine any work degrading which tends to the betterment of the home. Remember, 'Who sweeps a room as for Thy law, makes that and the action fine.'

"Our lives are all made up of such small, commonplace things and this is such a commonplace old world, Mary. 'The commonplace earth and the commonplace sky make up the commonplace day,' and 'God must have loved common people, or He would not have made so many of them.' And, what if we are commonplace? We cannot all be artists, poets and sculptors. Yet, how frequently we see people in commonplace surroundings, possessing the soul of an artist, handicapped by physical disability or lack of means! We are all necessary in the great, eternal plan. 'Tis not good deeds alone for which we receive our reward, but for the performance of duty well done, in however humble circumstances our lot is cast. Is it not Lord Houghton who says: 'Do not grasp at the stars, but do life's plain, common work as it comes, certain that daily duties and daily bread are the sweetest things of life.' I consider a happy home in the true sense of the word one of the greatest of blessings. How important is the work of the housemother and homemaker who creates the home! There can be no happiness there unless the wheels of the domestic machinery are oiled by loving care and kindness to make them run smoothly, and the noblest work a woman can do is training and rearing her children. Suffrage, the right of woman to vote; will it not take women from the home? I am afraid the home will then suffer in consequence. Will man accord woman the same reverence she has received in the past? Should she have equal political rights? A race lacking respect for women would never advance socially or politically. I think women could not have a more important part in the government of the land than in rearing and educating their children to be good, useful citizens. In what nobler work could women engage than in work to promote the comfort and well-being of the ones they love in the home? I say, allow men to make the laws, as God and nature planned. I think women should keep to the sphere God made them for — the home. Said Gladstone, 'Woman is the most perfect when most womanly.' There is nothing, I think, more despicable than a masculine, mannish woman, unless it be an effeminate, sissy man. Dr. Clarke voiced my sentiments when he said: 'Man is not superior to woman, nor woman to man. The relation of the sexes is one of equality, not of better or worse, of higher and lower. The loftiest ideal of humanity demands that each shall be perfect in its kind and not be hindered in its best work. The lily is not inferior to the rose, nor the oak superior to the clover; yet the glory of the lily is one and the glory of the oak is another, and the use of the oak is not the use of the clover.'

"This present-day generation demands of women greater efficiency in the home than ever before. And Mary, many of the old-time industries which I had been accustomed to as a girl have passed away. Electricity and numerous labor-saving devices make household tasks easier, eliminating some altogether. When housekeeping you will find time to devote to many important questions of the day which we old-time housekeepers never dreamed of having. Considerable thought should be given to studying to improve and simplify conditions of the home-life. It is your duty. Obtain books; study food values and provide those foods which nourish the body, instead of spending time uselessly preparing dainties to tempt a jaded appetite. Don't spoil Ralph when you marry him. Give him good, wholesome food, and plenty of it; but although the cooking of food takes up much of a housekeeper's time, it is not wise to allow it to take up one's time to the exclusion of everything else. Mary, perhaps my views are old-fashioned. I am not a 'new woman' in any sense of the word. The new woman may take her place beside man in the business world and prove equally as efficient, but I do not think woman should invade man's sphere any more than he should assume her duties."

"Aunt Sarah, I am surprised to hear you talk in that manner about woman's sphere," replied Mary, "knowing what a success you are in the home, and how beautifully you manage everything you undertake. I felt, once you recognized the injustice done woman in not allowing them to vote, you would feel differently, and since women are obliged to obey the laws, should they not have a voice in choosing the lawmakers? When you vote, it will not take you out of the home. You and Uncle John will merely stop on your way to the store, and instead of Uncle John going in to write and register what he thinks should be done and by whom it should be done, you too will express your opinion. This will likely be twice a year. By doing this, no woman loses her womanliness, goodness or social position, and to these influences the vote is but another influence. I know there are many things in connection with the right of equal suffrage with what you do not sympathize.

"Aunt Sarah, let me tell you about a dear friend of mine who taught school with me in the city. Emily taught a grammar grade, and did not get the same salary the men teachers received for doing the same work, which I think was unfair. Emily studied and frequently heard and read about what had been done in Colorado and other States where women vote. She got us all interested, and the more we learned about the cause the harder we worked for it. Emily married a nice, big, railroad man. They bought a pretty little house in a small town, had three lovely children and were very happy. More than ever as time passed Emily realized the need of woman's influence in the community. It is true, I'll admit, Aunt Sarah, house-keeping and especially home-making are the great duties of every woman, and to provide the most wholesome, nourishing food possible for the family is the duty of every mother, as the health, comfort and happiness of the family depend so largely on the *common sense* (only another name for efficiency) and skill of the homemaker, and the wise care and though she expends on the preparation of wholesome, nutritious food in the home, either the work of her own hands or prepared under her direction. You can *not* look after these duties without getting *outside* of your home, especially when you live like Emily, in a town where the conditions are so different from living as you do on a farm in the country. Milk, bread and water are no longer controlled by the woman in her home, living in cities and towns; and just because women want to look out for their families they should have a voice in the larger problems of municipal housekeeping. To return to Emily, she did not bake her own bread, as you do, neither did she keep a cow, but bought milk and bread to feed the children. Wasn't it her duty to leave the home and see where these products were produced, and if they were sanitary? And, knowing the problem outside the home would so materially affect the health, and perhaps lives, of her children, she felt it her distinctive duty to keep house in a larger sense. When the children became old enough to attend school, Emily again took up her old interest in schools. She began to realize how much more just it would be if an equal number of women were on the school board."

"But what did the husband think of all this?" inquired Aunt Sarah, dubiously.

"Oh, Tom studied the case, too, at first just to tease Emily, but he soon became as enthusiastic as Emily. He said, 'The first time you are privileged to vote, Emily, I will hire an automobile to take you to the polls in style.' But poor Emily was left alone with her children last winter. Tom died of typhoid fever. Contracted it from the bad drainage. They lived in a town not yet safeguarded with sewerage. Now Emily is a taxpayer as well as a mother, and she has no say as far as the town and schools are concerned. There are many cases like that, where widows and unmarried women own property, and they are in no way represented. And think of the thousands and thousands of women who have no home to stay in and no babies to look after."

"Mercy, Mary! Do stop to take breath. I never thought when I started this subject I would have an enthusiastic suffragist with whom to deal."

"I am glad you started the subject, Aunt Sarah, because there is so much to be said for the cause. I saw you glance at the clock and I see it is time to prepare supper. But some day I'm going to stop that old clock and bring down some of my books on 'Woman's Suffrage' and you'll he surprised to hear what they have done in States where equal privileges were theirs. I am sure 'twill not be many years before every State in the Union will give women the right of suffrage."

After Mary retired that evening Aunt Sarah had a talk with her John, whom she knew needed help on the farm. As a result of the conference, Mary wrote to Ralph the following day, asking him to spend his vacation on the farm as a "farm hand." Needless to say, the offer was gladly accepted by Ralph, if for no other reason than to be near the girl he loved.

Ralph came the following week—"a strapping big fellow," to quote Uncle John, being several inches over six feet.

"All you need, young chap," said Mary's Uncle, "is plenty of good, wholesome food of Sarah's and Mary's preparing, and I'll see that you get plenty of exercise in the fresh air to give you an appetite to enjoy it, and you'll get a healthy coat of tan on your pale cheeks before the Summer is ended."

Ralph Jackson, or "Jack," as he was usually called by his friends, an orphan like Mary, came of good, old Quaker stock, his mother having died immediately after giving birth to her son. His father, supposed to be a wealthy contractor, died when Ralph was seventeen, having lost his fortune through no fault of his own, leaving Ralph penniless.

Ralph Jackson possessed a good face, a square, determined jaw, sure sign of a strong will and quick temper; these Berserker traits he inherited from his father; rather unusual in a Quaker. He possessed a head of thick, coarse, straight brown hair, and big honest eyes. One never doubted his word, once it had been given. 'Twas good as his bond. This trait he inherited also from his father, noted for his truth and integrity. Ralph was generous to a fault. When a small boy he was known to take off his shoes and give them to a poor little Italian (who played a violin on the street for pennies) and go home barefoot.

Ralph loved Mary devotedly, not only because she fed him well at the farm, as were his forefathers, the "Cave Men," fed by their mates in years gone by, but he loved her first for her sweetness of disposition and lovable ways; later, for her quiet unselfishness and lack of temper over trifles — so different from himself.

When speaking to Mary of his other fine qualities, Aunt Sarah said: "Ralph is a manly young fellow; likeable, I'll admit, but his hasty temper is a grave fault in my eyes."

Mary replied, "Don't you think men are very queer, anyway, Aunt Sarah? I do, and none of us is perfect."

RALPH JACKSON

To Mary, Ralph's principal charm lay in his strong, forceful way of surmounting difficulties, she having a disposition so different. Mary had a sweet, motherly way, seldom met with in so young a girl, and this appealed to Ralph, he having never known "mother love," and although not at all inclined to be sentimental, he always called Mary his "Little Mother Girl" because of her motherly ways.

ROCKY VALLEY

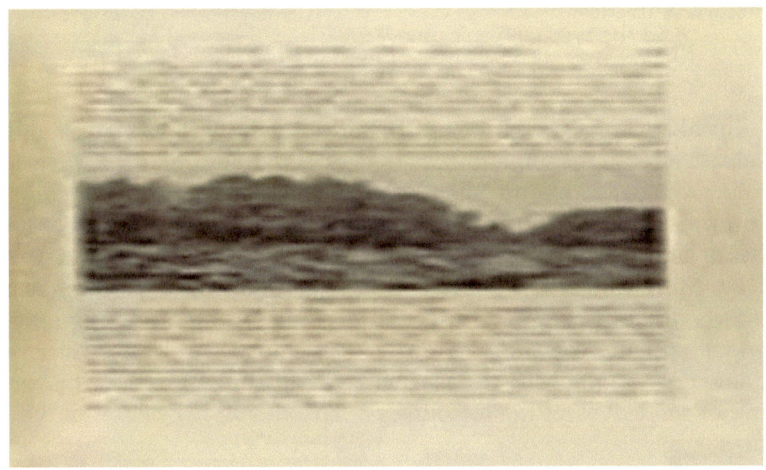

"Well," continued Mary's Aunt, "a quick temper is one of the most difficult faults to overcome that flesh is heir to, but Ralph, being a young man of uncommon good sense, may in time curb his temper and learn to control it, knowing that unless be does so it will handicap him in his career. Still, a young girl will overlook many faults in the man she loves. Mary, ere marrying, one should be sure that no love be lacking to those entering these sacred bonds. 'Tis not for a day, but for a lifetime, to the right thinking. Marriage, as a rule, is too lightly entered into in this Twentieth Century of easy divorces, and but few regard matrimony in its true holy relation, ordained by our Creator. If it be founded on the tower of enduring love and not ephemeral passion, it is unassailable, lasting in faith and honor until death breaks the sacred union and annuls the vows pledged at God's holy altar."

"Well," replied Mary, as her Aunt paused to take breath, "I am sure of my love for Ralph."

"God grant you may both be happy," responded her Aunt.

"Mary, did you ever hear this Persian proverb? You will understand why I have so much to say after hearing it."

"'Says a proverb of Persia provoking mirth;
When this world was created by order divine.
Ten measures of talk were put down on the earth,
And the woman took nine.'"

Speaking to Mary of life on the farm one day, Ralph laughingly said: "I am taught something new every day. Yesterday your Uncle told me it was 'time to plant corn when oak leaves were large as squirrels' ears.'" Ralph worked like a Trojan. In a short time both his hands and face took on a butternut hue. He became strong and robust. Mary called him her "Cave Man," and it taxed the combined efforts of Aunt Sarah and Mary to provide food to satisfy the ravenous appetite Mary's "Cave Man" developed. And often, after a busy day, tired but happy, Mary fell asleep at night to the whispering of the leaves of the Carolina poplar outside her bedroom window.

But country life on a farm has its diversions. One of Mary's and Ralph's greatest pleasures after a busy day at the farm was a drive about the surrounding country early Summer evenings, frequently accompanied by either Elizabeth or Pauline Schmidt, their nearest neighbors.

One of the first places visited by them was a freak of nature called "Rocky Valley," situated at no great distance from the farm.

PROFESSOR SCHMIDT

CHAPTER VII

PROFESSOR SCHMIDT.

A small country place named "Five Oaks," a short distance from "Clear Spring" farm, was owned by a very worthy and highly-educated, but rather eccentric, German professor. He came originally from Heidelberg, but had occupied the position of Professor of German for many years in a noted university in a near by town. A kind, warm-hearted, old-fashioned gentleman was the Professor; a perfect Lord Chesterfield in manners. Very tall, thin almost to emaciation, although possessed of excellent health; refined, scholarly looking: a rather long, hooked nose, faded, pale-blue eyes; snowy, flowing "Lord Dundreary" whiskers, usually parted in the centre and twisted to a point on either side with the exceedingly long, bony fingers of his well-kept, aristocratic-looking white hands. He had an abrupt, quick, nervous manner when speaking. A fringe of thin, white hair showed at the lower edge of the black silk skull cap which he invariably wore about home, and in the absence of this covering for his bald head, he would not have looked natural to his friends.

The Professor always wore a suit of well-brushed, "shiny" black broadcloth, and for comfort old-fashioned soft kid "gaiters," with elastic in the sides. He was a man with whom one did not easily become acquainted, having very decided opinions on most subjects. He possessed exquisite taste, a passionate love of music, flowers and all things beautiful; rather visionary, poetical and a dreamer; he was not practical, like his wife; warm-hearted, impulsive, energetic Frau Schmidt, who was noted for her executive abilities. I can imagine the old Professor saying as Mohammed has been quoted as saying, "Had I two loaves, I would sell one and buy hyacinths to feed my soul." Impulsive, generous to a fault, quick to take offense, withal warm-hearted, kind and loyal to his friends, he was beloved by the students, who declared that "Old Snitzy" always played fair when he was obliged to reprimand them for their numerous pranks, which ended sometimes, I am obliged to confess, with disastrous results. The dignified old Professor would have raised his mild, blue, spectacled eyes in astonishment had he been so unfortunate as to have overheard the boys, to whom he was greatly attached, call their dignified preceptor by such a nickname.

The Professor's little black-eyed German wife, many years younger than her husband, had been, before her marriage, teacher of domestic science in a female college in a large city. "She was a most excellent housekeeper," to quote the Professor, and "a good wife and mother."

The family consisted of "Fritz," a boy of sixteen, with big, innocent, baby-blue eyes like his father, who idolized his only son, who was alike a joy and a torment. Fritz attended the university in a near-by town, and was usually head of the football team. He was always at the front in any mischief whatever, was noted for getting into scrapes innumerable through his love of fun, yet he possessed such a good-natured, unselfish, happy-go-lucky disposition that one always forgave him.

Black-eyed, red-cheeked Elizabeth was quick and impulsive, like her mother. A very warm and lasting friendship sprung up between merry Elizabeth and serious Mary Midleton during Mary's Summer on the farm, although not at all alike in either looks or disposition, and Elizabeth was Mary's junior by several years.

The third, last and least of the Professor's children was Pauline, or "Pollykins," as she was always called by her brother Fritz, the seven-year-old pet and baby of the family. A second edition of Fritz, the same innocent, questioning, violet-blue eyes, fair complexion, a kissable little mouth and yellow, kinky hair, she won her way into every one's heart and became greatly attached to Mary, who was usually more patient with the little maid (who, I must confess, was sometimes very willful) than was her sister Elizabeth. Mary, who had never been blessed with a sister, dearly loved children, and thought small "Polly" adorable, and never wearied telling her marvelous fairy tales.

FRAU SCHMIDT

CHAPTER VIII.

USES OF AN OLD-FASHIONED WARDROBE.

Shortly after Mary's advent at the farm she one day said: "Aunt Sarah, the contents of this old trunk are absolutely worthless to me; perhaps they may be used by you for carpet rags."

"Mary Midleton!" exclaimed Aunt Sarah, in horrified tones, "you extravagant girl. I see greater possibilities in that trunk of partly-worn clothing than, I suppose, a less economically-inclined woman than I ever would have dreamed of."

Mary handed her Aunt two blue seersucker dresses, one plain, the other striped. "They have both shrunken, and are entirely too small for me," said Mary.

"Well," said her Aunt, considering, "they might be combined in one dress, but you need aprons for kitchen work more useful than those little frilly, embroidered affairs you are wearing. We should make them into serviceable aprons to protect your dresses. Mary, neatness is an attribute that every self-respecting housewife should assiduously cultivate, and no one can be neat in a kitchen without a suitable apron to protect one from grime, flour and dust."

"What a pretty challis dress; its cream-colored ground sprinkled over with pink rose buds!"

Mary sighed. "I always did love that dress, Aunt Sarah, 'Twas so becoming, and he—he—admired it so!"

"And HE, can do so still," replied Aunt Sarah, with a merry twinkle in her kind, clear, gray eyes, "for that pale-green suesine skirt, slightly faded, will make an excellent lining, with cotton for an interlining, and pale green Germantown yarn with which to tie the comfortable. At small cost you'll have a dainty, warm spread which will be extremely pretty in the home you are planning with HIM. I have several very pretty-old-style patchwork quilts in a box in the attic which I shall give you when you start housekeeping. That pretty dotted, ungored Swiss skirt will make dainty, ruffled sash curtains for bedroom windows. Mary, sometimes small beginnings make great endings; if you make the best of your small belongings, some day your homely surroundings will be metamorphosed into what, in your present circumstances, would seem like extravagant luxuries. An economical young couple, beginning life with a homely, home-made rag carpet, have achieved in middle age, by their own energy and industry, carpets of tapestry and rich velvet, and costly furniture in keeping; but, never—never, dear, are they so valued, I assure you, as those inexpensive articles, conceived by our inventive brain and manufactured by our own deft fingers during our happy Springtime of life when, with our young lover husband, we built our home nest on the foundation of pure, unselfish, self-sacrificing love."

Aunt Sarah sighed; memory led her far back to when she had planned her home with her lover, John Landis, still her lover, though both have grown gray together, and shared alike the joys and sorrows of the passing years. Aunt Sarah had always been the perfect "housemother" or "Haus Frau," as the Germans phrase it, and on every line of her matured face could be read an anxious care for the family welfare. Truly could it be said of her, in the language of Henry Ward Beecher: "Whoever makes home seem to the young dearer and happier is a public benefactor."

Aunt Sarah said earnestly to Mary, "I wish it were possible for me to impart to young, inexperienced girls, about to become housewives and housemothers, a knowledge of those small economics, so necessary to health and prosperity, taught me by many years of hard work, mental travail, experience and some failures. In this extravagant Twentieth Century economy is more imperative than formerly. We feel that we need so much more these days than our grandmothers needed; and what we need, or feel that we need, is so costly. The housemother has larger problems today than yesterday.

"Every husband should give his wife an allowance according to his income, so that she will be able to systematize her buying and occasionally obtain imperishable goods at less cost. Being encouraged thus to use her dormant economical powers; she will become a powerful factor in the problem of home-making along lines that will essentially aid her husband in acquiring a comfortable competency, if not a fortune. Then she will have her husband's interest truly at heart; will study to spend his money carefully, and to the best advantage; and she herself, even, will be surprised at the many economies which will suggest themselves to save his hard-earned money when she handles that money herself, which certainly teaches her the saving habit and the value of money.

"The majority of housewives of today aren't naturally inclined to be extravagant or careless. It is rather that they lack the knowledge and experience of spending money, and spending it to the best advantage for themselves and their household needs.

"'Tis a compulsory law in England, I have heard, to allow a wife pin money, according to a man's means. 'Tis a most wise law. To a loyal wife and mother it gives added force, dignity and usefulness to have a sufficient allowance and to be allowed unquestioningly to spend that money to her best ability. Her husband, be he a working or professional man, would find it greatly to his advantage in the home as well as in his business and less of a drain on his bank account should he give his wife a suitable allowance and trust her to spend it according to her own intelligence and thrift.

"Child, many a man is violently prejudiced against giving a young wife money; many allow her to run up bills, to her hurt and to his, rather than have her, even in her household expenditure, independent of his supervision. I sincerely hope, dear, that your intended, Ralph Jackson, will be superior to this male idiosyncrasy, to term it mildly, and allow you a stated sum monthly. The home is the woman's kingdom, and she should be allowed to think for it, to buy for it, and not to be cramped by lack of money to do as she thinks best for it."

"But, Aunt Sarah, some housewives are so silly that husbands cannot really be blamed for withholding money from them and preventing them from frittering it away in useless extravagance."

"Mary, wise wives should not suffer for those who are silly and extravagant. I don't like to be sarcastic, but with the majority of the men, silliness appeals to them more than common sense. Men like to feel their superiority to us. However, though inexperienced, Mary, you aren't silly or extravagant, and Ralph could safely trust you with his money. It makes a woman so self-respecting, puts her on her mettle, to have money to do as she pleases with, to be trusted, relied upon as a reasoning, responsible being. A man, especially a young husband, makes a grave mistake when he looks upon his wife as only a toy to amuse him in his leisure moments and not as one to be trusted to aid him in his life work. A trusted young housewife, with a reasonable and regular allowance at her command, be she ever so inexperienced, will soon plan to have wholesome, nutritious food at little cost, instead of not knowing until a half hour before meal time what she will serve. She would save money and the family would be better nourished; nevertheless, I would impress it on the young housewife not to be too saving or practice too close economy, especially when buying milk and eggs, as there is nothing more nutritious or valuable. A palatable macaroni and cheese; eggs or a combination of eggs and milk, are dishes which may be substituted occasionally, at less expense, for meat. A pound of macaroni and cheese equals a pound of steak in food value. Take time and trouble to see that all food be well cooked and served, both in an attractive and appetizing manner. Buy the cheaper cuts of stewing meats, and by long, slow simmering, they will become sweet and tender and of equal nutritive value as higher priced sirloins and tenderloins.

"But, Mary, I've not yet finished that trunk and its contents. That slightly-faded pink chambray I'll cut up into quilt blocks. Made up with white patches, and quilted nicely, a pretty quilt lined with white, will be evolved. I have such a pretty design of pink and white called the 'Winding Way,' very simple to make. The beauty of the quilt consists altogether in the manner in which the blocks are put together, or it might be made over the pattern called 'The Flying Dutchman.' From that tan linen skirt may be made a laundry bag, shoe pocket, twine bag, a collar bag and a table runner, the only expense being several skeins of green embroidery silk, and a couple yards of green cord to draw the bags up with, and a couple of the same-hued skirt braids for binding edges, and," teasingly, "Mary, you might embroider Ralph Jackson's initials on the collar and laundry bag."

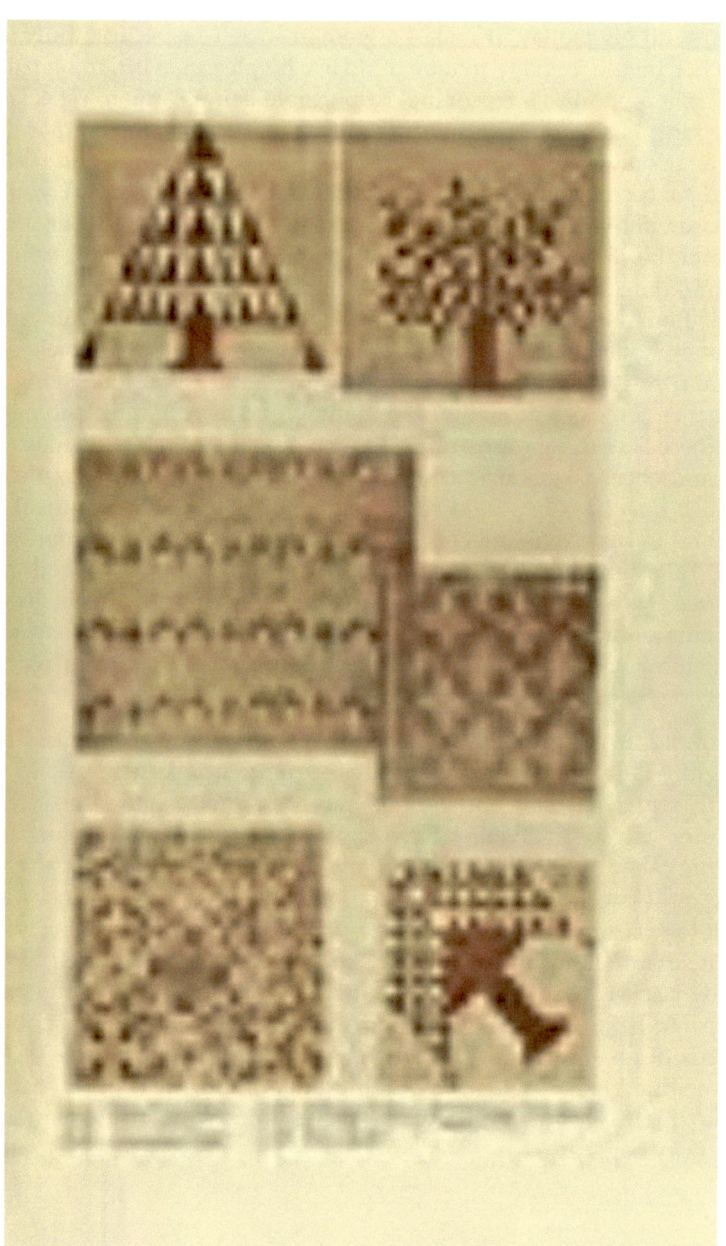

A-12 Pine Tree Quilt
A-13 Tree of Life
A-14 Pineapple
A-15 Enlarged Block of Winding Way Quilt
A-16 Lost Rose in the Wilderness
A-17 Tree Quilt

Mary blushed rosily red and exclaimed in an embarrassed manner, most bewitchingly, "Oh!"

Aunt Sarah laughed. She thought to have Mary look that way 'twas worth teasing her.

"Well, Mary, we can in leisure moments, from that coarse, white linen skirt which you have discarded, make bureau scarfs, sideboard cover, or a set of scalloped table mats to place under hot dishes on your dining-room table. I will give you pieces of asbestos to slip between the linen mats when finished. They are a great protection to the table. You could also make several small guest towels with deep, hemstitched ends with your initials on. You embroider so beautifully, and the drawn work you do is done as expertly as that of the Mexican women."

"Oh, Aunt Sarah, how ingenious you are."

"And, Mary, your rag carpet shall not be lacking. We shall tear up those partly-worn muslin skirts into strips one-half inch in width, and use the dyes left over from dyeing Easter eggs. I always save the dye for this purpose, they come in such pretty, bright colors. The rags, when sewed together with some I have in the attic, we'll have woven into a useful carpet for the home you are planning.'

"Oh! Aunt Sarah," exclaimed Mary, "do you mean a carpet like the one in the spare bedroom?"

"Yes, my dear, exactly like that, if you wish."

"Indeed I do, and I think one like that quite good enough to have in a dining-room. I think it so pretty. It does not look at all like a common rag carpet."

"No, my dear, it is nothing very uncommon. It is all in the way it is woven. Instead of having two gay rainbow stripes about three inches wide running through the length of the carpet, I had it woven with the ground work white and brown chain to form checks. Then about an inch apart were placed two threads of two shades of red woolen warp, alternating with two threads of two shades of green, across the whole width, running the length of the carpet. It has been greatly admired, as it is rather different from that usually woven. All the rag carpets I found in the house when we moved here, made by John's mother, possessed very wide stripes of rainbow colors, composed of shaded reds, yellows, blues and greens. You can imagine how very gorgeous they were, and so very heavy. Many of the country weavers use linen chain or warp instead of cotton, and always use wool warp for the stripes."

"Aunt Sarah, I want something so very much for the Colonial bedroom I should like to have when I have a home of my very own."

"What is it, dear? Anything, e'en to the half of my kingdom," laughingly replied her Aunt.

"Why, I'd love to have several rag rugs like those in your bedroom, which you call 'New Colonial' rugs."

"Certainly, my dear. They are easily made from carpet rags. I have already planned in my mind a pretty rag rug for you, to be made from your old, garnet merino shirtwaist, combined with your discarded cravenette stormcoat.

"And you'll need some pretty quilts, also," said her Aunt.

"I particularly admire the tree quilts," said Mary.

"You may have any one you choose; the one called 'Tree of Paradise,' another called 'Pineapple Design,' which was originally a border to 'Fleur de lis' quilt or 'Pine Tree,' and still another called 'Tree of Life,' and 'The Lost Rose in the Wilderness.'"

"They are all so odd," said Mary, "I scarcely know which one I think prettiest."

"All are old-fashioned quilts, which I prize highly," continued her Aunt. "Several I pieced together when a small girl, I think old-time patchwork too pretty and useful an accomplishment to have gone out of fashion.

"You shall have a small stand cover like the one you admired so greatly, given me by Aunt Cornelia. It is very simple, the materials required being a square of yard-wide unbleached muslin. In the centre of this baste a large, blue-flowered handkerchief with cream-colored ground, to match the muslin. Turn up a deep hem all around outside edge; cut out quarter circles of the handkerchief at each of four corners; baste neatly upon the muslin, leaving a space of muslin the same width as the hem around each quarter circle; briarstitch all turned-in edges with dark-blue embroidery silk, being washable, these do nicely as covers for small tables or stands on the veranda in Summertime."

"Aunt Sarah," ecstatically exclaimed Mary, "you are a wizard to plan so many useful things from a trunk of apparently useless rags. What a treasure Uncle has in you. I was fretting about having so little to make my home attractive, but I feel quite elated at the thought of having a carpet and rugs already planned, besides the numerous other things evolved from your fertile brain."

Aunt Sarah loved a joke. She held up an old broadcloth cape. "Here is a fine patch for Ralph Jackson's breeches, should he ever become sedentary and need one."

Mary reddened and looked almost offended and was at a loss for a reply.

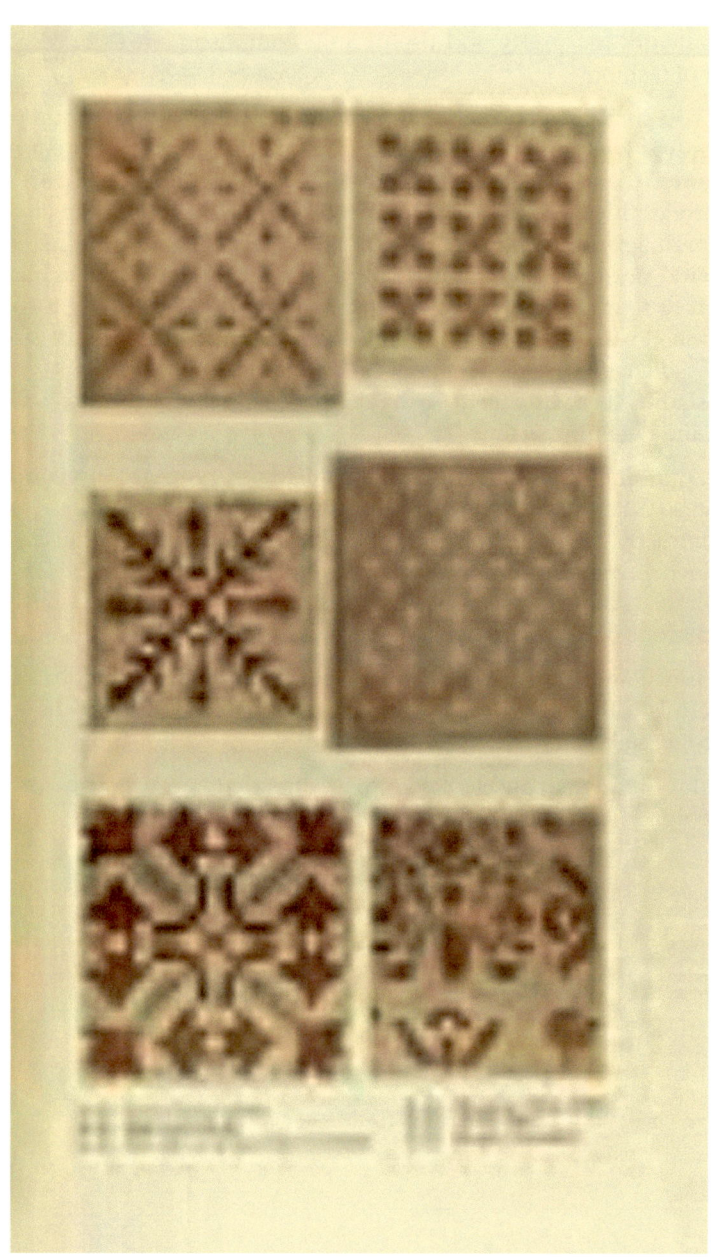

A-18 Fleur DeLys Quilt
A-19 Oak Leaf Quilt
A-20 One Block of Fleur DeLys Quilt
A-21 Winding Way Quilt
A-22 Tulip Quilt
A-23 Flower Pot Quilt

Greatly amused, Aunt Sarah quoted ex-President Roosevelt: "'Tis time for the man with the patch to come forward and the man with the dollar to step back,'" and added, "Never mind, Mary, your Ralph is such an industrious, hustling young man that he will never need a patch to step forward, I prophesy that with such a helpmeet and 'Haus Frau' as you, Mary, he'll always be most prosperous and happy. Kiss me, dear."

Mary did so, and her radiant smile at such praise from her honored relative was beautiful to behold.

OLD RAG CARPET

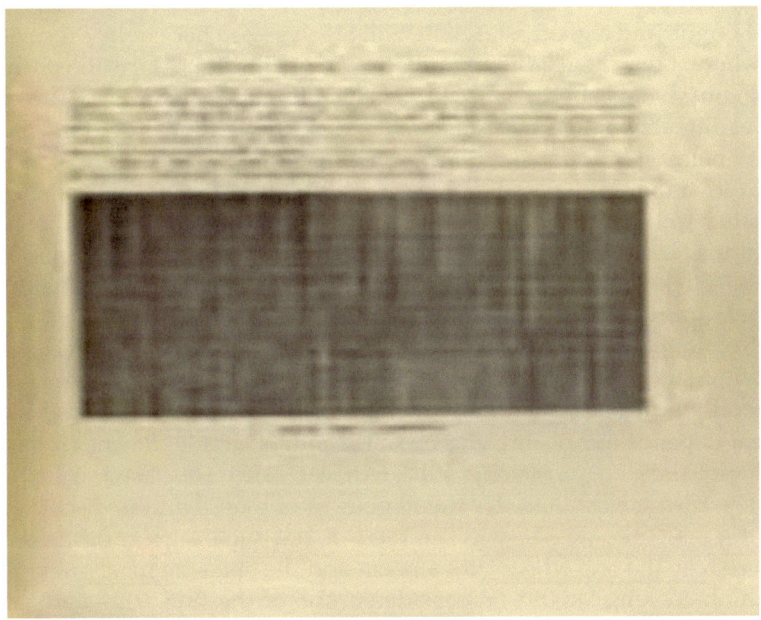

CHAPTER IX.

POETRY AND PIE.

"Aunt Sarah," questioned Mary one day, "do you mind if I copy some of your recipes?"

"Certainly not, my dear," replied her Aunt.

"And I'd like to copy some of the poems, also, I never saw any one else have so much poetry in a book of cooking recipes."

"Perhaps not," replied her Aunt, "but you know, Mary, I believe in combining pleasure with my work, and our lives are made up of poetry and prose, and some lives are so very prosy. Many times when too tired to look up a favorite volume of poems, it has rested me to turn the pages of my recipe book and find some helpful thought, and a good housewife will always keep her book of recipes where it may be readily found for reference. I think, Mary, the poem 'Pennsylvania,' by Lydia M.D. O'Neil, a fine one, and I never tire of reading it over and over again. I have always felt grateful to my old schoolmaster. Professor T— —, for teaching me, when a school girl, to love the writing of Longfellow, Whittier, Bryant, Tennyson and other well-known poets. I still, in memory, hear him repeat 'Thanatopsis,' by Bryant and 'The Builders,' by Longfellow. The rhymes of the 'Fireside Poet' are easily understood, and never fail to touch the heart of common folk. I know it appears odd to see so many of my favorite poems sandwiched in between old, valued cooking recipes, but, Mary, the happiness of the home life depends so largely on the food we consume. On the preparation and selection of the food we eat depends our health, and on our health is largely dependent our happiness and prosperity. Who is it has said, 'The discovery of a new dish makes more for the happiness of man than the discovery of a star'? So, dearie, you see there is not such a great difference between the one who writes a poem and the one who makes a pie. I think cooking should be considered one of the fine arts—and the woman who prepares a dainty, appetizing dish of food, which appeals to the sense of taste, should be considered as worthy of praise as the artist who paints a fine picture to gratify our sense of sight. I try to mix all the poetry possible in prosaic every-day life. We country farmers' wives, not having the opportunities of our more fortunate city sisters, such as witnessing plays from Shakespeare, listening to symphony concerts, etc., turn to 'The Friendship of Books,' of which Washington Irving writes: 'Cheer us with the true friendship, which never deceived hope nor deserted sorrow.'"

"Yes," said Mary, "but remember, Aunt Sarah, Chautauqua will be held next Summer in a near-by town, and, as Uncle John is one of the guarantors, you will wish to attend regularly and will, I know, enjoy hearing the excellent lectures, music and concerts."

"Yea," replied her Aunt, "Chautauqua meetings will commence the latter part of June, and I will expect you and Ralph to visit us then. I think Chautauqua a godsend to country women, especially farmers' wives; it takes them away from their monotonous daily toil and gives them new thoughts and ideas."

"I can readily understand, Aunt Sarah, why the poem, 'Life's Common Things,' appeals to you; it is because you see beauty in everything. Aunt Sarah, where did you get this very old poem, 'The Deserted City'?"

"Why, that was given me by John's Uncle, who thought the poem fine."

> "Sad is the sight, the city once so fair!
> An hundred palaces lie buried there;
> Her lofty towers are fallen, and creepers grow
> O'er marbled dome and shattered portico.
> "Once in the gardens, lovely girls at play,
> Culled the bright flowers, and gently touched the spray;
> But now wild creatures in their savage joy
> Tread down the flowers and the plants destroy.
> "By night no torches in the windows gleam;
> By day no women in their beauty beam;
> The smoke has ceased—the spider there has spread
> His snares in safety—and all else is dead."

"Indeed, it is a 'gem,'" said Mary, after slowly reading aloud parts of several stanzas.

"Yes," replied her Aunt, "Professor Schmidt tells me the poem was written by Kalidasa (the Shakespeare of Hindu literature), and was written 1800 years before Goldsmith gave us his immortal work, 'The Deserted Village.'"

"I like the poem, 'Abou Ben Adhem and the Angel,'" said Mary, "and I think this true by Henry Ward Beecher:"

> "'Do not be troubled because you have not great virtues,
> God made a million spears of grass where He made one tree;
> The earth is fringed and carpeted not with forests but with
> grasses,
> Only have enough of little virtues and common fidelities,

And you need not mourn because you are neither a hero nor a saint.'

"This is a favorite little poem of mine, Aunt Sarah. I'll just write it on this blank page in your book."

There's a little splash of sunshine and a little spot of shade, always somewhere near,
The wise bask in the sunshine, but the foolish choose the shade.
The wise are gay and happy, on the foolish, sorrow's laid,
And the fault's their own, I fear.
For the little splash of sunshine and the little spot of shade
Are here for joint consumption, for comparison are made;
We're all meant to be happy, not too foolish or too staid.
And the right dose to be taken is some sunshine mixed with shade.

"Aunt Sarah, I see there is still space on this page to write another poem, a favorite of mine. It is called, 'Be Strong,' by Maltbie Davenport."

Be Strong!
We are not here to play, to dream, to drift;
We have hard word to do, and loads to lift,
Shun not the struggle; face it, 'tis God's gift.
Be Strong!
Say not the days are evil — who's to blame?
And fold the hands and acquiesce — Oh, shame!
Stand up, speak out, and bravely, in God's name.
Be Strong!
It matters not how deep intrenched the wrong,
How hard the battle goes, the day how long;
Faint not, fight on! Tomorrow comes the song,

LIFE'S COMMON THINGS.

How lovely are life's common things.
When health flows in the veins;
The golden sunshine of the days
When Phoebus holds the reins;
The floating clouds against the blue;
The fragrance of the air;
The nodding flowers by the way;
The green grass everywhere;
The feathery beauty of the elm,
With graceful-swaying boughs.
Where nesting songbirds find a home
And the night wind sighs and soughs;
The hazy blue of distant hill,
With wooded slope and crest;
The crimson sky when low at night
The sun sinks in the West;
The thrilling grandeur of the storm,
The lightning's vivid flash,
The mighty rush of wind and rain,
The thunder's awful crash.
And then the calm that follows storm,
And rainbow in the sky;
The rain-washed freshness of the earth —
A singing bird near by.
And oh, the beauty of the night!
Its hush, its thrill, its charm;
The twinkling brilliance of its stars;
Its tranquil peace and calm.
Oh, loving fatherhood of God
To give us every day
The lovely common things of life
To brighten all the way!
(Susan M. Perkins, in the Boston Transcript)

ABOU BEN ADHEM AND THE ANGEL.

Abou Ben Adhem—may his tribe increase—
Awoke one night from a deep dream of peace
And saw, within the moonlight of his room,
Making it rich and like a lily in bloom,
An angel writing in a book of gold.
Exceeding peace had made Ben Adhem bold,
And to the presence in the room he said:
"What writest thou?" The vision raised his head,
And with a look made of all sweet accord,
Answered: "The names of those who love the Lord."
"And is mine one?" said Abou. "Nay, not so,"
Replied the angel. Abou spoke low,
But cheerily still, and Said, "I pray thee, then,
write me as one that loves his fellow-men."
The angel wrote and vanished. The next night
It came again, with a great, wakening light,
And showed the names whom love of God had blessed,
And, lo! Ben Adhem's name led all the rest.
LEIGH HUNT.

CHAPTER X.

SIBYLLA LINSABIGLER.

A very original character was Sibylla Linsabigler, who had been a member of the Landis household several years. She was Aunt Sarah's only maid servant, but she disliked being referred to as a servant, and when she overheard "Fritz" Schmidt, as he passed the Landis farm on his way to the creek for a days fishing, call to Mary: "Miss Midleton, will you please send the butter over with the servant today, as I shall not return home in time for dinner" Sibylla said, "I ain't no servant. I'm hired girl What does that make out if I do work here? Pop got mad with me 'cause I wouldn't work at home no more for him and Mom without they paid me. They got three more girls to home yet that can do the work. My Pop owns a big farm and sent our 'Chon' to the college, and it's mean 'fer' him not to give us girls money for dress, so I work out, 'Taint right the way us people what has to work are treated these days," said Sibylla to herself, as she applied the broom vigorously to the gay-flowered carpet in the Landis parlor. "Because us folks got to work ain't no reason why them tony people over to the Perfessor's should call me a 'servant.' I guess I know I milk the cows, wash dishes, scrub floors, and do the washin' and ir'nin' every week, but I'm no 'servant,' I'm just as good any day as that good-fer-nothin' Perfesser's son," continued Sibylla, growing red in the face with indignation. "Didn't I hear that worthless scamp, Fritz Schmidt, a-referrin' to me and a-sayin' to Miss Midleton fer the 'servant' to bring over the butter? Betch yer life this here 'servant' ain't a-goin' to allow eddicated people to make a fool of her. First chance I get I'll give that Perfesser a piece of my mind."

Sibylla's opportunity came rather unexpectedly. The gentle, mild-mannered Professor was on good terms with his sturdy, energetic neighbor, John Landis, and frequently visited him for a neighborly chat. On this particular day he called as usual and found Sibvlla in the mood described.

"Good afternoon, Sibylla," said the Professor, good-naturedly. "How are you today?"

"I'd be a whole lot better if some people weren't so smart," replied Sibylla, venting her feelings on the broom. "Should think a Perfesser would feel himself too big to talk to a 'servant'."

"On the contrary, my dear girl, I feel honored. I presume you are not feeling as well as usual. What makes you think it is condescension for me to address you?" asked the genial old man, kindly.

"Well, since you ask me, I don't mind a-tellin' you. Yesterday your son insulted me, I won't take no insult from nobody, I am just as good as what you are, even if I hain't got much book larnin'."

With this deliverance, Sibylla felt she had done full justice to the occasion and would have closed the interview abruptly had not the Professor, with a restraining hand, detained her.

"We must get to the bottom of this grievance, Sibylla. I am sure there is some mistake somewhere. What did my son say?"

"Well, if you want to know," replied the irate domestic, 'I'll tell you. He called me a 'servant.' I know I'm only a working girl, but your son nor nobody else ain't got no right to abuse me by callin' me a 'servant'."

"Ah! I see. You object to the term 'servant' being applied to you," said the Professor, comprehendingly. "The word 'servant' is distasteful to you. You feel it is a disgrace to be called a servant. I see! I see!" In a fatherly way, the old man resumed: "In a certain sense we are all servants. The history of human achievements is a record of service. The men and women who have helped the world most were all servants—servants to humanity. The happiest man is he who serves. God calls some men to sow and some to reap; some to work in wood and stone; to sing and speak. Work is honorable in all, regardless of the capacity in which we serve. There is no great difference, after all, between the ordinary laborer and the railroad president; both are servants, and the standard of measurement to be applied to each man is the same. It is not so much a question of station in life as it is the question of efficiency. Best of all, work is education. There is culture that comes without college and university. He who graduates from the college of hard work is as honorable as he who takes a degree at Yale or Harvard; for wisdom can be found in shop and foundry, field and factory, in the kitchen amid pots and kettles, as well as in office and school. The truly educated man is the man who has learned the duty and responsibility of doing something useful, something helpful, something to make this old world of ours better and a happier place in which to live. The word 'servant,' Sibylla, is a beautiful one, rightly understood. The greatest man who ever lived was a servant. All His earthly ministry was filled with worthy deeds. When man pleaded with Him to rest, He answered: 'My Father worketh hitherto, and I work.' When one of Christ's followers desired to express the true nature of his work and office, he called himself a servant. He used a word, 'doulos,' which means, in the Greek language, a slave or a bond-servant. By the word 'doulos' he meant to say that his mission in life was to work, to do good, to serve. This man was a great preacher, but it is possible for any one to become a 'doulos' in so far as he is willing to serve God and his fellowman. You see, Sibylla, the spirit of Christian work and brotherly love is the spirit of 'doulos.' The word has been transformed by service and unselfish devotion to duty. Great men who have blessed the world, and good and noble women who have helped to uplift humanity, have done it through service. It is just as honorable to bake well, and cook well, and to do the humblest daily tasks efficiently, as it is to play well on the piano and talk fluently about the latest books."

At the conclusion of the Professor's little talk on the dignity of labor, a new light shone in Sibylla's eyes and a new thought gripped her soul. The spirit of "doulos" had displaced her antipathy toward the word servant.

"I'll take that butter over to the Professer's home right away," she said, to herself.

Before leaving Sibylla, the Professor quoted from the "Toiling of Felix," by Henry Vandyke:

"Hewing wood and drawing water, splitting stones and cleaving sod, All the dusty ranks of labour, in the regiment of God, March together toward His triumph, do the task His hands prepare; Honest toil is holy service, faithful work is praise and prayer."

They who work without complaining, do the holy will of God.

Heaven is blest with perfect rest, but the blessing of earth is toil.

Sibylla Linsabigler was a healthy, large-boned, solidly-built, typical "Pennsylvania German" girl. Her clear, pinkish complexion looked as if freshly scrubbed with soap and water. A few large, brown freckles adorned the bridge of her rather broad, flat nose. She possessed red hair and laughing, red-brown eyes, a large mouth, which disclosed beautiful even, white teeth when she smiled, extraordinary large feet and hands, strong, willing and usually good-natured, although possessed of a quick temper, as her red hair indicated. Kind-hearted to a fault, she was of great assistance to Aunt Sarah, although she preferred any other work to that of cooking or baking. She kept the kitchen as well as other parts of the house, to quote Aunt Sarah, "neat as a pin," and did not object to any work, however hard or laborious, as long as she was not expected to do the thinking and planning. She was greatly attached to both Aunt Sarah and Mary, but stood rather in awe of John Landis, who had never spoken a cross word to her in the three years she had lived at the farm.

Sarah Landis, knowing Sibylla to be an honest, industrious girl, appreciated her good qualities, thought almost as much of Sibylla as if she had been her daughter, and treated her in like manner, and for this reason, if for no other, she received willing service from the girl.

Sibylla, a swift worker at all times, never finished work so quickly as on Wednesday and Saturday evenings, when she "kept company" with Jake Crouthamel. "Chake," as Sibylla called him, was a sturdy, red-faced young farmer, all legs and arms. He appeared to be put together loosely at the joints, like a jumping-jack, and never appeared at ease in his ill-fitting "store clothes." He usually wore gray corduroy trousers and big cowhide boots, a pink and white striped shirt and red necktie.

Sibylla did not notice his imperfections, and thought him handsome as a Greek god.

Jake, an honest, industrious young fellow, worked on a near-by farm, owned his own carriage, and had the privilege of using one of the farm horses when he wished, so he and Sibylla frequently took "choy rides," as Sibylla called them.

Jake Crouthamel was usually called "Boller-Yockel," this name having been accorded him on account of his having delivered to a purchaser a load of hay largely composed of rag-weed. The man called him an old "Boller-Yockel," and the name had clung to Jake for years.

CHAPTER XI.

"NEW COLONIAL" RAG RUGS.

Several days had elapsed since that on which Mary's Aunt had planned to use the contents of her trunk to such good advantage, when Mary, coming into the room where her Aunt was busily engaged sewing, exclaimed: "Don't forget, Auntie, you promised to teach me to crochet rag rugs!"

A "HIT-AND-MISS" RUG

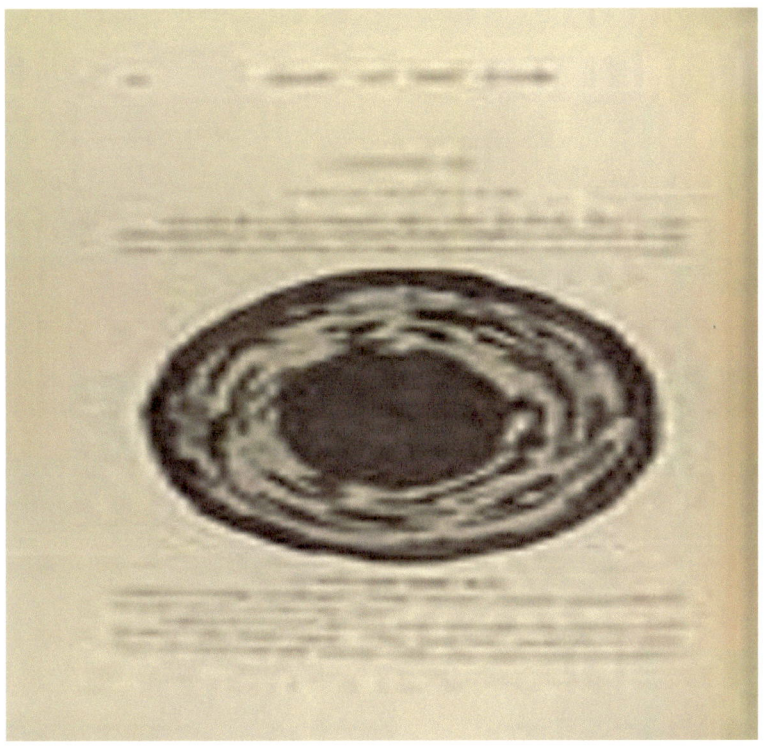

"Indeed, I've not forgotten, and will make my promise good at once," said Aunt Sarah. "We shall need quantities of carpet rags cut about one-half inch in width, the same as those used for making rag carpet. Of course, you are aware, Mary, that heavier materials should be cut in narrower strips than those of thinner materials. You will also require a long, wooden crochet needle, about as thick as an ordinary wooden lead pencil, having a hook at one end, similar to a common bone crochet needle, only larger. For a circular rug, crochet about twelve stitches (single crochet) over one end of a piece of candle wick or cable cord; or, lacking either of these, use a carpet rag of firm material; then draw the crocheted strip into as small a circle as possible, fasten and crochet round and round continuously until finished. The centre of a circular or oblong rug may be a plain color, with border of colored light and dark rags, sewed together promiscuously, called 'Hit and Miss.'

A BROWN AND TAN RUG

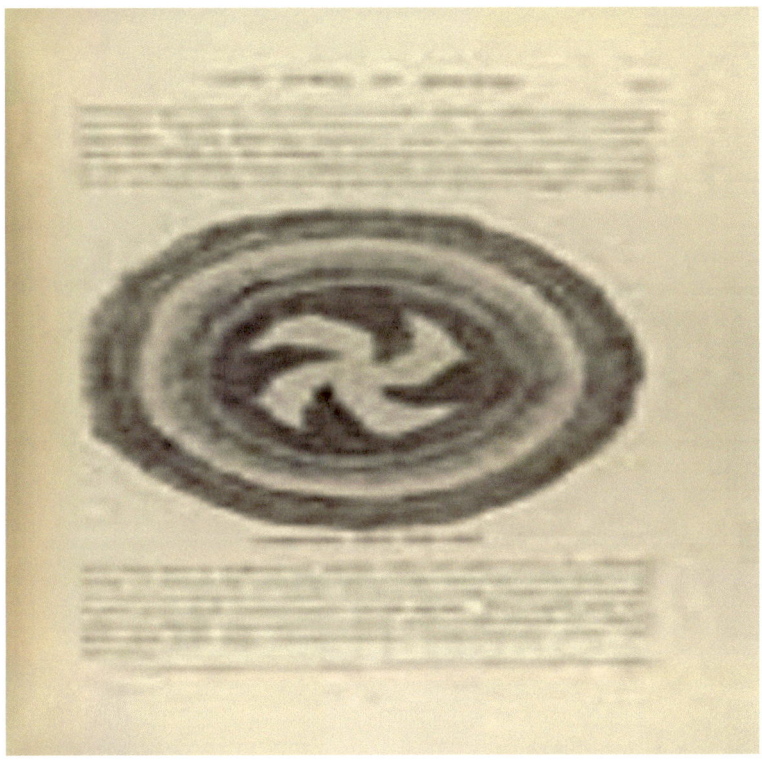

"Or you might have a design similar to a 'Pin-wheel' in centre of the circular rug, with alternate stripes, composed of dark and light-colored rags."

"I'd like one made in that manner from different shades that harmonize, browns and tans, for instance," said Mary.

"You may easily have a rug of that description," continued her Aunt. "With a package of brown dye, we can quickly transform some light, woolen carpet rags I possess into pretty shades of browns and tans."

RUG

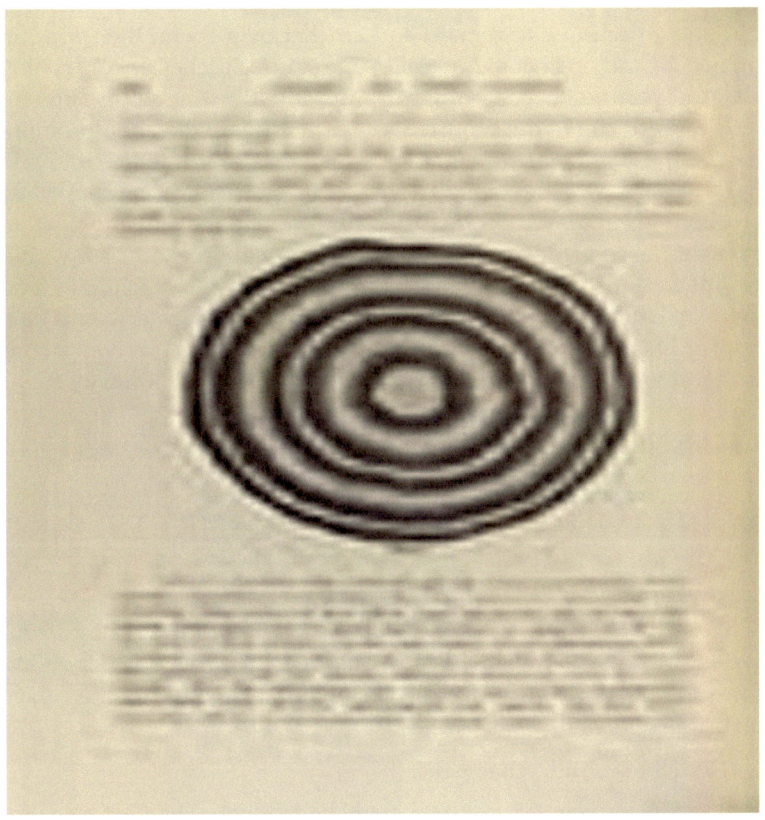

"For a circular rug, with design in centre resembling a pin-wheel, commence crocheting the rug same as preceding one. Crochet three rows of one color, then mark the rug off into four parts, placing a pin to mark each section or quarter of the rug. At each of four points crochet one stitch of a contrasting shade. Crochet once around the circle, using a shade similar to that of the centre of rug for design, filling in between with the other shade. For the following row, crochet two stitches beneath the one stitch (not directly underneath the stitch, but one stitch beyond), filling in between with the other color. The third row, add three stitches beneath the two stitches in same manner as preceding row, and continue, until design in centre is as large as desired, then crochet 'Hit or Miss' or stripes. Do not cut off the carpet rags at each of the four points after crocheting stitches, but allow each one to remain and crochet over them, then pick up on needle and crochet every time you require stitches of contrasting shade. Then crochet several rows around the rug with different shades until rug is the required size. The under side should be finished off as neatly as the right, or upper side. Mary, when not making a design, sew the rags together as if for weaving carpet. When crocheting circular rugs, occasionally stretch the outside row to prevent the rug from curling up at edges when finished, as it would be apt to do if too tightly crocheted. If necessary, occasionally add an extra stitch. Avoid also crocheting it too loosely, as it would then appear like a ruffle. The advantage of crocheting over a heavy cord is that the work may be easily drawn up more tightly if too lose."

CHAPTER XII.

MARY IMITATES NAVAJO BLANKETS.

On her return from an afternoon spent at Professor Schmidt's, Mary remarked to Aunt Sarah, "For the first time in my life I have an original idea!"

"Do tell me child, what it is!"

"The 'New Colonial' rag rugs we have lately finished are fine, but I'd just love to have a Navajo blanket like those owned by Professor Schmidt; and I intend to make a rag rug in imitation of his Navajo blanket."

"Yes," answered her Aunt, "I have always greatly admired them myself, especially the large gray one which covers the Professor's own chair in the library. The Professor brought them with him when he returned from 'Cutler's Ranch' at Rociada, near Las Vegas, New Mexico, where he visited his nephew, poor Raymond, or rather, I should say, fortunate Raymond, an only child of the Professor's sister. A quiet, studious boy, he graduated at the head of his class at an early age, but he inherited the weak lungs of his father, who died of consumption. Raymond was a lovable boy, with a fund of dry humor and wit—the idol of his mother, who, taking the advice of a specialist, accompanied her boy, as a last resort, to New Mexico, where, partly owing to his determination to get well, proper food and daily rides on the mesa, on the back of his little pinto pony, he regained perfect health, and today is well, happily married and living in Pasadena, California, so I have been told by Frau Schmidt, who dearly loves the boy."

"But Mary, forgive an old woman for rambling away from the subject in which you are interested—Navajo blankets. Ever since we planned to make a rug with a swastika in the centre, I nave been trying to evolve from my brain (and your Uncle John says my bump of inventiveness is abnormally large) a Navajo rag rug for the floor of the room you intend to furnish as Ralph's den, in the home you are planning. Well, my dear, a wooden crochet hook in your deft fingers will be the magic wand which will perform a miracle and transform into Navajo blankets such very commonplace articles as your discarded gray eiderdown kimona, and a pair of your Uncle's old gray trousers, which have already been washed and ripped by Sibylla, to be used for making carpet rags. These, combined with the gray skirt I heard you say had outlived its day of usefulness, will furnish the background of the rug. The six triangles in the centre of the rug, also lighter stripes at each end of the rug, we will make of that old linen chair-cover and your faded linen skirt, which you said I might use for carpet rags; and, should more material be needed, I have some old, gray woolen underwear in my patch bag, a gray-white, similar to the real Navajo. The rows of black with which we shall outline the triangles may be made from those old, black, silk-lisle hose you gave me, by cutting them round and round in one continuous strip. Heavy cloth should be cut in *very* narrow strips. Sibylla will do that nicely; her hands are more used to handling large, heavy shears than are yours. The linen-lawn skirt you may cut in strips about three-fourths of an inch in width, as that material is quite thin. I would sew rags of one color together like carpet rags, not lapping the ends more than necessary to hold them together. The rug will be reversible, both sides being exactly alike when finished. I should make the rug about fifty-three stitches across. This will require about six and one-fourth yards of carpet rags, when sewed together, to crochet once across. I think it would be wise to cut all rags of different weight materials before commencing to crochet the rug, so they may be well mixed through. I will assist you with the work at odd moments, and in a short time the rug will be finished."

The rug, when finished, was truly a work of art, and represented many hours of labor and thought. But Mary considered it very fascinating work, and was delighted with the result of her labor—a rug the exact imitation of one of the Professor's genuine Indian Navajo blankets, the work of her own hands, and without the expenditure of a penny.

Mary remarked: "I do not think all the triangles in my rug are the exact size of the paper pattern you made me, Aunt Sarah. The two in the centre appear larger than the others."

"Well," remarked her Aunt, "if you examine closely the blankets owned by Professor Schmidt, you will find the on the ones woven by Navajo Indians are not of an equal size."

'Tis said Navajo blankets and Serapes will become scarce and higher in price in the future, on account of the numerous young Indians who have been educated and who prefer other occupations to that of weaving blankets, as did their forefathers; and the present disturbance in Mexico will certainly interfere with the continuance of this industry for a time.

IMITATION OF NAVAJO BLANKET

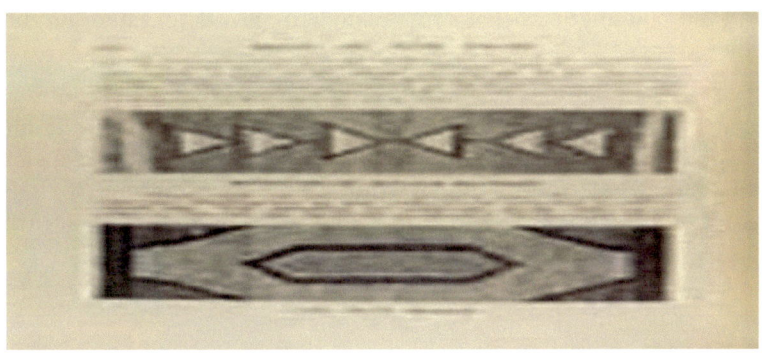

RUG WITH DESIGN

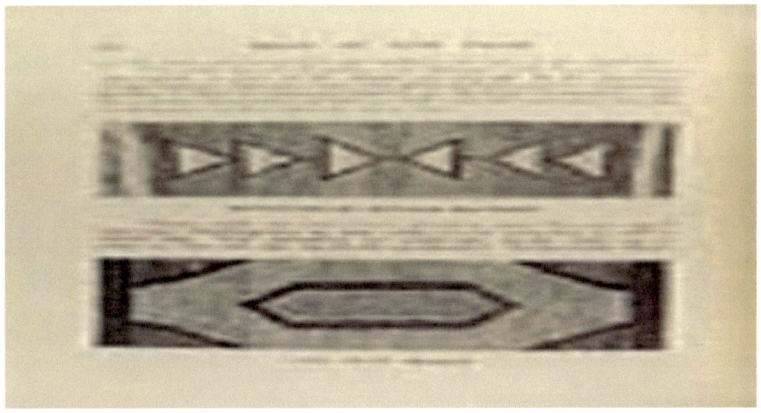

"Mary, while you have been planning your Navajo rug, I have been thinking how we may make a very attractive as well as useful rug. You remember, we could not decide what use to make of your old, tan cravenette stormcoat? I have been thinking we might use this, when cut into carpet rags, for the principal part of the rug, and that old, garnet merino blouse waist might be cut and used for the four corners of a rug, and we might have gay stripes in the centre of the rug to form a sort of design, and also put gay stripes at each end of the rug.

"And you might crochet a rug, plain 'Hit or Miss,' of rather bright-colored rags."

"Yes," said Mary, "I think I will crochet a swastika in the centre of a rug, as you suggest, of bright orange, outlined with black, and a stripe of orange edged with black at each end of the rug to match the centre. Don't you think that would be pretty, Aunt Sarah?"

"HIT-OR-MISS" RUG WITH SWASTIKA CENTRE

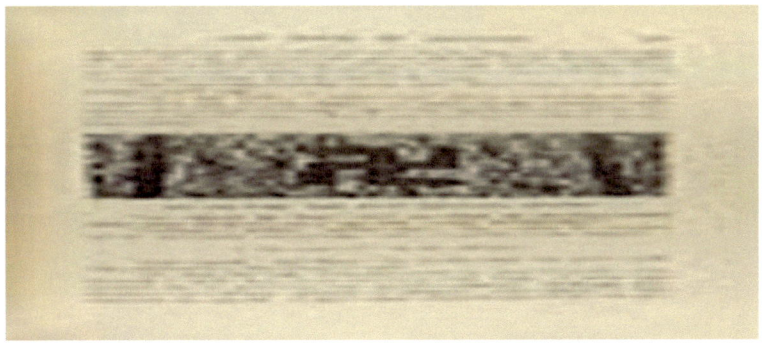

"Yes indeed, but Mary, don't you think the swastika would show more distinctly on a rug with a plain background?"

"Perhaps it would," replied Mary, "but I think I'll crochet one of very gayly-colored rags, with a swastika in the centre."

A "PRAYER RUG" OF SILK SCRAPS.

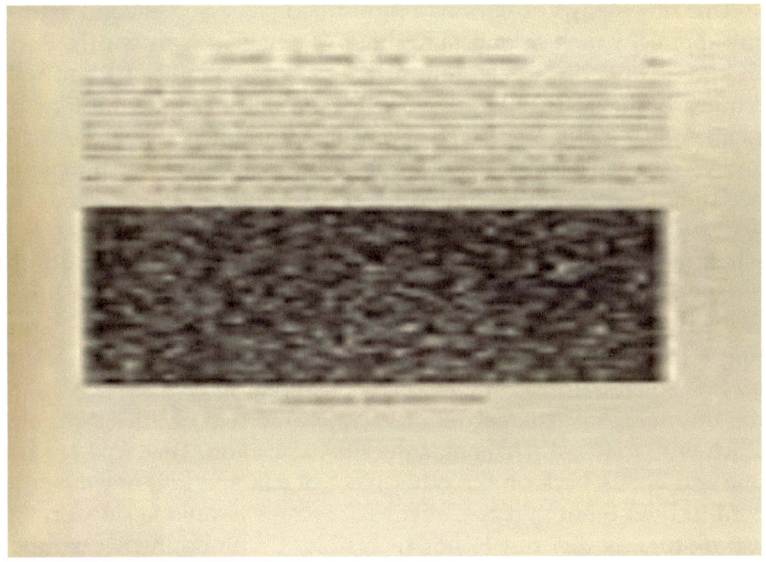

"Aunt Sarah," said Mary, "do tell me how that pretty little rug composed of silk scraps is made."

"Oh, that *silk* rug; 'twas given me by Aunt Cornelia, who finished it while here on a visit from New York. I never saw another like it, and it has been greatly admired. Although possessed of an ordinary amount of patience, I don't think I'll ever make one for myself. I don't admire knitted rugs of any description, neither do I care for braided rugs. I think the crocheted ones prettier. But, Mary, this small silk rug is easily made should you care to have one. I will commence knitting one for you at once. You will then find a use for the box of bright-colored silks you possess, many of which are quite too small to be used in any other manner. Professor Schmidt calls this a 'Prayer Rug.' He said: 'This rug, fashioned of various bright-hued silks of orange, purple and crimson, a bright maze of rich colors, without any recognizable figure or design, reminds me of the description of the 'Prayer Carpet' or rugs of the Mohammedans. They are composed of rich-hued silks of purple, ruby and amber. 'Tis said their delicacy of shade is marvelous and was suggested by the meadows of variegated flowers.' But this is a digression; you wished directions for making the rug.

"Use tiny scraps of various bright-hued silks, velvets and satins, cut about 3½ inches long and about one-half inch in width. Ends should always be cut slanting or bias; never straight. All you will require besides the silk scraps, will be a ball of common cord or twine, or save all cord which comes tied around packages, as I do, and use that and two ordinary steel knitting needles. When making her rug, Aunt Cornelia knitted several strips a couple of inches in width and the length she wished the finished rug to be. The strips when finished she sewed together with strong linen thread on the wrong side of the rug. She commenced the rug by knitting two rows of the twine or cord. (When I was a girl we called this common knitting 'garter stitch.') Then, when commencing to knit third row, slip off first stitch onto your other needle; knit one stitch, then lay one of the tiny scraps of silk across or between the two needles; knit one stitch with the cord. This holds the silk in position. Then fold or turn one end of silk back on the other piece of silk and knit one stitch of cord to hold them in place, always keeping silk on one side, on the top of rug, as this rug is not reversible. Continue in this manner until one row is finished. Then knit once across plain with cord, and for next row lay silk scraps in and knit as before. Always knit one row of the cord across plain after knitting in scraps of silk, as doing this holds them firmly in position. Of course, Mary, you will use judgment and taste in combining light and dark, bright and dull colors. Also, do not use several scraps of velvet together. Use velvet, silk and satin alternately. Should any scraps of silk be longer than others after knitting, trim off evenly so all will be of uniform size. When her rug was finished, Aunt Cornelia spread it, wrong side uppermost, on an unused table, covered it with a thick boiled paste, composed of flour and water, allowed it to dry thoroughly, then lined the rug with a heavy piece of denim. This was done to prevent the rug from curling up at edges, and caused it to lie flat on floor; but I think I should prefer just a firm lining or foundation of heavy burlap or denim."

"Thank you, Aunt Sarah, for your explicit directions. I cannot fail to know just how to knit a silk rug, should I ever care to do so. I think the work would be simply fascinating."

CHAPTER XIII.

THE GIRLS' CAMPFIRE, ORGANIZED BY MARY.

One day in early June, when all nature seemed aglow with happiness, we find Mary earnestly discussing with Elizabeth Schmidt the prosaic, humdrum life of many of the country girls, daughters of well-to-do farmers in the vicinity.

"I wish," said Mary, wrinkling her forehead thoughtfully, "I could think of some new interest to introduce into their lives; some way of broadening their outlook; anything to bring more happiness into their commonplace daily toil; something good and helpful for them to think about."

All at once Mary, who was not usually demonstrative, clapped her hands, laughed gleefully and said: "I have it, Elizabeth. The very thing! Suppose we start a 'girls' campfire,' right here in the country? I don't think we shall have any trouble to organize."

"And you, because you understand all about it, will be the Guardian," said Elizabeth.

At first Mary demurred, but, overcome by Elizabeth's pleading, finally gave a reluctant consent. They then made out a list of the girls they thought might be willing to join, Mary promising to write at once for a handbook. They separated, Elizabeth to call to see the girls, and Mary to interview their parents. Their efforts were rewarded with surprisingly gratifying results, for many of the girls had read about the "Campfire Girls" and were anxious to become members.

One afternoon, several weeks later, had you gone into the old apple orchard, at the farm, you would have seen thirteen eager young girls, ranging in age from fourteen to sixteen, listening intently to Mary, who was telling them about the "Campfire Girls." What she told them was something like this:

"Now girls, we are going to have a good time. Some of our good times will be play and some work. When you join, you will become a 'Wood Gatherer,' and after three months' successful work, if you have met certain qualifications, you will be promoted to the rank of 'Fire Maker.' Later on, when you come to realize what it means to be a 'Torch Bearer,' you will be put in that rank. The first law which you learn to follow is one which you must apply to your daily life. It is: Seek beauty, give service, pursue knowledge, be trustworthy, hold on to health, glorify work, be happy. 'The Camp Fire' has meant so much to girls I have known, for their betterment, and has been so helpful in many ways, you surely will never regret becoming a member of the organization, or be anything but happy if you keep their laws. There will be no dues, except what is collected for good times, and no expense except the cost of your ceremonial costume, epaulettes and honor beads. The latter are quite inexpensive. The honors are divided into several classes, and for each honor a bead is given as a symbol of your work. A special colored bead is given for each class. We shall meet about once every week. The monthly meeting is called the 'Council Fire.' I will tell you later about the 'Wohelo' ceremony. By the way, girls, 'Wohelo' stands for work, health and love. You see, the word is composed of the first two letters of each word."

The girls appeared to be greatly interested, and Mary felt very much encouraged. Some of the girls left to talk it over with the homefolks, while others, wishing to learn more of the organisation, plied Mary with numerous questions. Finally, in desperation, she said: "Girls, I will read you the following from the 'Camp Fire Girls' Handbook, which I received this morning:"

'The purpose of this organization is to show that the common things of daily life are the chief means of beauty, romance and adventure; to aid in the forming of habits making for health and vigor, the out-of-door habit and the out-of-door spirit; to devise ways of measuring and creating standards to woman's work, and to give girls the opportunity to learn how to "keep step," to learn team work, through doing it; to help girls and women serve the community, the larger home, in the same way they have always served the individual home; to give status and social recognition to the knowledge of the mother, and thus restore the intimate relationship of mothers and daughters to each other.'

"Well, girls," said Mary, as she laid aside the book, "I think you all understand what a benefit this will be to you, and I will do all in my power to help you girls, while I am at the farm this summer. It is too late to tell you any more today. The information I have given you will suffice for the present. Three cheers for our Camp Fire! which will be under way in two months, I trust."

The members of "Shawnee" Camp Fire held their first Council Fire, or Ceremonial Meeting, the second week in July. The girls, all deeply interested, worked hard to secure honors which were awarded for engaging in domestic duties well known to the home, for studying and observing the rules of hygiene and sanitation, and for learning and achievements in various ways. They held weekly meetings and studied diligently to win the rank of Fire Maker.

A girl, when she joins, becomes a Wood Gatherer; she then receives a silver ring.

The weeks pass swiftly by, and it is time for another Camp Fire. The girls selected as their meeting place for this occasion farmer Druckenmuller's peach orchard, to which they walked, a distance of about three miles from the home of Elizabeth Schmidt. They left about two o'clock in the afternoon, intending to return home before nightfall, a good time being anticipated, as they took with them lunch and materials for a corn-roast.

The peach orchard in question, covering many acres, was situated at the foot of a low hill. Between the two flowed an enchanting, fairy-like stream, the cultivated peach orchard on one side, and on the opposite side the forest-like hill, covered with an abundance of wild flowers.

When the afternoon set for the Council Fire arrived, had you happened to meet the fifteen merry, chattering girls, accompanied by two older girls, Mary and Lucy Robbins (the country school teacher), as chaperones, wending their way to the orchard, you, without a doubt, would have smiled and a question might naturally have arisen regarding their sanity. They certainly possessed intelligent faces, but why those queer-shaped Indian dresses? And such an awkward length for a young girl's dress! And why was their hair all worn hanging in one braid over each shoulder, with a band over the forehead? Why so many strings of gaudy beads around their necks? These questions may all be answered in one single sentence: The girls are dressed in Ceremonial Costume.

ELIZABETH SCHMIDT "LAUGHING WATER"

A great many delays along the way were caused by girls asking the names of the different wild flowers and weeds they noticed in passing. One of the girls stopped to examine a prickly-looking plant about two feet high, with little, blue flowers growing along the stem, and asked if any one knew the name of it. They were about to look it up in a small "Flower Guide" owned by one of the girls, when some one said: "Why, that is a weed called 'Vipers Bougloss,'" They also found cardinal flower, thorn apple, monkey flower and jewel-weed in abundance, wild sunflower, ginseng, early golden rod, "Joe-pie-weed," marshallow, black cohosh and purple loose-trifle. The girls also noticed various birds.

On a tall tree one of the girls espied a rose-breasted Grosbeak, rare in this part of Bucks County. They all stopped and watched for a short time a white-bellied Nut-hatch. The girls were startled as a Scarlet Tanger flew past to join his mate, and they at last reached their rendezvous, the orchard.

By half-past three they were all seated in a circle waiting for the ceremonies to begin. Mary Midleton, their Guardian, stepped to the front, saying: "Sunflower, light the fire." Sunflower, through several months of daily attainment, had become a Fire-maker and was very proud of the Fire-maker's bracelet she was entitled to wear. Sunflower was given that name because she always looked on the bright side of everything; she looked like a sunflower, too, with her tanned face and light, curly hair.

All the girls had symbolical names given them. "Lark" was so named because of her sweet voice and because she loved to sing; "Sweet Tooth," on account of her love for candy; "Quick Silver," because she was quick, bright and witty; "Great Buffalo," a girl who was very strong; Elizabeth Schmidt, "Laughing Water," so named because she laughed and giggled at everybody and everything; "Babbling Brook," because it seemed an utter impossibility for her to stop talking; "Burr," because she sticks to ideas and friends; "Faith," quiet and reserved; "Comet," comes suddenly and brings a lot of light; "Black Hawk," always eager at first, but inclined to let her eagerness wear off: "Pocahontas," because she never can hurry; "Ginger Foot," a fiery temper, "Gypsy," so named on account of her black hair; "Bright Eyes," for her bright, blue eyes; "Rainbow," for her many ways, and because she is pretty.

As "Sunflower" took the matches and knelt by the pile of wood and lighted the fire, she recited the Ode to the Fire:

"Oh, Fire! Long years ago, when our fathers fought with great beasts, you were their protector. From the cruel cold of winter you saved. When they needed food, you changed the flesh of beasts into savory meat for them. Through all ages your mysterious flame has been a symbol of the Great Spirit to them. Tonight we light this fire in remembrance of the Great Spirit Who gave you to us."

Then the girls sang the chant or chanted:

Wohelo for aye,
Wohelo for aye,
Wohelo for aye,
Wohelo for work,
Wohelo for health,
Wohelo,
Wohelo for love.

Then they recited the Wood-gatherer's Desire:

"It is my desire to be a Campfire Girl and keep the law of the Camp Fire, which is 'To Seek Beauty, Give Service, Pursue Knowledge, Be Trustworthy, Hold onto Health, Glorify Work, Be Happy,'"

None had yet attained the highest rank, that of Torch Bearer, won by still greater achievement, the Camp having been organized so recently. Their motto was "The light which has been given to me, I desire to pass undimmed to others."

"Gypsy," the secretary, then read the "Count" for the last meeting and called the roll, and the girls handed in the list of honors they had won in the last month. Some amused themselves playing games, while others gathered more wood.

At five o'clock the corn and white and sweet potatoes were in the fire roasting. A jolly circle of girls around the fire were busily engaged toasting "Weiners" for the feast, which was finally pronounced ready to be partaken of. The hungry girls "fell to" and everything eatable disappeared as if by magic; and last, but not least, was the toasting of marshmallows, speared on the points of long, two-pronged sticks (broken from near-by trees), which were held over the fire until the marshmallows turned a delicate color. When everything had been eaten, with the exception of several cardboard boxes, corn cobs and husks, the girls quickly cleared up. Then, seated around the fire, told what they knew of Indian legends and folklore.

Noticing the sun slowly sinking in the West, they quickly gathered together their belongings and started homeward singing, "My Country, 'tis of Thee, Sweet Land of Liberty."

Thus broke up the second Council Fire, and in the heart of each girl was the thought of how much the Campfire was helping them to love God and His works.

CHAPTER XIV.

MARY MAKES "VIOLET" AND "ROSE LEAF" BEADS.

"Aunt Sarah," exclaimed Mary one day, "you promised to tell me exactly how you made those 'Rose Beads' you have."

"Yes, my dear, and you must make the beads before the June roses are gone. The process is very simple. If you would have them very sweet, get the petals of the most fragrant roses. I used petals of the old-fashioned, pink 'hundred leaf' and 'blush roses.' Gather a quantity, for you will need them all. Grind them to a pulp in the food chopper, repeat several times and place the pulp and juice into an *iron* kettle or pan. This turns the pulp black, which nothing but an iron kettle will do; cook, and when the consistency of dough it is ready to mold into beads. Take a bit of the dough, again as large as the size you wish your beads to be when finished, as they shrink in size when dried, and make them of uniform size, or larger ones for the centre of the necklace, as you prefer. Roll in the palms of your hands, until perfectly round, stick a pin through each bead, then stick the pins into a bake board. Be careful the bead does not touch the board, as that would spoil its shape. Allow the beads to remain until perfectly dry. If they are to have a dull finish, leave as they are. If you wish to polish them, take a tiny piece of vaseline on the palm of the hand and rub them between the palms until the vaseline is absorbed. Then string them on a linen thread. Keep in a closed box to preserve their fragrance. Those I showed you, Mary, I made many years ago, and the scent of the roses clings 'round them still.'"

"Did you know, Mary, that beads may be made from the petals of the common wild blue violet in exactly the same manner as they are made from rose leaves?"

"No, indeed, but I don't think the making of beads from the petals of roses and violets as wonderful as the beads which you raise in the garden. Those shiny, pearl-like seeds or beads of silvery-gray, called 'Job's Tears,' which grow on a stalk resembling growing corn; and to think Professor Schmidt raised those which Elizabeth strung on linen thread, alternately with beads, for a portiere in their sitting-room."

"Yes, my dear, the beads must be pierced before they become hard; later they should be polished. Did you ever see them grow, Mary? The beads or 'tears' grow on a stalk about fifteen inches high and from the bead or 'tear' grows a tiny, green spear resembling oats. They are odd and with very little care may he grown in a small garden."

"They certainly are a curiosity," said Mary.

CHAPTER XV.

MARY AND ELIZABETH VISIT SADIE SINGMASTER.

Farmer Landis, happening to mention at the breakfast table his intention of driving over to the "Ax Handle Factory" to obtain wood ashes to use as a fertilizer, his wife remarked, "Why not take Mary with you, John? She can stop at Singmaster's with a basket of carpet rags for Sadie. I've been wanting to send them over for some time." Turning to Mary, she said: "Poor little, crippled Sadie! On account of a fall, which injured her spine, when a small child, she has been unable to walk for years. She cuts and sews carpet rags, given her by friends and neighbors, and from their sale to a carpet weaver in a near-by town, helps her widowed mother eke out her small income."

"I'd love to go see her," said Mary. Elizabeth Schmidt also expressed her willingness to go, when asked, saying: "I am positive mother will add her contribution to the carpet rags for Sadie, I do pity her so very much."

"Yes," said Mary's Aunt, "she is poor and proud. She will not accept charity, so we persuade her to take carpet rags, as we have more than we can possibly use."

On reaching the Singmaster cottage, the girls alighted with their well-filled baskets, Mary's Uncle driving on to the "Ax Handle Factory," promising to call for the girls on his return. The sad, brown eyes of Sadie, too large for her pinched, sallow face, shone with pleasure at sight of the two young girls so near her own age, and she smiled her delight on examining the numerous bright-colored patches brought by them. Thinking the pleasure she so plainly showed might appear childish to the two girls, she explained: "I do get so dreadfully tired sewing together so many dull homely rags. I shall enjoy making balls of these pretty, bright colors."

"Sadie," Mary inquired, "will you think me inquisitive should I ask what the carpet weaver pays you for the rags when you have sewed and wound them into balls?"

"Certainly not," replied Sadie. "Four cents a pound is what he pays me. It takes two of these balls to make a pound," and she held up a ball she had just finished winding.

"Is *that all* you get?" exclaimed Elizabeth.

"Have you ever made rag rugs?" inquired Mary.

"No, I have never even seen one. Are they anything like braided mats?"

"Yes, they are somewhat similar to them, but I crochet mine and think them prettier. I have made several, with Aunt Sarah's assistance. I'll come over and teach you to make them one of these days, should you care to learn, and I'm positive you will find ready sale for them. In fact, I've several friends in the city who have admired the ones I have, and would like to buy rugs for the Colonial rooms they are furnishing. Sadie, can you crochet?"

"Oh, yes. I can do the plain stitch very well."

"That is all that will be necessary. You will become very much interested in inventing new designs, it is very fascinating work, and it will be more remunerative than sewing carpet rags. Aunt Sarah will send you more carpet rags if you require them, and should you wish dull colors of blue or pink, a small package of dye will transform white or light-colored rags into any desired shade, to match the furnishings of different rooms. I think the crocheted rugs much prettier than the braided ones, which are so popular in the 'Nutting' pictures, and the same pretty shades may be used when rugs are crocheted."

When Farmer Landis came for the girls, he found them too busily engaged talking to hear his knock at the door. During the drive home Mary could think and talk of nothing but Sadie Singmaster, and the rugs she had promised to teach her to make at an early day. Elizabeth, scarcely less enthusiastic, said: "I've a lot of old things I'll give her to cut up for carpet rags."

Reaching home, Mary could scarcely wait an opportunity to tell Aunt Sarah all her plans for Sadie's betterment. When she finally did tell her Aunt, she smiled and said: "Mary, I'm not surprised. You are always planning to do a kind act for some one. You remind me of the lines, 'If I Can Live,' by Helen Hunt Jackson." And she repeated the following for Mary:

IF I CAN LIVE.

If I can live
To make some pale face brighter and to give
A second luster to some tear-dimmed eye,

Or e'en impart
One throb of comfort to an aching heart,
Or cheer some wayworn soul in passing by;
If I can lend
A strong hand to the fallen, or defend
The right against a single envious strain,
My life, though bare,
Perhaps, of much that seemeth dear and fair
To us of earth, will not have been in vain.
The purest joy,
Most near to heaven, far from earth's alloy,
Is bidding cloud give way to sun and shine;
And 'twill be well
If on that day of days the angels tell
Of me, she did her best for one of Thine.

CHAPTER XVI.

OLD PARLOR MADE BEAUTIFUL (MODERNIZED).

When John Landis came into possession of "Clear Spring" Farm, where his mother had lived during her lifetime, she having inherited it from her father, the rooms of the old farm house were filled with quaint, old-fashioned furniture of every description. "Aunt Sarah," on coming to the farm to live, had given a personal touch and cheery, homelike look to every room in the house, with one exception, the large, gloomy, old-fashioned parlor, which was cold, cheerless and damp. She confessed to Mary she always felt as if John's dead-and-gone ancestors' ghostly presences inhabited the silent room. The windows were seldom opened to allow a ray of sunlight to penetrate the dusk with which the room was always enveloped, except when the regular weekly sweeping day arrived; when, after being carefully swept and dusted, it was promptly closed. A room every one avoided, Aunt Sarah was very particular about always having fresh air and sunlight in every other part of the house but his one room. The old fireplace had been boarded up many years before Aunt Sarah's advent to the farm, so it could not be used. One day Mary noticed, while dusting the room (after it had been given a thorough sweeping by Sibylla, Aunt Sarah's one maid servant), that the small, many-paned windows facing the East, at one end of the parlor, when opened, let in a flood of sunshine; and in the evening those at the opposite end of the long room gave one a lovely view of the setting sun—a finer picture than any painted by the hand of a master. Mary easily persuaded her Aunt to make some changes in the unlivable room. She suggested that they consult her Uncle about repapering and painting the room and surprise him with the result when finished.

Aunt Sarah, who never did things by halves, said: "Mary, I have long intended 'doing over' this room, but thought it such a great undertaking. Now, with your assistance, I shall make a sweep of these old, antiquated heirlooms of a past generation. This green carpet, with its gorgeous bouquets of roses, we shall have combined with one of brown and tan in the attic. Your Uncle shall take them with him when he drives to town and have them woven into pretty, serviceable rugs for the floor."

"And, oh! Aunt Sarah," cried Mary, "do let's have an open fireplace. It makes a room so cheery and 'comfy' when the weather gets colder, on long winter evenings, to have a fire in the grate. I saw some lovely, old brass andirons and fender in the attic, and some brass candlesticks there also, which will do nicely for the mantel shelf over the fireplace. I'll shine 'em up, and instead of this hideously-ugly old wall paper with gay-colored scrawley figures, Aunt Sarah, suppose we get an inexpensive, plain, tan felt paper for drop ceiling and separate it from the paper on the side wall, which should be a warm, yellow-brown, with a narrow chestnut wood molding. Then this dull, dark, gray-blue painted woodwork; could any one imagine anything more hideously ugly? It gives me the 'blues' simply to look at it. Could we not have it painted to imitate chestnut wood? And don't you think we might paint the floor around the edges of the rug to imitate the woodwork? Just think of those centre panels of the door painted a contrasting shade of pale pink. The painter who did this work certainly was an artist. A friend of mine in the city, wishing to use rugs instead of carpets on her floors, and not caring to go to the expense of laying hardwood floors, gave the old floors a couple of coats of light lemon, or straw-colored paint, then stained and grained them a perfect imitation of chestnut, at small expense. The floors were greatly admired when finished, and having been allowed to dry thoroughly after being varnished, proved quite durable. I will write to my friend at once and ask her exactly how her floors were treated."

"Now, Mary, about this old-style furniture. The old grandfather clock standing in the corner, at the upper end of the room, I should like to have remain. It is one hundred and fifty years old and belonged to my folks, and, although old-fashioned, is highly valued by me."

"Of course," said Mary, "we'll certainly leave that in the room."

"Also," said Aunt Sarah, "allow the old cottage organ and large, old-fashioned bookcase belonging to your Uncle to remain. He has frequently spoken of moving his bookcase into the next room, when he was obliged to come in here for books, of which he has quite a valuable collection."

A-24 Seed Wreath
A-25 Wax Fruit
A-26 Old Parlor Mantel
A-27 Old Clock
A-28 Boquet of Hair Flowers

"Oh," said Mary, "no need of that. We will move Uncle John in here, near the bookcase, when we get our room fixed up. Aunt Sarah, we will leave that old-fashioned table, also, with one leaf up against the wall, and this quaint, little, rush-bottomed rocker, which I just dote on."

"Why, dear," exclaimed Aunt Sarah, "there are several chairs to match it in the attic, which you may have when you start housekeeping for your very own. And," laughingly, said her Aunt, "there is another old, oval, marble-topped table in the attic, containing a large glass case covering a basket of wax fruit, which you may have."

"No, Aunt Sarah," said Mary, "I don't believe I want the fruit, but I will accept your offer of the table. Well, Aunt Sarah, I know you won't have this old, black what-not standing in the corner of the room. I do believe it is made of spools, strung on wire, as supports for the shelves; then all painted black, imitation of ebony, I suppose. It must have been made in the Black Age, at the same time the old corner cupboard was painted, as Uncle John told me he scraped off three different layers of paint before doing it over, and one was black. It was originally made of cherry. It certainly looks fine now, with those new brass hinges and pretty, old-fashioned glass knobs."

"Yes, Mary," replied her Aunt, "and there is an old corner cupboard in the attic which belonged to my father, that you may have, and, with a very little labor and expense, Ralph can make it look as well as mine. It has only one door and mine possesses two."

"Aunt Sarah," exclaimed Mary, "you are a dear! How will I ever repay you for all your kindness to me?"

"By passing it on to some one else when you find some one needing help," said Aunt Sarah.

"Such a collection of odd things, Aunt Sarah, as are on this what-not I never saw. Old ambrotypes and daguerreotypes of gone and forgotten members of the 'freinshoft,' as you sometimes say. I don't believe you know any of them."

"Yes, the red plush frame on the mantel shelf contains a picture of John's Uncle, a fine-looking man, but he possessed 'Wanderlust' and has lived in California for many years.

"Oh, you mean the picture on the mantel standing near those twin gilded china vases, gay with red and blue paint?"

"Yes; and that small china and gilt stand with little bowl and pitcher was given me when a small child."

"Suppose I bring a basket and we will fill it with articles from the mantel and what-not," said Mary, "and carry them all to the attic, until you have a rummage sale some day. We'll burn these 'everlasting' and 'straw' flowers, and pampas grass, and this large apple stuck full of cloves. Here is a small china dog and a little china basket with a plaited china handle decorated with gilt, and tiny, pink-tinted china roses. And these large, glass marbles containing little silver eagles inside; also this small, spun-glass ship and blue-and-pink-striped glass pipe. Aunt Sarah, some of your ancestors must have attended a glass blowers' exhibition in years past."

"This branch of white coral, these large snail shells (when a child I remember holding them to my ear to hear a noise resembling the roar of the ocean), and this small basket, fashioned of twigs and tendrils of grape vine, then dipped in red sealing wax, certainly is a good imitation of coral, and this plate, containing a miniature ship composed of green postage stamps, we will place in your corner cupboard."

"And, Aunt Sarah, I suppose this deep, glass-covered picture frame containing a bouquet of hair flowers, most wonderfully and fearfully made, was considered a work of art in days past and gone, as was also the crescent in a frame on the opposite side of the room, composed of flowers made of various seeds of grain and garden vegetables. Those daisies, made of cucumber seeds with grains of red corn for centres, and those made of tiny grains of popcorn with a watermelon seed in centre, are cute. The latter look like breastpins with a circle of pearls around the edge. And this glass case on the table, containing a white cross, covered with wax tube roses, ivy leaves and fuchsias drooping from the arms of the cross, sparkling with diamond dust! The band of green chenille around its base matches the mat underneath, composed of green zephyr of different shades, knitted, then raveled to imitate moss, I suppose; and, no doubt, this marble-topped table has stood here for fifty years, in this same spot, for the express purpose of holding this beautiful (?) work of art."

"The hair flowers and the seed wreath were made by John's sister," replied Aunt Sarah.

"Aunt Sarah," exclaimed Mary, "I've an original idea. This oval, marble-topped table has such strong, solid legs of black walnut, suppose we remove the marble slab and have a large, circular top made of wood at the planing mill? Wait; I'll get my tape measure. About thirty-two inches in diameter will do. The new top we shall stain to match the walnut frame, and it could be easily fastened to the table with a couple of screws; and, after the marble top has been well scoured, we'll use it in the kitchen as a bake board on which to roll out pie crust."

Her Aunt as usual acquiesced to all Mary's suggestions.

"You're a dear, Aunt Sarah!" exclaimed Mary, as she gave her a hug, "and I'll embroider big, yellow daisies with brown centres of French knots on gray linen for a new table cover. Won't they look just sweet?"

"Yes, Mary, and I'll buy a large, new lamp with a pretty shade, as I feel sure your Uncle will like to sit here evenings to read his papers and farm journals."

"And don't forget the Shriners' little magazine, *The Crescent*, which amuses him so greatly. Aunt Sarah, I do wish those stiff, starchy-looking, blue-white Nottingham lace curtains at the windows had grown yellow with age. They would be ever so much prettier and softer looking, and they are such a pretty, neat design, too."

"Oh!" replied her Aunt, "that may be easily remedied. I'll just dip them into a little weak liquid coffee and that will give them a creamy tint, and take out the stiffness."

"Now," said Mary, "what shall we do with these stiff, ugly, hair-cloth-covered chairs and sofa?"

"Why," replied Aunt Sarah, "we shall buy cretonne or art cloth, in pretty shades of brown and tan or green, to harmonize with the wall paper, and make slip covers for them all. We could never think of dispensing with the sofa. It is a very important article of furniture in German households. The hostess usually gives the person of greatest distinction among her guests the place of honor beside her on the sofa."

"These chairs have such strong, well-made, mahogany frames it would be a pity not to use them. Now," continued Mary, "about the pictures on the wall. Can't we consign them all to the attic? We might use some of the frames. I'll contribute unframed copies of 'The Angelus' and 'The Gleaners,' by Millet; and I think they would fit into these plain mahogany frames which contain the very old-fashioned set of pictures named respectively 'The Lovers,' 'The Declaration,' 'The Lovers' Quarrel' and 'The Marriage.' They constitute a regular art gallery. I'll use a couple of the frames for some small Colonial and apple blossom pictures I have, that I just love, by Wallace Nutting. Mine are all unframed; 'Maiden Reveries,' 'A Canopied Roof' and a 'Ton of Bloom,' I think are sweet. Those branches of apple trees, covered with a mass of natural-looking pink blossoms, are exquisite."

"Yes," remarked Aunt Sarah, "they look exactly like our old Baldwin, Winesap and Cider apple trees in the old, south meadow in the Spring. And, Mary, we'll discard those two chromos, popular a half century ago, of two beautiful cherubs called respectively, 'Wide Awake' and 'Fast Asleep,' given as premiums to a popular magazine. I don't remember if the magazine was 'Godey's,' 'Peterson's' or 'Home Queen'; they have good, plain, mahogany frames which we can use."

"And, Aunt Sarah," said Mary, "we can cut out the partition in this large, black-walnut frame, containing lithograph pictures of General George Washington, 'the Father of his Country' (we are informed in small letters at the bottom of the picture), and of General Andrew Jackson, 'the hero of New Orleans.' Both men are pictured on horseback, on gayly-caparisoned, prancing white steeds, with scarlet saddle cloth, edged with gold bullion fringe. The Generals are pictured clad in blue velvet coats with white facings of cloth or satin vest and tight-fitting knee breeches, also white and long boots reaching to the knee. Gold epaulettes are on their shoulders, and both are in the act of lifting their old-fashioned Continental hats, the advancing army showing faintly in the background. How gorgeously they are arrayed! We will use this frame for the excellent, large copy you have of 'The Doctor' and the pictured faces of the German composers — Beethoven, Wagner, Mendelssohn, Haydn, Schubert and Mozart, which I have on a card with a shaded brown background, will exactly fit into this plain frame of narrow molding, from which I have just removed the old cardboard motto, 'No place like home,' done with green-shaded zephyr in cross-stitch."

A-29 An Old Sampler
A-30 Old Woven Basket
A-31 Wax Cross
A-32 Old Spinning Wheel

"Now, Mary, with the couple of comfortable rockers which I intend purchasing, I think we have about finished planning our room."

"If you are willing, Aunt Sarah, I should like to make some pretty green and brown cretonne slips to cover those square sofa pillows in place of the ones made of small pieces of puffed silk and the one of colored pieces of cashmere in log cabin design, I do admire big, fat, plain, comfortable pillows, for use instead of show. And we must have a waste paper basket near the table beside Uncle John's chair. I shall contribute green satin ribbon for an immense bow on the side of the basket. Oh! Aunt Sarah! You've forgotten all about this odd, woven basket, beside the what-not, filled with sea shells. I don't care for the shells, but the basket would make a lovely sewing basket."

"You may have the basket, Mary, if you like it. It came from Panama, or perhaps it was bought at Aspinwall by John's Uncle, many years ago, when he came home on a visit from California, by way of the Isthmus, to visit old friends and relatives. John's Mother always kept it standing on the floor in one corner of the room beside the what-not."

"Aunt Sarah, why was straw ever put under this carpet?"

"The straw was put there, my dear, to save the carpet, should the boards on the floor be uneven. My Mother was always particular about having *cut rye straw*, because it was softer and finer than any other. It was always used in those days instead of the carpet linings we now have. I remember sometimes, when the carpet had been newly laid, in our home, immediately after house cleaning time, the surface of the floor looked very odd; full of bumps and raised places in spots, until frequent walking over it flattened down the straw. This room happens to have a particularly good, even floor, as this part of the house was built many years later than the original, old farm house, else it would not do to have it painted."

"Aunt Sarah, may I have the old spinning wheel in the attic? I'd love to furnish an old Colonial bedroom when I have a home of my very own. I'll use the rag carpet you made me for the floor, the old-fashioned, high-post bed Uncle John said I might have, and the 'New Colonial' rugs you taught me to make.

"Yes, my dear, and there is another old grandfather's clock in the attic which you may have; and a high-boy also, for which I have no particular use."

"Aunt Sarah, we shall not put away this really beautiful old sampler worked in silk by Uncle John's grandmother when a girl of nine years. It is beautifully done, and is wonderful, I think. And what is this small frame containing a yellowed piece of paper cut in intricate designs, presumably with scissors?"

"Look on the back of the picture and see what is written there, my dear," said her Aunt.

Mary slowly read: "'This is the only picture I owned before my marriage. I earned the money to buy it by gathering wheat heads.'"

"It belonged to my grandmother," said Aunt Sarah. "In old times, after the reapers had left the field, the children were allowed to gather up the wheat remaining, and, I suppose, grandmother bought this picture with the money she earned herself, and considered it quite a work of art in her day. It is over one hundred years old."

CHAPTER XVII.

AN OLD SONG EVENING.

Aunt Sarah and Mary spent few idle moments while carrying out their plans for "doing over" the old parlor. Finally, 'twas finished. Mary breathed a sigh of satisfaction as the last picture was hung on the wall. She turned to her Aunt, saying, "Don't you think the room looks bright, cheery and livable?"

"Yes," replied her Aunt, "and what is more essential, homey, I have read somewhere, 'A woman's house should be as personal a matter as a spider's web or a snail's shell; and all the thought, toil and love she puts into it should be preserved a part of its comeliness and homelikeness forever, and be her monument to the generations.'"

"Well, Aunt Sarah," replied Mary, "I guess we've earned our monument. The air that blows over the fields, wafted in from the open window, is sweet with the scent of grain and clover, and certainly is refreshing. I'm dreadfully tired, but so delighted with the result of our labors. Now we will go and 'make ready,' as Sibylla says, before the arrival of Ralph from the city. I do hope the ice cream will be frozen hard. The Sunshine Sponge Cake, which I baked from a recipe the Professor's wife gave me, is light as a feather. 'Tis Ralph's favorite cake. Let's see; besides Ralph there are coming all the Schmidts, Lucy Robbins, the school teacher, and Sibylla entertains her Jake in the kitchen. I promised to treat him to ice cream; Sibylla was so good about helping me crack the ice to use for freezing the cream. We shall have an 'Old Song Evening' that will amuse every one."

Quite early, as is the custom in the country, the guests for the evening arrived; and both Mary and Aunt Sarah felt fully repaid for their hard work of the past weeks by the pleasure John Landis evinced at the changed appearance of the room.

The Professor's wife said, "It scarcely seems possible to have changed the old room so completely."

Aunt Sarah replied, "Paint and paper do wonders when combined with good taste, furnished by Mary."

During the evening one might have been forgiven for thinking Professor Schmidt disloyal to the Mother Country (he having been born and educated in Heidelberg) had you overheard him speaking to Ralph on his favorite subject, the "Pennsylvania German." During a lull in the general conversation in the room Mary heard the Professor remark to Ralph: "The Pennsylvania Germans are a thrifty, honest and industrious class of people, many of whom have held high offices. The first Germans to come to America as colonists in Pennsylvania were, as a rule, well to do. Experts, when examining old documents of Colonial days, after counting thousands of signatures, found the New York 'Dutch' and the Pennsylvania 'Germans' were above the average in education in those days. Their dialect, the so-called 'Pennsylvania German' or 'Dutch,' as it is erroneously called by many, is a dialect which we find from the Tauber Grund to Frankfurt, A.M. As the German language preponderated among the early settlers, the language of different elements, becoming amalgamated, formed a class of people frequently called 'Pennsylvania Dutch'."

Professor Harbaugh, D.D., has written some beautiful poems in Pennsylvania German which an eminent authority, Professor Kluge, a member of the Freiburg University, Germany, has thought worthy to be included among the classics. They are almost identical with the poems written by Nadler in Heidelberger Mundart, or dialect.

Mary, who had been listening intently to the Professor, said, when he finished talking to Ralph: "Oh, please, do repeat one of Professor Harbaugh's poems for us."

He replied, "I think I can recall several stanzas of 'Das Alt Schulhaus an der Krick.' Another of Professor Harbaugh's poems, and I think one of the sweetest I have ever read, is 'Heemweeh.' Both poems are published in his book entitled 'Harbaugh's Harfe,' in Pennsylvania German dialect, and possess additional interest from the fact that the translations of these poems, in the latter part of the same book, were made by the author himself."

"Oh, do repeat all that you remember of both the poems," begged Mary.

The Professor consented, saying: "As neither you nor Mr. Jackson understand the Pennsylvania German dialect, I shall translate them for you, after repeating what I remember. 'Heemweeh' means Homesickness, but first I shall give you 'Das Alt Schulhaus an der Krick'."

[A]DAS ALT SCHULHAUS AN DER KRICK.

Heit is 's 'xactly zwansig Johr,
Dass ich bin owwe naus;
Nau bin ich widder lewig z'rick
Un schteh am Schulhaus an d'r Krick,
Juscht neekscht an's Dady's Haus.
Ich bin in hunnert Heiser g'west,
Vun Marbelstee' un Brick,
Un alles was sie hen, die Leit,
Dhet ich verschwappe eenig Zeit
For's Schulhaus an der Krick.
* * * * * * *
Der Weisseech schteht noch an der Dhier —
Macht Schatte iwwer's Dach:
Die Drauwerank is ah noch grie' —
Un's Amschel-Nescht — guk juscht mol hi' —
O was is dess en Sach!
* * * * * * *
Do bin ich gange in die Schul,
Wo ich noch war gans klee';
Dort war der Meeschter in seim Schtuhl,
Dort war sei' Wip, un dort sei' Ruhl, —
Ich kann's noch Alles sch'.
Die lange Desks rings an der Wand —
Die grose Schieler drum;
Uf eener Seit die grose Mad,
Un dort die Buwe net so bleed —
Guk, wie sie piepe rum!
* * * * * * *
Oh horcht, ihr Leit, wu nooch mir lebt,
Ich schreib eich noch des Schtick:
Ich warn eich, droll eich, gebt doch Acht,
Un memmt uf immer gut enacht,
Des Schulhaus an der Krick!

[A]

From "Harbaugh's Harfe." Published by the Publication and Sunday School Board of the Reformed Church, Philadelphia, Pa. Used by permission.

THE OLD SCHOOL-HOUSE AT THE CREEK.

Today it is just twenty years,
Since I began to roam;
Now, safely back, I stand once more,
Before the quaint old school-house door,
Close by my father's home.
I've been in many houses since,
Of marble built, and brick;
Though grander far, their aim they miss,
To lure heart's old love from this
Old school-house at the creek.
*   *   *   *   *   *   *

The white-oak stands before the door,
And shades the roof at noon;
The grape-vine, too, is fresh and green;
The robin's nest! — Ah, hark! — I ween
That is the same old tune!
*   *   *   *   *   *   *

'Twas here I first attended school,
When I was very small;
There was the Master on his stool,
There was his whip and there his rule —
I seem to see it all.
The long desks ranged along the walls,
With books and inkstands crowned;
Here on this side the large girls sat,
And there the tricky boys on that —
See! how they peep around!
*   *   *   *   *   *   *

Ye, who shall live when I am dead —
Write down my wishes quick —
Protect it, love it, let it stand,
A way-mark in this changing land —
That school-house at the creek.

HEEMWEH.

126

Ich wees net was die Ursach is—
Wees net, warum ich's dhu:
'N jedes Johr mach ich der Weg
Der alte Heemet zu;
Hab weiter nix zu suche dort—
Kee' Erbschaft un kee' Geld;
Un doch treibt mich des Heemgefiehl
So schtark wie alle Welt;
Nor'd schtart ich ewe ab un geh,
Wie owe schun gemeldt.
Wie nacher dass ich kumm zum Ziel,
Wie schtarker will ich geh,
For eppes in mei'm Herz werd letz
Un dhut m'r kreislich weh.
Der letschte Hiwel schpring ich nuf;
Un ep ich drowe bin,
Schtreck ich mich uf so hoch ich kann
Un guk mit Luschte hin;
Ich seh's alt Schtee'haus dorch die Beem,
Un wott ich war schunm drin.
* * * * * * *
Wie gleich ich selle Babble Beem,
Sie schtehn wie Brieder dar;
Un uf'm Gippel—g'wiss ich leb!
Hockt alleweil 'n Schtaar!
'S Gippel biegt sich—guk, wie's gaunscht—
'R hebt sich awer fescht;
Ich seh sei' rothe Fliegle plehn,
Wann er sei' Feddere wescht;
Will wette, dass sei' Fraale hot
Uf sellem Baam 'n Nescht!
* * * * * * *
Guk! werklich, ich bin schier am Haus!—
Wie schnell geht doch die Zeit!
Wann m'r so in Gedanke geht.
So wees m'r net wie weit.
Dort is d'r Schhap, die Walschkornkrip,
Die Seiderpress dort draus;
Dort is die Scheier, un dort die Schpring—

Frisch quellt des Wasser raus;
Un guk! die sehm alt Klapbord-Fens,
Un's Dheerle vor'm Haus.
* * * * * * *
Zwee Blatz sin do uf dare Bortsch,
Die halt ich hoch in Acht,
Bis meines Lebens Sonn versinkt
In schtiller Dodtes-Nacht!
Wo ich vum alte Vaterhaus
'S erscht mol bin gange fort.
Schtand mei' Mammi weinend da,
An sellem Rigel dort:
Un nix is mir so heilig nau
Als grade seller Ort.
* * * * * * *
Was macht's dass ich so dort hi' guk,
An sell End vun der Bank!
Weescht du's? Mei' Herz is noch net dodt,
Ich wees es, Got sei Dank!
Wie manchmal sass mai Dady dort,
Am Summer-Nochmiddag,
Die Hande uf der Schoos gekreizt,
Sei Schtock bei Seite lag.
Was hot er dort im Schtille g'denkt?
Wer mecht es wisse — sag?

HOME-SICK NESS.

I know not what the reason is:
Where'er I dwell or roam,
I make a pilgrimage each year,
To my old childhood home.
Have nothing there to give or get —
No legacy, no gold —
Yet by some home-attracting power
I'm evermore controlled;
This is the way the homesick do,
I often have been told.

* * * * * * *

As nearer to the spot I come
More sweetly am I drawn;
And something in my heart begins
To urge me faster on.
Ere quite I've reached the last hilltop—
You'll smile at me, I ween!—
I stretch myself high as I can,
To catch the view serene—
The dear old stone house through the trees
With shutters painted green!

* * * * * * *

How do I love those poplar trees;
What tall and stalely things!
See! on the top of one just now
A starling sits and sings.
He'll fall!—the twig bends with his weight!
He likes that danger best.
I see the red upon his wings,—
Dark shining is the rest.
I ween his little wife has built
On that same tree her nest.

* * * * * * *

See! really I am near the house;
How short the distance seems!
There is no sense of time when one
Goes musing in his dreams.
There is the shop—the corn-crib, too—
The cider-press—just see!
The barn—the spring with drinking cup
Hung up against the tree.
The yard-fence—and the little gate
Just where it used to be.

* * * * * * *

Two spots on this old friendly porch
I love, nor can forget,
Till dimly in the night of death
My life's last sun shall set!
When first I left my father's house,

One summer morning bright,
My mother at that railing wept
Till I was out of sight!
Now like a holy star that spot
Shines in this world's dull night.
*   *   *   *   *   *   *

What draws my eye to yonder spot —
That bench against the wall?
What holy mem'ries cluster there,
My heart still knows them all!
How often sat my father there
On summer afternoon;
Hands meekly crossed upon his lap,
He looked so lost and lone,
As if he saw an empty world,
And hoped to leave it soon.

At the conclusion of his recital, Mary heartily thanked the Professor, and, at his request, obediently seated herself at the old, but still sweet-toned cottage organ, and expressed her willingness to play any old-time songs or hymns requested, and saying, "I know Aunt Sarah's favorite," commenced playing, "My Latest Sun is Sinking Fast," followed by "This Old-Time Religion," "Jesus, Lover of My Soul," "One of the Sweet Old Chapters," "Silver Threads Among the Gold" and the sweet old hymn, "In the Summer Land of Song," by Fanny Crosby.

At John Landis' request, she played and sang "Auld Lang Syne." "When You and I Were Young, Maggie," "Old Folks at Home" and "Old Black Joe."

Lucy Robbins, when asked for her favorites, replied; "In the Gloaming," "The Old, Old Home'" "The Lost Chord" and "Better Bide a Wee."

The Professor then asked his daughter Elizabeth to give them the music of a song from German Volkslied, or Folk Song, with the words of which all except Mary and Ralph were familiar. Professor

Schmidt sang in his high, cracked voice to Elizabeth's accompaniment the words of the German song, beginning:

Du, Du liegest mir in Herzen
Du, Du liegst mir in Sinn
Du, Du machst mir viel Schmerzen
Weist nicht wie gut ich Dir binn
Ja, ja, ja, ja, Du weist nicht wie gut ich Dir bin.

The young folks all joined in the chorus. Fritz Schmidt asked Elizabeth to play "Polly Wolly Doodle" for little Pollykins, which Frit sang with gusto. Fritz then sang the rollicking German song, "Lauderbach," to an accompaniment played by Mary, and followed by singing "Johnny Schmoker," with appropriate gestures in the chorus commencing "My Pilly, Willy Wink, das is mein fifa," etc., ending with "My fal, lal, lal, my whach, whach, das ist mein doodle soch," which he emphasised by shrugging his shoulders, to the no small enjoyment of the young folks, who thought the silly, old German song no end of fun. This was followed by a favorite college song, "Mandalay," by Fritz.

Then Elizabeth Schmidt played and sang a pretty little German song called "Meuhlen Rad," meaning The Mill Wheel, taught her by her mother.

MEUHLEN RAD.

In einen kuhlen grunde
Da steht ein meuhlen rad;
Mein libste ist versch wunden,
Die dort gewhoned hat;
Sie sat mir treu versprochen,
Gab ihr ein ring dabei;
Sie hat die treu gebrochen,
Das ringlein sprang entzwei.

She translated it for the benefit of Ralph and Mary: "In a cool, pleasant spot, stands a mill. My loved one, who lived there, has disappeared. She promised to be true to me, and I gave her a ring. She broke her promise and the ring broke in two."

Fritz then caught his little sister Pauline around the waist and waltzed her to one end of the long room, saying: "Mary, play the piece, 'Put On Your Old Gray Bonnet,' and Pollykins and I will do the cakewalk for you."

Polly, who had become quite a proficient little dancer under her sister's teaching, was very willing to do her share in the evening's entertainment, and it was pronounced a decided success.

Mary then said, "I'll play my favorite schottische, composed by our old friend, the Professor. I have not yet procured a copy of his latest piece of music, 'The Passing of the Dahlias.' I think it is still with the publishers."

Mary, after playing "Rock of Ages," left the room to see about serving refreshments, when Elizabeth Schmidt took her place at the instrument. After playing "The Rosary," she turned to Ralph, who had been greatly amused by the German songs on the program, all of which were quite new to him, and said: "What shall I play for you?"

He replied, "'My Little Irish Rose'—no, I mean 'The River Shannon.'"

"Don't you mean 'That Grand Old Name Called Mary?'" mischievously inquired Fritz Schmidt, who could not refrain from teasing Ralph, which caused a laugh at his expense, as all present were aware of his love for Mary. Elizabeth, to cover Ralph's confusion, quickly replied: "I'll play my favorite, 'The End of a Perfect Day.'"

The party was pronounced a success, and broke up at a late hour for country folks. Before leaving, Mary's Uncle said: "Now, let's sing 'Home, Sweet Home,' and then all join in singing that grand old hymn, 'My Country, 'Tis of Thee,' to the new tune by our friend, the Bucks County Editor."

PALASADES OR NARROWS OF NOCKAMIXON

134

A VISIT TO THE "PENNSYLVANIA PALISADES," AS THE "NARROWS" OF THE DELAWARE RIVER ARE CALLED.

All hailed with delight Aunt Sarah's proposal that the Schmidt and Landis families, on the Fourth of July, drive over to the Narrows, visit Aunt Sarah's old home at Nockamixon, and see the "Ringing Rocks" and "High Falls," situated a short distance from the rocks, near which place picnics were frequently held. John Landis readily agreed to the proposed plan, saying, "The meadow hay and clover are cut, and I'll not cut the wheat until the fifth day of July."

The third of July was a busy day at both farm houses, preparing savory food of every description with which to fill hampers for the next day's outing. Small Polly Schmidt was so perfectly happy, at the thought of a proposed picnic, she could scarcely contain herself, and as her sister Elizabeth said, "did nothing but get in every one's way." Little Polly, being easily offended, trudged over to the Landis farm to see Mary, with whom she knew she was a great favorite.

The morning of the Fourth dawned bright and clear. Quite early, while the earth was still enveloped in a silvery mist, and on the lattice work of filmy cobwebs, spun over weeds and grass, dewdrops, like tiny diamonds, sparkled and glistened, until dissolved by the sun's warm rays, the gay party left home, for the "Palisades" were quite a distance from the farm, to drive being the only way of reaching the place, unless one boarded the gasoline motorcar, called the "Cornfield Express" by farmers living in the vicinity of Schuggenhaus Township.

There is something indescribably exhilarating about starting for an early drive in the country before sunrise on a bright, clear morning in midsummer, when "the earth is awaking, the sky and the ocean, the river and forest, the mountain and plain." Who has not felt the sweet freshness of early morning before "the sunshine is all on the wing" or the birds awaken and begin to chatter and to sing? There is a hush over everything; later is heard the lowing of cattle, the twitter of birds and hum of insect life, proclaiming the birth of the new day. Passing an uncultivated field, overgrown with burdock, wild carrots, mullein, thistle and milk weed, Mary alighted and gathered some of the pods of the latter, inclosing imitation of softest down, which she used later for filling sofa pillows.

"Look at those pretty wild canaries!" exclaimed Aunt Sarah, "yellow as gold, swinging on the stem of a tall weed."

"Professor Schmidt, can you tell me the name of that weed?" questioned Mary. "I have always admired the plant, with its large leaves and long, drooping racemes of crimson seeds.

"That," replied the Professor, "is a foreign plant, a weed called Equisetum from 'Equi,' a horse, and 'Setum'—tail. The country folk hereabout call it 'Horsetail.' It belongs to the Crptogamous or flowerless plants. There are only four specimens of this plant in America. I, too, have always greatly admired the plant."

The Professor was quite a noted botanist. There were few flowers, plants or weeds of which he was ignorant of the name or medicinal value. Another bird lazily picked seeds from the thistle blossoms. "See," exclaimed Aunt Sarah, "one bird has a spear of grass in its mouth!"

"Yellow star grass," said the Professor, "with which to make a nest. They never mate until the last of June, or first part of July. The tiny, little robbers ate up nearly all my sunflower seeds in the garden last summer."

"Well," replied Mary, "you know, Professor, the birds must have food. They are the farmer's best friend. I hope you don't begrudge them a few sunflower seeds, I love birds. I particularly admire the 'Baltimore Oriole,' with their brilliant, orange-colored plumage; they usually make their appearance simultaneously with the blossoms in the orchard in the south meadow; or so Aunt Sarah tells me. I love to watch them lazily swinging on the high branches of tall trees. On the limb of a pear tree in the orchard one day, I saw firmly fastened, a long, pouch-like nest, woven with rare skill. Securely fastened to the nest by various colored pieces of twine and thread was one of smaller size, like a lean-to added to a house, as if the original nest had been found too small to accommodate the family of young birds when hatched. The oriole possesses a peculiar, sweet, high-whistled trill, similar to this — 'La-la-la-la,' which always ends with the rising inflection."

Fritz Schmidt, who had been listening intently to Mary, gravely remarked, "An oriole built a nest on a tall tree outside my bedroom window, and early every morning, before the family arise, I hear it sing over and over again what sounds exactly like 'Lais Die Beevil!' which translated means 'Read your Bible'."

"Even the birds are 'Dutch,' I believe, in Bucks County," said Fritz. "I think these must be German Mennonites, there being quite a settlement of these honest, God-fearing people living on farms at no great distance from our place."

THE CANAL AT THE NARROWS

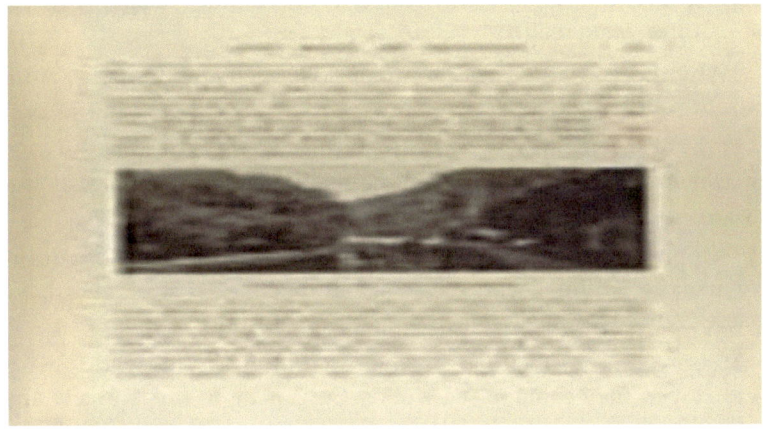

As they drove along the country road, parallel with the Delaware River, just before reaching the Narrows. Mary was greatly attracted by the large quantities of yellow-white "sweet clover," a weed-like plant found along the Delaware River, growing luxuriantly, with tall, waving stems two to four feet high. The clover-like flowers, in long, loose racemes, terminating the branches, were so fragrant that, like the yellow evening primrose, the scent was noticeable long before one perceived the flowers. And, strange to tell, sweet clover was never known to grow in this locality until the seed was washed up on the bank of the river some ten or twelve years previous to the date of my story, when the Delaware River was higher than it was ever before known to be.

"The first place we shall visit," said Aunt Sarah, "will be my grandmother's old home, or rather, the ruins of the old home. It passed out of our family many years ago; doors and windows are missing and walls ready to tumble down. You see that old locust tree against one side the ruined wall of the house?" and with difficulty she broke a branch from the tree saying, "Look, see the sharp, needle-shaped thorns growing on the branch! They were used by me when a child to pin my dolls' dresses together. In those days, pins were too costly to use; and look at that large, flat rock not far distant from the house! At the foot of that rock, when a child of ten,

I buried the 'Schild Krote Family' dolls, made from punk (when told I was too big a girl to play with dolls). I shed bitter tears, I remember. Alas! The sorrows of childhood are sometimes deeper than we of maturer years realize."

"Why did you give your family of dolls such an odd name, Aunt Sarah?" questioned Mary.

"I do not remember," replied her Aunt. "Schild Krote is the German name for turtle. I presume the name pleased my childish fancy."

"Suppose we visit my great-great-grandfather's grave in the nearby woods. I think I can locate it, although so many years have passed since I last visited it."

Passing through fields overgrown with high grass, wild flowers and clover, they came to the woods. Surprising to say, scarcely any underbrush was seen, but trees everywhere—stately Lebanon cedars, spruce and spreading hemlock, pin oaks, juniper trees which later would be covered with spicy, aromatic berries; also beech trees. Witch hazel and hazel nut bushes grew in profusion. John Landis cut a large branch from a sassafras tree to make a new spindle on which to wind flax, for Aunt Sarah's old spinning wheel (hers having been broken), remarking as he did so, "My mother always used a branch of sassafras wood, having five, prong-like branches for this purpose, when I was a boy, and she always placed a piece of sassafras root with her dried fruit."

The Professor's wife gathered an armful of yarrow, saying, "This is an excellent tonic and should always be gathered before the flowers bloom. I wonder if there is any boneset growing anywhere around here."

Boneset, a white, flowering, bitter herb, dearly beloved and used by the Professor's wife as one of the commonest home remedies in case of sickness, and equally detested by both Fritz and Pauline.

THE NARROWS OR PENNSYLVANIA PALISA-DES

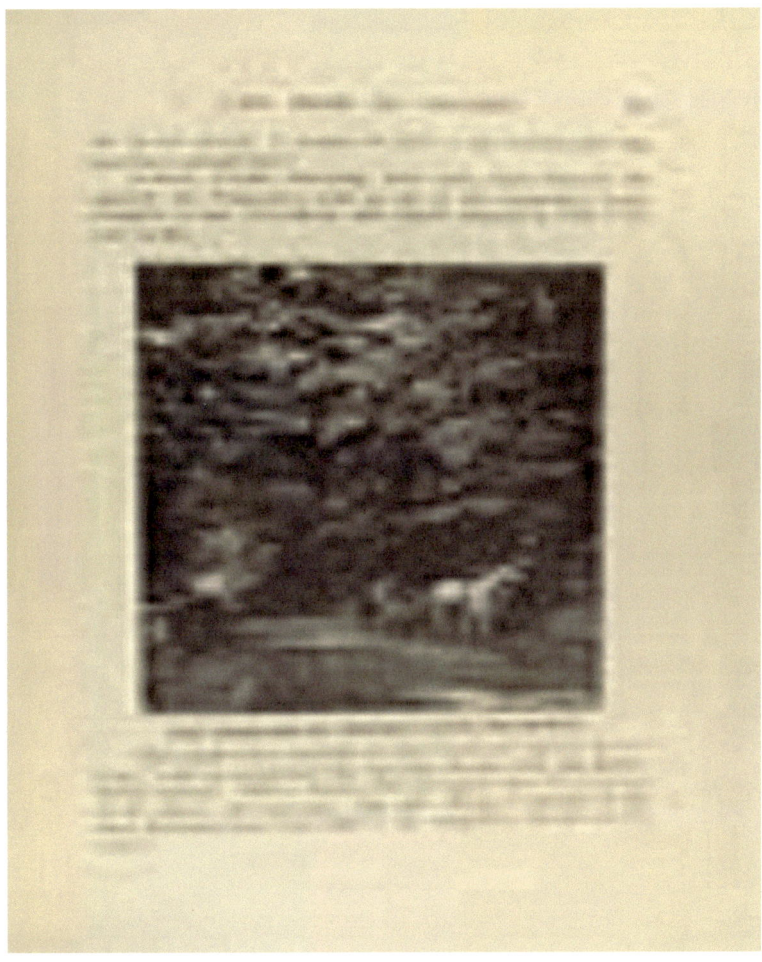

Mary gathered a bouquet of wild carrot, or "Queen Anne's Lace," with its exquisitely fine, lace-like flowers with pale green-tinted centres. Mary's Uncle could not agree with her in praise of the dainty wild blossoms. He said: "Mary, I consider it the most detested weed with which I am obliged to contend on the farm."

TOP ROCK

After quite a long, tiresome walk in the hot sun, they discovered the lonely grave, covered with a slab of granite surrounded by a small iron railing and read the almost illegible date—"Seventeen Hundred and Forty." Ralph said, "If he ever sighed for a home in some vast wilderness, his wish is granted." It certainly was a lonely grave in the deep woods, and gave all the members of the party a sad and eerie feeling as they wended their way out into the sunlight again, to the waiting carriages, and were soon driving swiftly along the Narrows, as they have been called from time immemorial by the inhabitants, although I prefer the name of Pennsylvania Palisades, as they are sometimes called.

Said Professor Schmidt: "Numerous tourists visit the Narrows every year. The Narrows are said to resemble somewhat the Palisades on the Hudson. I have seen, the latter and think these greatly resemble them and are quite as interesting and picturesque."

"The name Narrows is derived from the fact that at this place the Delaware River has forced itself through the rocky barrier," continued the Professor, "hedged in on one side by cliffs of perpendicular rock, three hundred feet high, extending some distance along the river, leaving scarcely room at some places for the river and the canal. Some quite rare plants grow here, said to be found in few other localities in the United States. You see the highest flat rock along the Narrows? It is called 'Top Rock' and rises to a height of more than three hundred feet. We shall drive around within a short distance of it; then, after passing a small house, we are obliged to walk across a field of ploughed ground; follow the well-beaten path between trees and undergrowth, and 'Top Rock' is before us. Stepping upon the high ledge of rock projecting out over the road beneath, we discover it may also be reached by following a precipitous path and clinging to bushes and trees, but none of the party venture. Recently the body of a man who had been searching for rare birds' eggs on the side of this self-same rock was found dead on the path below the rocks. What caused his fall is not known. No wonder Aunt Sarah says it makes her dizzy when you boys skip stones across the river while standing on the rock."

The beautiful view of the Delaware River and the scenery on the opposite side was something long to be remembered. While the party were going into raptures over the beautiful sight, Professor Schmidt turned to Mary and remarked: "In those rocks which rise in perpendicular bluffs, several hundred feet above the level of the river, are evidence that prehistoric man may have inhabited the caves in these same walls of rock along the Delaware. From implements and weapons found, it does not require any great effort of imagination to believe the 'Cave Man' dwelt here many centuries ago."

Fritz Schmidt was much interred in his father's conversation, and from that time on called Ralph Jackson Mary's "Cave Man."

Leaving Top Rock, the party wended their way back to the waiting carriages in the road, and drove to the "Ringing or Musical Rocks." They had been informed that their nearest approach to the rocks was to drive into the woods to reach them. Passing a small shanty at the roadside, where a sign informed the passerby that soft drinks were to be obtained, the party dismounted and found, to their surprise, a small pavilion had been erected with bench, table and numerous seats composed of boards laid across logs, where camp meetings had formerly been held. As the large trees furnished shade, and a spring of fresh water was near by, they decided to "strike" camp and have lunch before going farther into the woods.

Aunt Sarah and the Professor's wife spread a snowy cloth over the rough wooden table, quickly unpacked the hampers, and both were soon busily engaged preparing sandwiches of bread, thinly sliced, pink cold ham and ground peanuts, fried chicken and beef omelette; opening jars of home-made pickles, raspberry jam and orange marmalade.

"Oh!" said Pauline, "I'm so hungry for a piece of chocolate cake. Let me help shell the eggs, so we can soon have dinner."

"Here's your fresh spring water," called Fritz, as he joined the party, a tin pail in his hand, "We had such an early breakfast, I'm as hungry as a bear."

The party certainly did full justice to the good things provided with a lavish hand by Frau Schmidt and Aunt Sarah. All were in high spirits. The Professor quoted from the Rubaiyat of Omar Khayyam —

Here with a loaf of bread beneath the bough.
A flask of wine, a book of verse and thou,
Beside me singing in the wilderness,
And wilderness is Paradise enow.

Ralph cast a look at Marry, unnoticed by any one else, as much as to say, "The old tentmaker voiced my sentiments."

RINGING ROCKS OF BRIDGTON TOWNSHIP BUCKS COUNTY. PA.

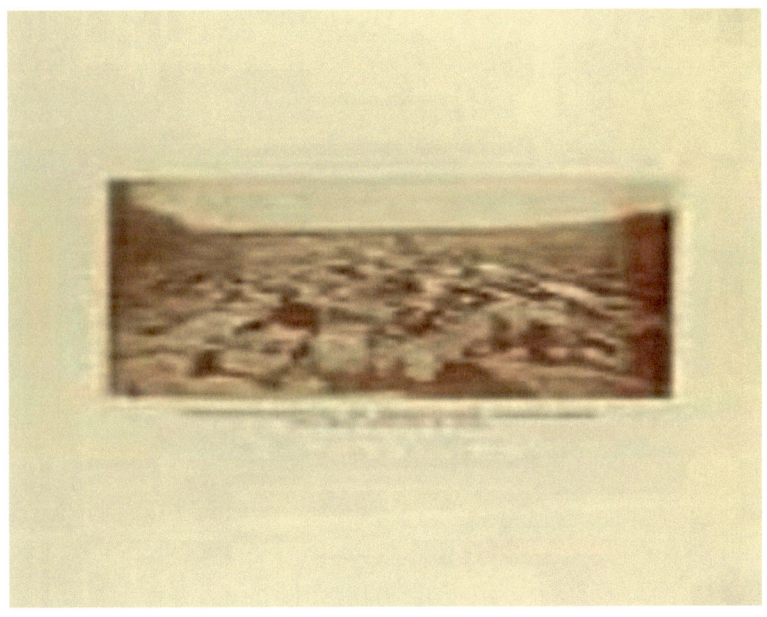

HIGH FALLS

After the hampers had been repacked and stowed away in the carriages, they with the horses were left in the shade while the party walked to "High Falls," at no great distance from the camp. "High Falls," a beautiful waterfall about thirty feet high and fifty feet wide, is situated several hundred feet east of the Ringing Rocks. The water, before dashing below, passes over a large, solid, level floor of rock. After gazing at the Falls and picturesque surroundings, they searched through the woods for the Ringing Rocks, a peculiar formation of rocks of irregular shape and size, branching out from a common centre in four directions. The rocks vary in size from a few pounds to several tons in weight. Arriving there, Aunt Sarah said: "Ralph, you will now find use for the hammer which I asked you to bring." Ralph struck different rocks with the hammer, and Fritz Schmidt struck rocks with other pieces of rock, and all gave a peculiar metallic sound, the tones of each being different. The rocks are piled upon each other to an unknown depth, not a particle of earth being found between them, and not a bush or spear of grass to be seen. They occupy a space of about four and a half acres and are a natural curiosity well worth seeing. The young folks scrambled over the rocks for a time, and, having made them ring to their hearts' content, were satisfied to return to camp and supper.

BIG ROCK AT ROCKY DALE

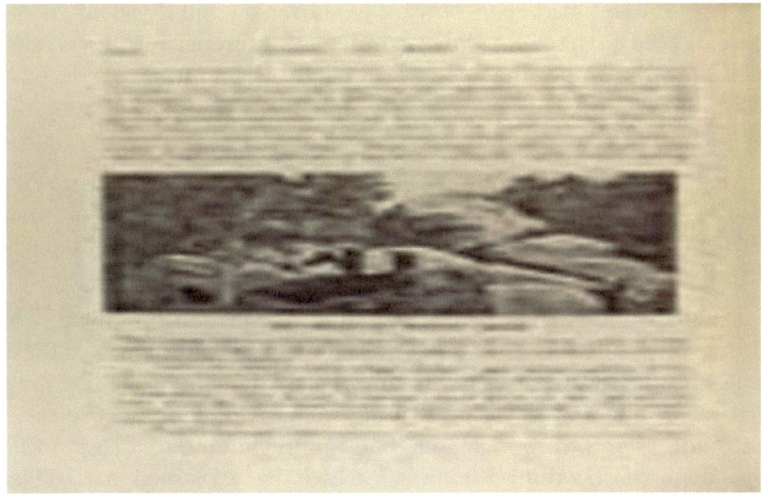

"Not far distant from High Falls," said John Landis, when all were comfortably seated near the table, with a sandwich in hand, "is a place called Roaring Rocks, also a freak of nature. I remember, when a boy, I always went there in the fall of the year, after the first hard frost, to pick persimmons. The water could he distinctly heard running underneath the rocks at a considerable depth."

Ralph Jackson remarked to Aunt Sarah: "I never imagined there were so many interesting, natural features right here in Bucks County."

"Oh, yes," exclaimed the impressible Fritz Schmidt, "we have a few things besides pigs and potatoes."

"Yes, Ralph," said the Professor, "there are still several places of interest you will like to see. 'Stony Garden' is another very interesting freak of nature. It is about two and a half miles from the small town of 'Snitzbachsville,' as Fritz calls the hamlet, and 'tis a wild spot. About an acre is covered with trap rock. The stones are of odd shapes and sizes and appear as if thrown into the forest in the wildest confusion. No earth or vegetation is found about them. 'Tis said the rocks are similar to those found at Fingal's Cave, Ireland, and also at the Palisades on the Hudson, and are not found anywhere else in this section of the country."

"And Ralph," said Fritz, "I want to show you 'Big Rock,' at Avondale, where a party of us boys camped one summer for two weeks. Oh! but I remember the good pies given us by a farmer's wife who sold us milk and eggs, and who lived just across the fields from our camp."

"I think," said John Landis, "it is time we began hitching up our horses and starting for home. We have a long drive before us, and, therefore, must make an early start. Sarah, get the rest of the party together and pack up your traps."

At that moment the Professor came in sight with an armful of ferns, the rich loam adhering to their roots, and said: "I'm sure these will grow." Later he planted them on a shady side of the old farm house at "Five Oaks," where they are growing today. Professor Schmidt, after a diligent search, had found clinging to a rock a fine specimen of "Seedum Rhodiola," which he explained had never been found growing in any locality in the United States except Maine. Little Pauline, with a handful of flowers and weeds, came trotting after Mary, who carried an armful of creeping evergreen called partridge berry, which bears numerous small, bright, scarlet berries later in the season. Ralph walked by her side with a basket filled to overflowing with quantities of small ferns and rock moss, with which to border the edge of the waiter on which Mary intended planting ferns; tree moss or lichens, hepaticas, wild violets, pipsissewa or false wintergreen, with dark green, waxy leaves veined with a lighter shade of green; and wild pink geraniums, the foliage of which is prettier than the pink blossoms seen later, and they grow readily when transplanted.

Aunt Sarah had taught Mary how to make a beautiful little home-made fernery. By planting these all on a large waiter, banking moss around the edges to keep them moist and by planting them early, they would be growing finely when taken by her to the city in the fall of the year—a pleasant reminder of her trip to the "Narrows" of the Delaware River. Frau Schmidt brought up the rear, carrying huge bunches of mint, pennyroyal and the useful herb called "Quaker Bonnet."

THE OLD TOWPATH AT THE NARROWS

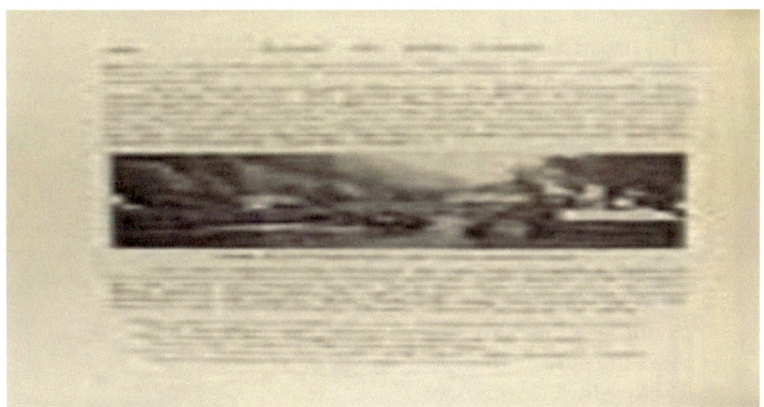

Driving home at the close of the day, the twinkling lights in farm house windows they swiftly passed, were hailed with delight by the tired but happy party, knowing that each one brought them nearer home than the one before. To enliven the drowsy members of the party, Fritz Schmidt sang the following to the tune of "My Old Kentucky Home," improvising as he sang:

> The moon shines bright on our "old Bucks County home,"
> The meadows with daisies are gay,
> The song of the whipporwill is borne on the breeze,
> With the scent of the new mown hay.
> Oh! the Narrows are great with their high granite peaks,
> And Ringing Rocks for ages the same;
> But when daylight fades and we're tired and cold,
> There's no place like "hame, clear alt hame."

The last lingering rays of the sun idealized the surrounding fields and woods with that wonderful afterglow seen only at the close of day. The saffron moon appeared to rise slowly from behind the distant tree-tops, and rolled on parallel with them, and then ahead, as if to guide them on their way, and the stars twinkled one by one from out the mantle of darkness which slowly enveloped the earth.

155

The trees they swiftly passed, when the moonbeams touched them, assumed gigantic, grotesque shapes in the darkness. Mary quoted from a favorite poem, "The Huskers," by Whittier:

'Till broad and red as when he rose, the sun
Sank down at last,
And, like a merry guest's farewell, the day
In brightness passed.
And lo! as through the western pines,
On meadow, stream and pond,
Flamed the red radiance of a sky,
Set all afire beyond.
Slowly o'er the eastern sea bluffs,
A milder glory shone,
And the sunset and the moon-rise
Were mingled into one!
As thus into the quiet night,
The twilight lapsed away,
And deeper in the brightening moon
The tranquil shadows lay.
From many a brown, old farm house
And hamlet without name,
Their milking and their home tasks done,
The merry huskers came.

"You mean 'The Merry Picknickers Came,'" said Fritz Schmidt, as Mary finished, "and here we are at home. Good night, all."

CHAPTER XIX.

MARY IS TAUGHT TO MAKE PASTRY, PATTIES AND "ROSEN KUCHEN."

Mary's Aunt taught her to make light, flaky pastry and pies of every description. In this part of Bucks County a young girl's education was considered incomplete without a knowledge of pie-making. Some of the commonest varieties of pies made at the farm were "Rivel Kuchen," a pie crust covered with a mixture of sugar, butter and flour crumbled together; "Snitz Pie," composed of either stewed dried apples or peaches, finely mashed through a colander, sweetened, spread over a crust and this covered with a lattice-work of narrow strips of pastry laid diamond-wise over the top of the pie; "Crumb" pies, very popular when served for breakfast, made with the addition of molasses or without it; Cheese pies, made of "Smier Kase;" Egg Custard, Pumpkin and Molasses pie.

Pies were made of all the different fruits and berries which grew on the farm. When fresh fruits were not obtainable, dried fruits and berries were used. Pie made from dried, sour cherries was an especial favorite of Farmer Landis, and raisin or "Rosina" pie, as it was usually called at the farm, also known as "Funeral" pie, was a standby at all seasons of the year, as it was invariably served at funerals, where, in old times, sumptuous feasts were provided for relatives and friends, a regular custom for years among the "Pennsylvania Germans," and I have heard Aunt Sarah say, "In old times, the wives of the grave-diggers were always expected to assist with the extra baking at the house where a funeral was to be held."

It would seem as if Bucks County German housewives did not like a dessert without a crust surrounding it.

The Pennsylvania German farmers' wives, with few exceptions, serve the greatest variety of pies at a meal of any class of people I know; not alone as a dessert at twelve o'clock dinner, but frequently serve several different varieties of pie at breakfast and at each meal during the day. No ill effects following the frequent eating of pie I attribute to their active life, the greater part of which, during the day, was usually spent in the open air, and some credit may he due the housewife for having acquired the knack of making *good* pie crust, which was neither very rich nor indigestible, if such a thing be possible.

The combination of fruit and pastry called pie is thought to be of American invention. Material for pies at a trifling cost were furnished the early settlers in Bucks County by the large supply of fruit and vegetables which their fertile farms produced, and these were utilized by the thrifty German housewives, noted for their wise management and economy.

The Professor's wife taught Mary to make superior pastry, so flaky and tender as to fairly melt in one's mouth; but Mary never could learn from her the knack of making a dainty, crimped edge to her pies with thumb and forefinger, although it looked so very simple when she watched "Frau Schmidt" deftly roll over a tiny edge as a finish to the pie.

Mary laughingly told the Professor's wife (when speaking of pies) of the brilliant remark she made about lard, on first coming to the farm. Her Aunt Sarah, when baking pies one day, said to her, "Look, Mary, see this can of snowy lard, rendered from pork, obtained from our fat pigs last winter!"

"Why, Aunt Sarah!" exclaimed Mary, "is lard made from pork fat? I always thought lard was made from milk and butter was made from cream."

The Professor's wife possessed, besides a liking for pies, the German's fondness for anything pertaining to fritters. She used a set of "wafer and cup irons" for making "Rosen Kuchen," as she called the flat, saucer-like wafer; and the cup used for serving creamed vegetables, salads, etc., was similar to pattie cases.

"The 'Wafer and Cup Irons,'" said Frau Schmidt, "were invented by a friend of mine, also a teacher and an excellent cook, besides; she gave me several of her original recipes, all to be served on wafers or in patties. You shall have a set of the irons when you start housekeeping. Mary. You will be surprised at the many uses you will find for them. They are somewhat similar to Rosette Irons, but I think them an improvement. They are pieces of fluted steel fastened to a long handle and one is cup-shaped. This latter is particularly fine for making patties. Then the cup may be filled and served on saucer-like wafers, which I call 'Rosen Kuchen,' or the 'Rosen Kuchen' may be simply dusted with a mixture consisting of one cup of sugar, one teaspoonful of cinnamon and a quarter teaspoonful of powdered cardamon seed, and served on a plate, as dainty cakes or wafers."

Aunt Sarah, when cooking fritters, always used two-thirds lard and one-third suet for deep frying, but "Frau Schmidt" taught Mary to use a good brand of oil for this purpose, as she thought food fried in oil more digestible and wholesome than when fried in lard. The patties or wafers were easily made. "Frau Schmidt" placed the long-handled iron in hot fat, the right temperature for frying fritters. When the iron was heated she quickly and carefully wiped off any surplus fat, then at once dipped the hot wafer iron into a bowl containing the batter she had prepared (the recipe for which she gave Mary), then dipped the iron into the hot fat; when the batter had lightly browned she gently dropped it from the iron onto brown paper, to absorb any fat which might remain. These are quickly and easily prepared and, after a few trials, one acquires proficiency. Pattie cases or cup-shapes are made in a similar manner. They are not expensive and may be kept several weeks in a cool, dry place. When wanted for table use, place in a hot oven a few minutes to reheat. They make a dainty addition to a luncheon by simply dusting the "Rosen Kuchen" with pulverised sugar. Creamed vegetables of any variety may be served on them by placing a spoon of cream dressing on top of each, over which grate yolk of hard boiled egg; or use as a foundation on which to serve salads; or serve fruit on them with whipped cream. The patties or cups may be used to serve creamed chicken, oysters, or sweetbreads if no sugar be used in the batter. These pattie cases are exactly like those sold at delicatessen counters, in city stores, and are considered quite an addition to a dainty luncheon. They are rather expensive to buy, and we country housewives cannot always procure them when wanted, and they may be made at home with a small amount of labor and less expense.

"The Germans make fritters of almost everything imaginable," continued the Professor's wife. "One day in early Spring I saw a German neighbor gathering elderberry blossoms, of which she said she intended making fritters. I asked her how they were made, being curious, I will confess. She sent me a plate of the fritters and they were delicious. I will give you her recipe should you care for it. Mary, have you ever eaten a small, sweet wafer called 'Zimmet Waffle?' My mother made them at Christmas time, in Germany. Should I be able to procure a small 'waffle,' or I should call it wafer, iron, in the city, I will teach you how they are made. I think them excellent. My mother made a cake dough similar to that of pound cake. To one portion she added cinnamon, to the other chocolate, and the last portion was flavored with vanilla. A piece of dough the size of a small marble was placed in the wafer iron, which was then pressed together and held over the fire in the range, by a long handle, until the wafer was crisp and brown. They are delicious and will keep indefinitely."

The Professor's wife finished speaking to Mary, and turned to her daughter Elizabeth, saying, "It is time I mix the dough if we are to have 'Boova Shenkel' for dinner today. I see the potatoes have steamed tender."

"Oh, goody!" said Pauline, "I just love 'Boova Shenkel!'"

"Then," said her Mother, "run down into the cellar and get me three eggs for them, and Mary, I'll write off the recipe for you, if you wish it, as I feel sure you'll like them as well as Pauline. And Elizabeth, dust powdered sugar over this plate of 'Rosen Kuchen,' and you, Mary and Pauline, leave this hot kitchen and have lunch out in the 'Espalier,' as your Father calls it."

"I think," said Mary to Elizabeth, after they were seated in the shade, prepared to enjoy the "Rosen Kuchen," "this little, natural, home-grown summer-house is the oddest and prettiest little place I've ever seen."

"Yes," assented Elizabeth, "Father said he made it as nearly like as possible to a large one at Weisbaden, no great distance from his old home in Germany. He says the 'Frauer Esche,' meaning Weeping Ash, at Weisbaden, had tables and benches placed beneath spreading branches of the tree, and picnics were frequently held there. This one was made by the larger branches of the Weeping Ash, turning downward, fastened by pieces of leather to a framework nailed to the top of posts in the ground, about two yards apart, surrounding the tree. The posts, you notice, are just a little higher than an ordinary man, and when the leaves thickly cover the tops and sides, protecting one from the sun's rays, it is an ideal Summer-house. We frequently sit here evenings and afternoons; Mother brings her sewing and Pauline her doll family, which, you know, is quite numerous."

"I never saw a Summer-house at all like it," said Mary.

The Professor's wife not only taught Mary the making of superior pastry and the cooking of German dishes, but what was of still greater importance, taught her the value of different foods; that cereals of every description, flour and potatoes, are starchy foods; that cream, butter, oil, etc., are fat foods; that all fruits and vegetables contain mineral matter; and that lean meat, eggs, beans, peas and milk are muscle-forming foods. These are things every young housekeeper should have a knowledge of to be able to plan nourishing, wholesome, well-balanced meals for her family. And not to serve at one time a dish of rice, cheese and macaroni, baked beans and potatoes. Serve instead with one of these dishes fruit, a vegetable or salad. She said, "beans have a large percentage of nutriment and should be more commonly used." She also said graham and corn bread are much more nutritious than bread made from fine white flour, which lacks the nutritious elements. Indian corn is said to contain the largest amount of fat of any cereal. It is one of our most important cereal foods and should be more commonly used by housewives; especially should it be used by working men whose occupation requires a great amount of physical exercise. Particularly in cold weather should it be frequently served, being both cheap and wholesome.

The Professor's wife laughingly remarked to Mary, "When I fry fritters or 'Fast Nacht' cakes, Fritz and Pauline usually assist such a large number of them in disappearing before I have finished baking, I am reminded of 'Doughnutting Time,' by J.W. Foley. Have you never read the poem? I sometimes feel that it must have been written by me."

[A]"DOUGHNUTTING TIME."

Wunst w'en our girl wuz makin' pies an' doughnuts — 'ist a
Lot —
We stood around with great, big eyes, 'cuz we boys like 'em
hot;
And w'en she dropped 'em in the lard, they sizzled 'ist like
fun,
And w'en she takes 'em out, it's hard to keep from takin' one.

And 'en she says: "You boys'll get all spattered up with grease."
And by-um-by she says she'll let us have 'ist one apiece;
So I took one for me, and one for little James McBride,
The widow's only orfunt son, 'ats waitin' there outside.
An' Henry, he took one 'ist for himself an' Nellie Flynn,
'At's waitin' at the kitchen door and dassent to come in,
Becuz her Mother told her not; and Johnny, he took two,
'Cus Amey Brennan likes 'em hot, 'ist like we chinnern do.
'En Henry happened 'ist to think he didn't get a one
For little Ebenezer Brink, the carpet beater's son,
Who never gets 'em home, becuz he says, he ain't quite sure,
But thinks perhaps the reason wuz, his folkeses are too poor.
An 'en I give my own away to little Willie Biggs
'At fell down his stairs one day, an' give him crooked legs,
'Cuz Willie always seems to know w'en our girl's goin' to bake.
He wouldn't ast for none. Oh, no! But, my! he's fond of cake.
So I went back an' 'en I got another one for me,
Right out the kittle smokin' hot, an' brown as it could be;
An' John he got one, too, becuz he give his own to Clare,
An' w'en our girl she looked, there wuz 'ist two small doughnuts
there.
My! she wuz angry w'en she looked an' saw 'ist them two there,
An' says she knew 'at she had cooked a crock full an' to spare;
She says it's awful 'scouragin' to bake and fret an' fuss,
An' w'en she thinks she's got 'em in the crock, they're all in us.

[A]

The poem "Doughnutting Time," from "Boys and Girls," published by E.P. Dutton, by permission of the author, James W. Foley.

The Professor's wife gave Mary what she called her most useful recipe. She said, "Mary, this recipe was almost invaluable to me when I was a young housekeeper and the strictest economy was necessary. Sift into a bowl, one cup of flour, one even teaspoonful of baking powder (I use other baking powders occasionally, but prefer 'Royal'), then cut through the flour either one tablespoonful of butter or lard, add a pinch of salt, and mix into a soft dough with about one-half cup of sweet milk. Mix dough quickly and lightly, handling as little as possible. Drop large spoonfuls of the batter in muffin pans and bake in a quick oven for tea biscuits; or, sift flour thickly over the bread board, turn out the dough, roll several times in the flour, give one quick turn with the rolling-pin to flatten out dough, and cut out with small cake cutter, (I prefer using a small, empty tin, ½ pound baking powder can, to cut out cakes.) Place close together in an agate pan and bake, or bake in one cake in a pie tin and for shortcake; or place spoonfuls of the dough over veal or beef stew and potatoes or stewed chicken, and cook, closely covered, about fifteen minutes. Of course, you will have sufficient water in the stew pan to prevent its boiling away before the pot-pie dumplings are cooked, and, of course, you know, Mary, the meat and potatoes must be almost ready to serve when this dough is added. Then I frequently add one teaspoonful of sugar to the batter and place spoonfuls over either freshly stewed or canned sour cherries, plums, rhubarb or apples. In fact, any tart fruit may be used, and steam, closely covered, or place large tablespoonful of any fruit, either canned or stewed, in small custard cups, place tablespoonfuls of batter on top and steam or bake, and serve with either some of the stewed fruit and fruit juice, sugar and cream, or any sauce preferred."

"The varieties of puddings which may be evolved from this one formula," continued the Professor's wife, "are endless, and, Mary, I should advise you to make a note of it. This quantity of flour will make enough to serve two at a meal, and the proportions may be easily doubled if you wish to serve a large family."

"Then, Mary, I have a recipe taken from the 'Farmers' Bulletin' for dumplings, which I think fine. You must try it some time. Your Aunt Sarah thinks them 'dreadfully extravagant.' They call for four teaspoonfuls of baking powder to two cups of flour, but they are perfect puff balls, and this is such a fast age, why not use more baking powder if an advantage? I am always ready to try anything new I hear about."

"Yes," replied Mary, "I just love to try new recipes, I will experiment with the dumplings one of these days. Aunt Sarah says I will never use half the recipes I have; but so many of them have been given me by excellent and reliable old Bucks County cooks, I intend to copy them all in a book, and keep for reference after I leave the farm."

CHAPTER XX.

OLD POTTERIES AND DECORATED DISHES.

One day, looking through the old corner cupboard, Mary exclaimed, "Aunt Sarah, you certainly possess the finest collection of quaint old china dishes I have ever seen. I just love those small saucers and cups without handles; yes, and you have plates to match decorated with pinkish, lavender peacock feathers, and those dear little cups and saucers, decorated inside with pink and outside with green flowers, are certainly odd; and this queerly-shaped cream jug, sugar bowl and teapot, with pale green figures, and those homely plates, with dabs of bright red and green, they surely must be very old!"

Old Earthenware Dish

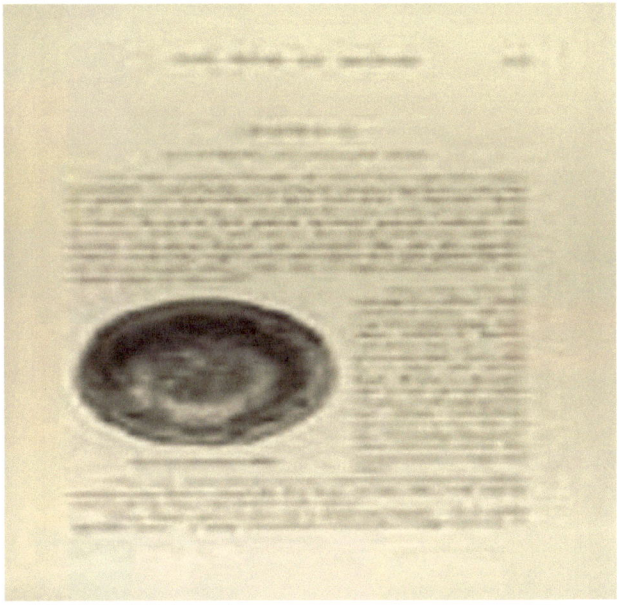

"Yes, dear, they all belonged to either John's mother or mine. All except this one large, blue plate, which is greatly valued by me, as it was given me many years ago by a dear old friend, Mary Butler, a descendant of one of the oldest families in Wyoming Valley, whose, forefathers date back to the time of the 'Wyoming Massacre,' about which so much has been written in song and story.

"The very oddest plates in your collection are those two large earthenware dishes, especially that large circular dish, with sloping sides and flat base, decorated with tulips."

SGRAFFITO PLATE Manufactured by One of the Oldest Pennsylvania German Potterers in 1786

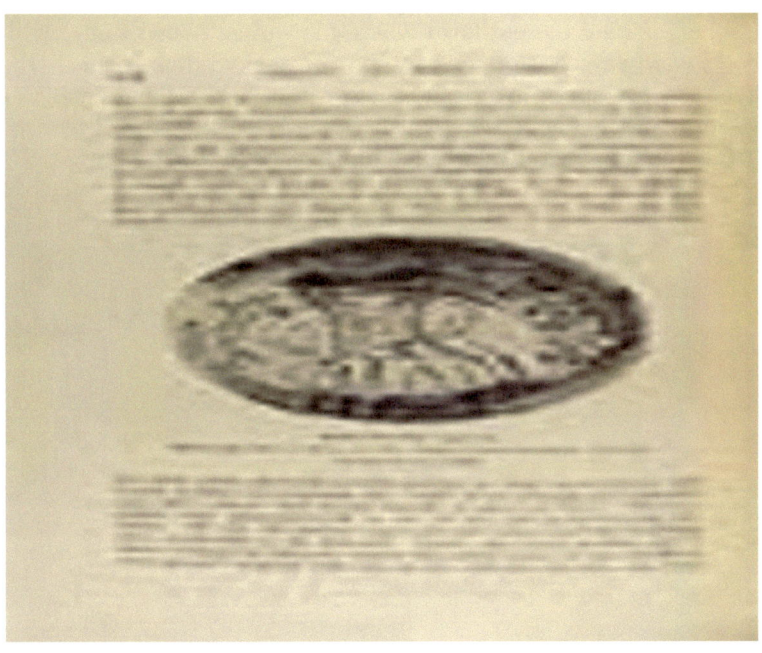

"Yes, Mary, and it is the one I value most highly. It is called sgraffito ware. A tulip decoration surrounds a large red star in the centre of the plate. This belonged to my mother, who said it came from the Headman pottery at Rockhill Township, about the year 1808. I know of only two others in existence at the present time; one is in a museum in the city of Philadelphia and the other one is in the Bucks County Historical Society at Doylestown, Pa. The other earthenware plate you admire, containing marginal inscription in German which when translated is 'This plate is made of earth, when it breaks the potter laughs,' is the very oldest in my collection, the date on it, you see, is 1786. Those curved, shallow earthenware pie plates, or 'Poi Schissel,' as they are frequently called in this part of Bucks County, I value, even if they are quite plain and without decoration, as they were always used by my mother when baking pies, and I never thought pies baked in any other shaped dish tasted equally as good as hers. These pie plates were manufactured at one of the old potteries near her home. All the old potters have passed away, and the buildings have crumbled to the ground. Years ago, your mother and I, when visiting the old farm where the earlier years of our childhood were passed, stopped with one of our old-time friends, who lived directly opposite the old Herstine pottery, which was then in a very dilapidated condition; it had formerly been operated by Cornelius Herstine (we always called him 'Neal' Herstine)."

OLD PLATES FOUND IN AUNT SARAH'S COR-
NER CUPBOARD

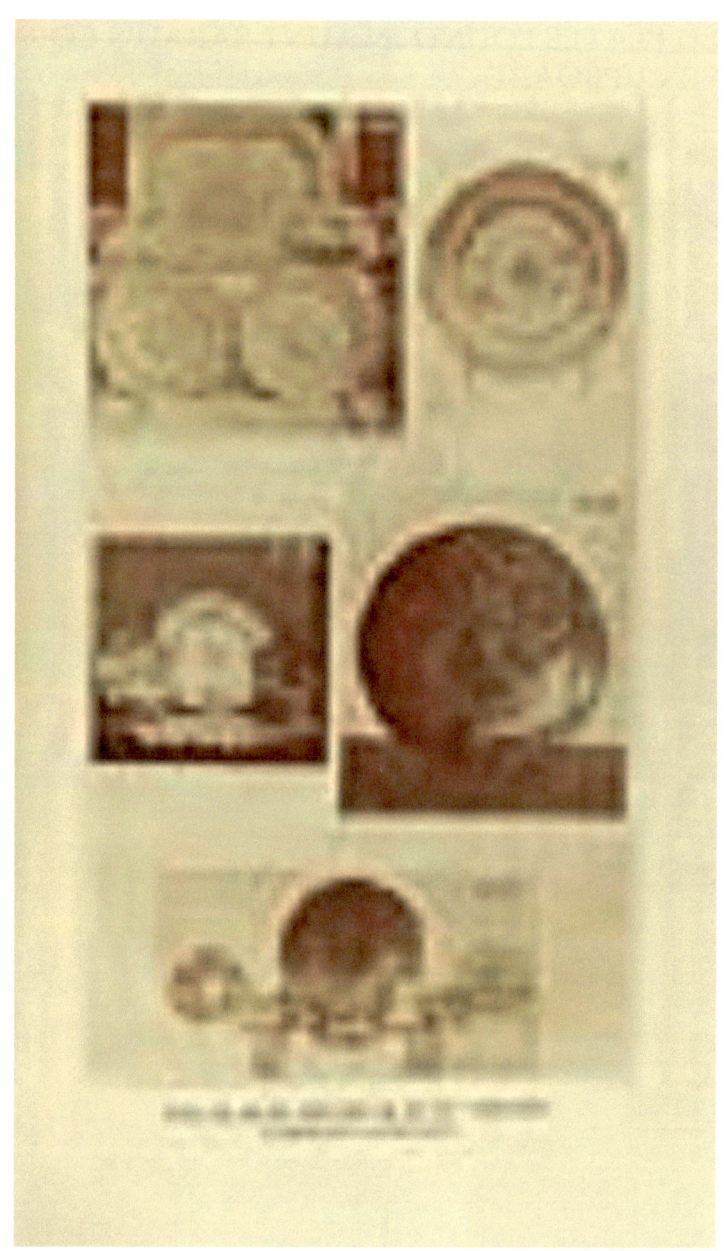

"Together we crossed the road, forced our way through tangled vines and underbrush, and, peering through windows guiltless of glass, we saw partly-finished work of the old potters crumbling on the ground. The sight was a sad one. We realized the hand of time had crumbled to dust both the potter and his clay. Still nearer my old home was the McEntee pottery. From earliest childhood our families were friends. We all attended the 'Crossroads' School, where years later a more modern brick structure was built, under the hill; not far distant from 'The Narrows' and the 'Ringing Rocks.' Yes, Mary, my memory goes back to the time when the McEntee pottery was a flourishing industry, operated by three brothers, John, Patrick and Michael. When last I visited them but few landmarks remained."

"Was there a pottery on your father's farm, Aunt Sarah?" inquired Mary.

"No. The nearest one was the McEntee pottery, but the grandson of the old man who purchased our old farm at my father's death had a limekiln for the purpose of burning lime, and several miles distant, at the home of my uncle, was found clay suitable for the manufacture of bricks. Only a few years ago this plant was still in operation. My father's farm was situated in the upper part of Bucks County, in what was then known as the Nockamixon Swamp, and at one time there were in that neighborhood no less than seven potteries within two miles of each other."

"Why," exclaimed Mary, "were there so many potteries in that locality?"

"'Twas due, no doubt, to the large deposits of clay found there, well suited to the manufacture of earthenware. The soil is a clayey loam, underlaid with potter's clay. The old German potters, on coming to this country, settled mostly in Eastern Pennsylvania, in the counties of Bucks and Montgomery. The numerous small potteries erected by the early settlers were for the manufacture of earthenware dishes, also pots of graded sizes. These were called nests, and were used principally on the farm for holding milk, cream and apple-butter. Jugs and pie plates were also manufactured. The plates were visually quite plain, but they produced occasionally plates decorated with conventionalized tulips, and some, more elaborate, contained besides figures of animals, birds and flowers. Marginal inscriptions in English and German decorate many of the old plates, from which may be learned many interesting facts concerning the life and habits of the early settlers. I think, judging from the inscriptions I have seen on some old plates, it must have taxed the ingenuity of the old German potters to think up odd, original inscriptions for their plates."

"Aunt Sarah, how was sgraffito ware made? Is it the same as slip-decorated pottery?"

"No, my dear, the two are quite different. The large plate you so greatly admired is called sgraffito or scratched work, sometimes called slip engraving. It usually consists of dark designs on a cream-colored ground. After the plates had been shaped over the mold by the potter, the upper surface was covered by a coating of white slip, and designs were cut through this slip to show the earthenware underneath. This decoration was more commonly used by the old potters than slip decorating, which consisted in mixing white clay and water until the consistency of cream. The liquid clay was then allowed to run slowly through a quill attached to a small cup, over the earthenware (before burning it in a kiln) to produce different designs. The process is similar to that used when icing a cake, when you allow the icing to run slowly from a pastry tube to form fanciful designs. I have watched the old potters at their work many a time when a child. The process employed in the manufacture of earthenware is almost the same today as it was a century ago, but the appliances of the present day workmen are not so primitive as were those of the old German potters. Mary, a new pottery works has been started quite lately in the exact locality where, over one hundred years ago, were situated the Dichl and Headman potteries, where my highly-prized, old sgraffito plate was manufactured. I hear the new pottery has improved machinery for the manufacture of vases, flower pots, tiles, etc. They intend manufacturing principally 'Spanish tiles' from the many acres of fine clay found at that place. The clay, it is said, burns a beautiful dark, creamy red. As you are so much interested in this subject, Mary, we shall visit this new pottery some day in the near future, in company with your Uncle John. It is no great distance from the farm. Quite an interesting story I have heard in connection with a pottery owned by a very worthy

Quaker in a near-by town may interest you, as your father was a Philadelphia Quaker and Ralph's parents were Quakers also."

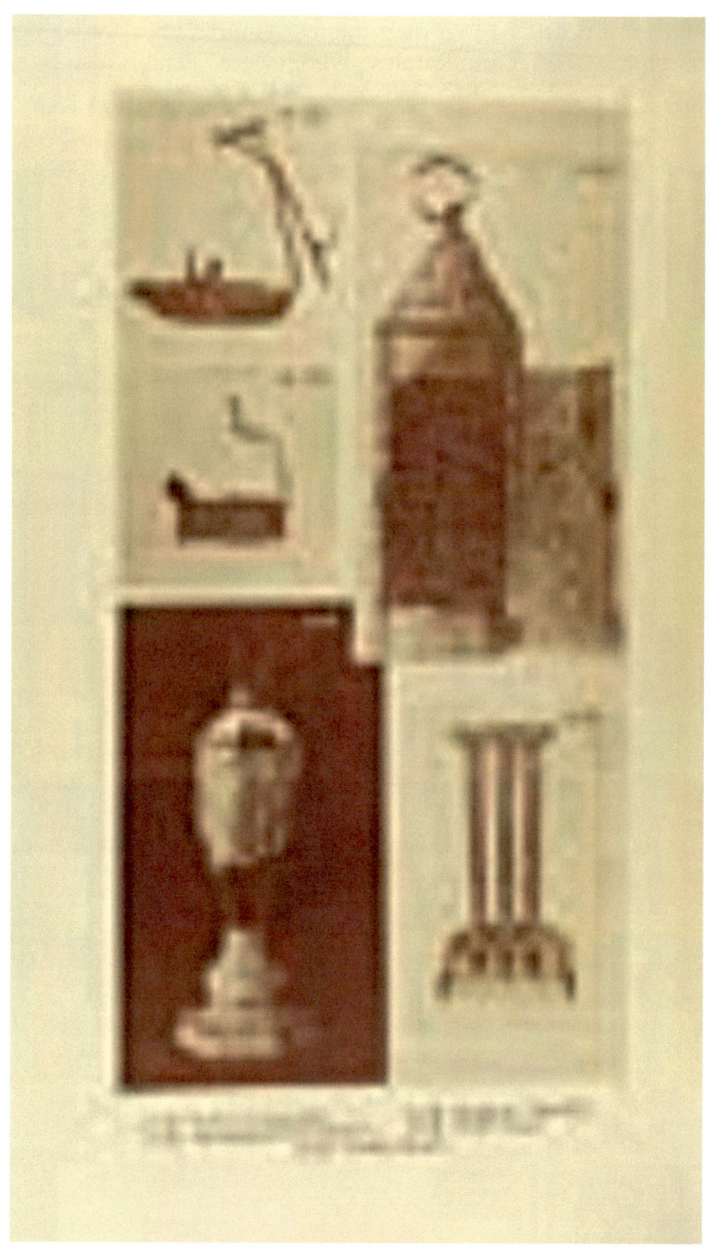

A-38 Schmutz Amschel
A-39 Antiquated Tin Lantern
A-40 Schmutz Amschel
A-41 Fluid Lamp
A-42 Candle Mould

"Yes, indeed, Aunt Sarah! I'd love to hear the story."

"This Quaker sympathized with the colored race, or negroes, in the South. This was, of course, before slavery was abolished. You don't remember that time, Mary, You are too young. It is only history to you, but I lived it, and when the slaves ran away from their owners and came North to Philadelphia they were sent from there, by sympathizers, to this Quaker, who kept an underground station. The slaves were then placed, under his direction, in a high 'pot wagon,' covered with layers or nests of earthenware pots of graduated sizes. I heard the driver of one of these pot wagons remark one time that when going down a steep hill, he put on the brake and always held his breath until the bottom of the hill was reached, fearing the pots might all be broken. The wagon-load containing earthenware and slaves was driven to Stroudsburg, where the pots were delivered to a wholesale customer. Here the runaways were released from their cramped quarters and turned over to sympathizing friends, who assisted them in reaching Canada and safety. I have frequently met the fine-looking, courtly old gentleman who owned the pottery, and old Zacariah Mast, the skilled German potter whom he employed. They were for many years familiar figures in the little Quaker town, not many miles distant. Both passed away many years ago."

Mary, who still continued her explorations of the corner cupboard, exclaimed: "Oh! Aunt Sarah! Here is another odd, old plate, way back on the lop shelf, out of sight."

"Yes, dear, that belonged to your Uncle John's mother. It has never been used and was manufactured over one hundred years ago at an old pottery in Bedminister Township, Bucks County. Some of those other quaint, old-fashioned plates also belonged to John's mother. Your Uncle loves old dishes and especially old furniture; he was so anxious to possess his grandfather's old 'Solliday' clock. In the centre of the face of the clock a hand indicated the day of the month and pictures of two large, round moons on the upper part of the clock's face (resembling nothing so much as large, ripe peaches) represented the different phases of the moon. If new moon, or the first or last quarter, it appeared, then disappeared from sight. It was valued highly, being the last clock made by the old clockmaker; but John never came into possession of it, as it was claimed by an elder sister. I value the old clock which stands in the parlor because 'twas my mother's, although it is very plain. This old cherry, corner cupboard was made for my grandmother by her father, a cabinetmaker, as a wedding gift, and was given me by my mother. Did you notice the strong, substantial manner in which it is made? It resembles mission furniture."

"Do tell me, Aunt, what this small iron boat, on the top shelf, was ever used for? It must be of value, else 'twould not occupy a place in the cupboard with all your pretty dishes."

"Yes, dearie, 'twas my grandmother's lamp, called in old times a 'Schmutz Amschel' which, translated, means a grease robin, or bird. I have two of them. I remember seeing my grandmother many a time, when the 'Amschel' was partly filled with melted lard or liquid fat, light a piece of lamp wick hanging over the little pointed end or snout of the lamp. The lamp was usually suspended from a chain fastened to either side. A spike on the chain was stuck into the wall, which was composed of logs. This light, by the way, was not particularly brilliant, even when one sat close beside it, and could not be compared with the gas and electric lights of our present day and generation. That was a very primitive manner of illumination used by our forefathers.

"Mary, did you notice the gayly-decorated, old-fashioned coffee pot and tea caddy in the corner cupboard? They belonged to my grandmother; also that old-fashioned fluid lamp, used before coal-oil or kerosene came into use; and that old, perforated tin lantern also is very ancient.

"Mary, have you ever read the poem, The Potter and the Clay?' No? Then read it to me, dear, I like it well; 'tis a particular favorite of mine, I do not remember by whom it was written."

THE POTTER AND THE CLAY.

(Jeremiah xviii 2-6.)

> The potter wrought a work in clay, upon his wheel;
> He moulded it and fashioned it, and made it feel,
> In every part, his forming hand, his magic skill,
> Until it grew in beauty fair beneath his will.
> When lo! through some defect, 'twas marred and broken lay,
> Its fair proportions spoiled, and it but crumbling clay;
> Oh, wondrous patience, care and love, what did he do?
> He stooped and gathered up the parts and formed anew.
> He might have chosen then a lump of other clay
> On which to show his skill and care another day,
> But no; he formed it o'er again, as seemed him good;
> And who has yet his purpose scanned, his will withstood?
> Learn thou from this a parable of God's great grace
> Toward the house of Israel, His chosen race;
> He formed them for His praise; they fell and grieved Him sore,
> But He will yet restore and bless them evermore.
> And what He'll do for Israel, He'll do for thee;
> Oh soul, so marred and spoiled by sin, thou yet shall see
> That He has power to restore, He will receive,
> And thou shall know His saving grace, only believe.
> Despair not, He will form anew thy scattered life,
> And gather up the broken parts, make peace from strife;
> Only submit thou to His will of perfect love,
> And thou shall see His fair design in Heaven above.

CHAPTER XXI.

THE VALUE OF WHOLESOME, NUTRITIOUS FOOD.

"Yes, my dear," said Frau Schmidt (continuing a conversation which had occurred several days previously between herself and Mary), "we will have more healthful living when the young housewife of the present day possesses a knowledge of different food values (those food products from which a well-balanced meal may be prepared) for the different members of her household. She should endeavor to buy foods which are most nourishing and wholesome; these need not necessarily consist of the more expensive food products. Cheaper food, if properly cooked, may have as fine a flavor and be equally as nutritious as that of higher price.

"And, Mary, when you marry and have a house to manage, if possible, do your own marketing, and do not make the mistake common to so many young, inexperienced housewives, of buying more expensive food than, your income will allow. Some think economy in purchasing food detrimental to their dignity and to the well-being of their families; often the ones most extravagant in this respect are those least able to afford it. Frequently the cause of this is a lack of knowledge of the value of different foods. The housewife with a large family and limited means should purchase cheaper cuts of meat, which become tender and palatable by long simmering. Combine them with different vegetables, cooked in the broth, and serve as the principal dish at a meal, or occasionally serve dumplings composed of a mixture of flour and milk, cooked in the broth, to extend the meat flavor. Frequently serve a dish of rice, hominy, cornmeal and oatmeal, dried beans and peas. These are all nutritious, nourishing foods when properly cooked and attractively served. And remember, Mary, to always serve food well seasoned. Many a well-cooked meal owes its failure to please to a lack of proper seasoning. This is a lesson a young cook must learn. Neither go to the other extreme and salt food too liberally. Speaking of salt, my dear, have you read the poem, 'The King's Daughters,' by Margaret Vandegrift? If not, read it, and then copy it in your book of recipes."

"THE KING'S DAUGHTERS."

The King's three little daughters, 'neath the palace window straying,

Had fallen into earnest talk that put an end to playing;
And the weary King smiled once again to hear what they were saying;
"It is I who love our father best," the eldest daughter said;
"I am the oldest princess," and her pretty face grew red;
"What is there none can do without? I love him more than bread."
Then said the second princess, with her bright blue eyes aflame;
"Than bread, a common thing like bread! Thou hast not any shame!
Glad am I, it is I, not thou, called by our mother's name;
I love him with a better love than one so tame as thine,
More than—Oh! what then shall I say that is both bright and fine?
And is not common? Yes, I know. I love him more than wine."
Then the little youngest daughter, whose speech would sometimes halt,
For her dreamy way of thinking, said, "Nay, you are both in fault.
'Tis I who love our father best, I love him more than salt."
Shrill little shrieks of laughter greeted her latest word,
As the two joined hands exclaiming. "But this is most absurd!"
And the King, no longer smiling, was grieved that he had heard,
For the little youngest daughter, with her eyes of steadfast grey,
Could always move his tenderness, and charm his care away;
"She grows more like her mother dead," he whispered day by day,
"But she is very little and I will find no fault,
That while her sisters strive to see who most shall me exalt,
She holds me nothing dearer than a common thing like salt."
The portly cook was standing in the courtyard by the spring,
He winked and nodded to himself, "That little quiet thing
Knows more than both the others, as I will show the King."

That afternoon, at dinner, there was nothing fit to eat.
The King turned angrily away from soup and fish and meat,
And he found a cloying sweetness in the dishes that were sweet;
"And yet," he muttered, musing, "I cannot find the fault;
Not a thing has tasted like itself but this honest cup of malt."
Said the youngest princess, shyly: "Dear father, they want salt."
A sudden look of tenderness shone on the King's dark face,
As he sat his little daughter in the dead queen's vacant place,
And he thought: "She has her mother's heart; Ay, and her mother's grace;
Great love through channels will find its surest way.
It waits not state occasions, which may not come or may;
It comforts and it blesses, hour by hour, and day by day."

CHAPTER XXII.

A VARIETY OF CAKES EVOLVED FROM ONE

"Aunt Sarah," questioned Mary one day, "will you tell me how it is possible to evolve a number of cakes from one recipe?"

"Certainly I will, my dear," said her Aunt. "For instance, take the simple recipe from which I have for years baked layer cake. You may have other recipes given you, equally as good, but I feel positive none better. The cake made from this recipe is not rich enough to be unwholesome, but a good, reliable, inexpensive, easily-made cake, with which I have never had a failure.

"The recipe, as you know, consists of 1¼ cups of granulated sugar, ½ cup of a mixture of butter and sweet lard (or use all butter), ½ cup sweet milk, 2 cups flour and 2 teaspoonfuls baking powder. 3 eggs.

"The simplest manner of baking this cake is in two square cake pans. When baked, take from pans and ice each cake with a boiled chocolate icing and put together as a layer cake, or ice each cake with a plain, boiled white icing and, when this is cold, you may spread over top of each cake unsweetened chocolate, which has been melted over steam after being grated. When cake is to be served, cut in diamonds or squares. Or add to the batter 1 cup of chopped hickory nut meats, bake in 2 layers and cut in squares.

"For a chocolate loaf cake, add two generous tablespoonfuls of unsweetened melted chocolate to the batter just before baking. If you wish a chocolate layer cake, use the same batter as for the chocolate loaf cake, bake in two layer pans and put together with white boiled icing.

"Or, add to this same batter one scant teaspoonful of cinnamon, ginger, ½ teaspoonful of grated nutmeg and cloves, a cup of raisins or dried currants, and you have a small fruit cake.

"Or, add a small quantity of thinly-shaved citron to the original recipe, flavor with lemon, bake in a loaf and spread a white icing flavored with lemon extract over top of cake, and you have a lemon cake.

"Or, add chocolate and spices to one-half the batter (about one-half as much chocolate and spices as were used in batter for fruit cake) and place spoonfuls of the light and dark batter alternately in a cake pan, until all batter has been used, and you will have a cheap, old-fashioned Marble cake.

"Or, bake cake over original recipe, in two-layer pans, placing between layers either tart jelly, a creamy cornstarch filling, grated cocoanut, apple cream filling, or you might even use half the recipe given for the delicious icing or filling for Lady Baltimore cake.

"Lastly, bake small cakes from this same recipe. Mary, you should have small pans for baking these delicious little cakes, similar to those I possess, which I ordered made at the tinsmith's. I took for a pattern one Frau Schmidt loaned me. They are the exact size of one-quarter pound boxes of Royal baking powder. Cut the box in three pieces of equal height, and your cakes will be equally as large in diameter as the baking powder box, but only one-third as high. I think I improved on Frau Schmidt's cake tins, as hers were all separate, I ordered twelve tins, similar to hers, to be fastened to a piece of sheet iron. I had two of these iron sheets made, containing twenty-four little pans. I place a generous tablespoonful of the batter in each of the twenty-four small pans, and cakes rise to the top of pans. Usually I have batter remaining after these are filled. Ice all the cake except the top with a white boiled icing or chocolate icing. These small cakes keep exceedingly well, and are always liked by young folks and are particularly nice for children's parties".

"Speaking of cakes, Aunt Sarah," said Mary, "have you ever used Swansdown cake flour? I have a friend in the city who uses it for making the most delicious Angel cake, and she gave me a piece of Gold cake made over a recipe in 'Cake Secrets,' which comes with the flour, and it was fine. I'll get a package of the flour for you the first time I go to the city. The flour resembles a mixture of ordinary flour and cornstarch. It is not a prepared flour, to be used without baking powder, and you use it principally for baking cakes. I have the recipe for both the Gold and Angel cakes, with the instructions for baking same. They are as follows:"

ANGEL CAKE.

"For the Angel cake, use one even cupful of the whites of egg (whites of either eight large or nine small eggs); a pinch of salt, if added when beating eggs, hastens the work. One and one-quarter cups granulated sugar, 1 cup of Iglehart's Swansdown cake flour. Sift flour once, then measure and sift three times. Beat whites of eggs about half, add ½ teaspoonful of cream of tartar then beat whites of eggs until they will stand of their own weight. Add sugar, then flour, not by stirring, but by folding over and over, until thoroughly mixed. Flavor with ½ teaspoonful of vanilla or a few drops of almond extract. As much care should be taken in baking an Angel Food cake as in mixing. Bake in an ungreased patent pan. Place the cake in an oven that is just warm enough to know there is a fire inside the range. Let the oven stay just warm through until the batter has raised to the top of the cake pan, then increase the heat gradually until the cake is well browned over. If by pressing the top of the cake with the finger it will spring back without leaving the impression of the finger, the cake is done through. Great care should be taken that the oven is not too hot to begin with, as the cake will rise too fast and settle or fall in the baking. It should bake in from 35 to 40 minutes' time. When done, invert the pan and let stand until cold before removing it. Should you see cake browning before it rises to top of pan, throw your oven door open and let cold air rush in and cool your oven instantly. Be not afraid. The cold air will not hurt the cake. Two minutes will cool any oven. Watch cake closely. Don't be afraid to open oven door every three or four minutes. This is the only way to properly bake this cake. When cake has raised above top of pan, increase your heat and finish baking rapidly. Baking too long dries out the moisture, makes it tough and dry. When cake is done it begins to shrink. Let it shrink back to level of pan. Watch carefully at this stage and take out of oven and invert immediately. Rest on centre tube of pan. Let hang until perfectly cold, then take cake carefully from pan. When baking Angel cake always be sure the oven bakes good brown under bottom of cake. If cake does not crust under bottom it will fall out when inverted and shrink in the fall."

"I never invert my pans of Angel cake on taking them from oven," said Mary's Aunt, "as the cakes are liable to fall out even if the pan is not greased. I think it safer to allow the pans containing the cakes to stand on a rack and cool without inverting the pan.

"Suppose, Mary, we bake a Gold cake over the recipe from 'Cake Secrets,' as eggs are plentiful; but we haven't any Swansdown flour. I think we will wait until we get it from the city."

GOLD CAKE.

Yolks of 8 eggs; 1¼ cups granulated sugar, ¾ cup of butter, ¾ cup water, 2½ cups of Swansdown cake flour, 2 heaping teaspoonfuls of baking powder, ½ teaspoonful lemon extract. Sift flour once, then measure. Add baking powder and sift three times. Cream butter and sugar thoroughly; beat yolks to a stiff froth; add this to creamed butter and sugar, and stir thoroughly through. Add flavor, add water, then flour. Stir very hard. Place in a slow oven at once. Will bake in from 30 to 40 minutes. Invert pan immediately it is taken from oven.

Mary, this batter may also be baked in layers with any kind of filling desired. The Angel cake receipt is very similar to an original recipe Frau Schmidt gave me; she uses cornstarch instead of Swansdown flour and she measures the eggs in a cup instead of taking a certain number; she thinks it more exact.

"Aunt Sarah, did you know Frau Schmidt, instead of using flour alone when baking cakes, frequently uses a mixture of flour and cornstarch? She sifts together, several times, six cups of flour and one cup of cornstarch, and uses this instead of using flour alone.

"I dearly love the Professor's wife—she's been so very good to me," exclaimed Mary.

"Yes," replied her Aunt, "she has very many lovable qualities."

Mary's liking for bright, energetic Frau Schmidt was not greater than the affection bestowed on Mary by the Professor's wife, who frequently entertained Mary with tales of her life when a girl in Germany, to all of which Mary never tired listening. One Aunt, a most estimable woman, held the position of valued and respected housekeeper and cook for the Lord Mayor of the city wherein she resided. Another relative, known as "Schone Anna," for many years kept an inn named "The Four Seasons," noted for the excellent fare served by the fair chatelaine to her patrons. The inn was made famous by members of the King's household stopping there while in the town during the Summer months, which was certainly a compliment to her good cooking. One of the things in which she particularly excelled was potato cakes raised with yeast. Frau Schmidt had been given a number of these valuable recipes by her mother, all of which she offered to Mary. One recipe she particularly liked was "Fast Nacht Cakes," which the Professor's wife baked always without fail on Shrove Tuesday (or "Fast Nacht" day), the day before the beginning of Lent. This rule was as "unchangeable as the law of Medes and Persians," and it would have been a very important event, indeed, which would have prevented the baking of these toothsome delicacies on that day.

CHAPTER XXIII.

THE OLD "TAUFSCHIEN."

BIRTH AND CHRISTENING CERTIFICATE

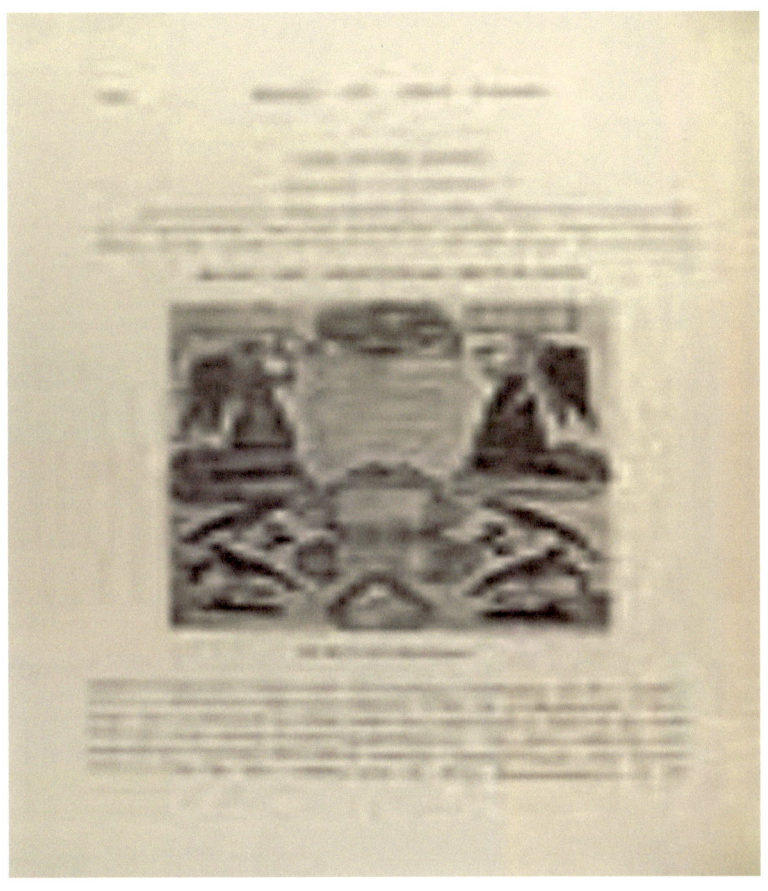

OLD TAUFSCHIEN

Aunt Sarah had long promised to show Mary her Grandmother's "Taufschien," and she reverently handled the large old family Bible, which contained between its sacred pages the yellowed paper, being the birth and christening certificate of her grandmother, whom we read was born in 1785, in Nockamixon Township, was confirmed in 1802, and was married in 1805 to the man who was later Aunt Sarah's grandfather. The old certificate was signed by a German Reformed minister named Wack, who history tells us was the first young man of that denomination to be ordained to the ministry in America. Folded with this "Taufschien" is another which has never been filled out. This is printed in German. Pictures of women, perhaps they are intended to represent angels, with golden wings, clothed in loose-flowing crimson drapery and holding harps in their hands; birds with gayly-colored plumage of bluish green, crimson and yellow, perched on branches of what presumably represent cherry trees, also decorate the page. Religious hymns printed on the "Taufschiens," encircled with gay stripes of light blue and yellow, dotted with green, further embellish them. On one we read:

"Infinite joy or endless woe,
Attend on every breath;
And yet, how unconcerned we go
Upon the brink of death."

"Mary, this old 'Taufschien' of my grandmother's is one of my most cherished possessions. Would you like to see your Uncle's old deed, which he came into possession of when he inherited the farm from his father?"

Carefully unfolding the stiff old parchment or pigskin deed, yellowed and brown spotted with age, Mary could faintly decipher the writing wherein, beautifully written, old-fashioned penmanship of two hundred years ago stated that a certain piece of land in Bucks County, Beginning at a Chestnut Oak, North to a post; then East to a large rock, and on the South unsettled land, which in later years was conveyed to John Landis.

"This deed," said Mary's Aunt, "was given in 1738, nearly two hundred years ago, by John, Thomas and Richard Penn, sons of William Penn by his second marriage, which occurred in America. His eldest son, John Penn, you have no doubt heard, was called 'The American,' he having been born in this country before William Penn's return to Europe, where he remained fifteen years, as you've no doubt heard."

At the bottom of the deed a blue ribbon has been slipped through cuts in the parchment, forming a diamond which incloses what is supposed to be the signature of Thomas Penn.

"Aunt Sarah, I am not surprised that you value this old deed of the farm and these 'Taufschiens' of your grandmother I should frame them, so they may be preserved by future generations."

CHAPTER XXIV.

THE OLD STORE ON THE RIDGE ROAD.

Aunt Sarah found in Mary a willing listener when talking of the time in years past when her grandfather kept a small "Country Store" on the Ridge Road in Bucks County. She also remembered, when a child of ten, accompanying her grandfather on one of his trips when he drove to Philadelphia to purchase goods for his store.

"They had no trolley cars in those days?" asked Mary.

"No, my dear, neither did they have steam cars between the different towns and cities as we have now."

"At grandfather's store could be bought both groceries and dry goods. The surrounding farmers' wives brought to the store weekly fresh print butter, eggs, pot cheese and hand-case, crocks of apple-butter, dried sweet corn, beans, cherries, peach and apple 'Snitz,' taking in exchange sugar, starch, coffee, molasses, etc. My father tapped his sugar maples and mother cooked down the syrup until thick, and we used that in place of molasses. They also took in exchange shaker flannel, nankeen, indigo blue and 'Simpson' gray calico, which mother considered superior to any other, both for its washing and wearing qualities. The farmers who came occasionally to the store to shop for different members of the family frequently bought whole pieces of calico of one pattern, and," affirmed Aunt Sarah, "I knew of one farmer who bought several whole pieces of one pattern with rather large figures on a dark wine ground, resembling somewhat the gay figures on an old paisley shawl. He said 'twas a good, serviceable color, and more economical to buy it all alike, and remarked: 'What's the difference, anyway? Calico is calico.' From the same piece of calico his wife made dresses, aprons and sunbonnets for herself and daughters, shirts for the farmer and his sons (the boys were young, fortunately), and patchwork quilts and comfortables from the remainder."

"Rather monotonous, I should think," said Mary. "I am surprised his wife did not make him wear coat and trousers made from the same piece of calico."

THE OLD STORE ON RIDGE ROAD

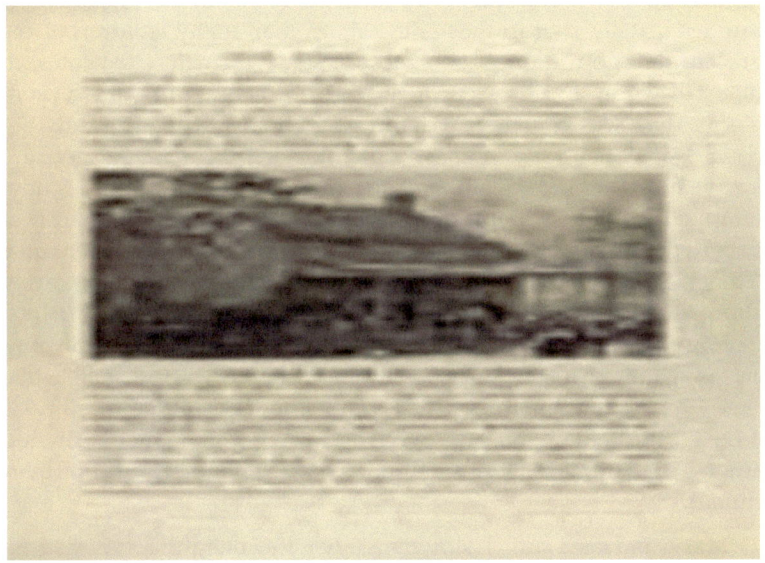

"The dry goods," continued Aunt Sarah, "retained the scent of coffee, cheese and dried fruits some time after being purchased but no one minded that in those days. I still remember how perfectly wonderful to me when a child appeared the large, wide-mouthed glass jars containing candy. There were red and white striped mint sticks, striped yellow and white lemon sticks and hoarhound and clear, wine-colored sticks striped with lines of white, flavored with aniseseed. One jar contained clear lemon-colored 'Sour Balls,' preferred by us children on account of their lasting qualities, as also were the jujubees, which resembled nothing so much as gutta percha, and possessed equally as fine flavor; also pink and yellow sugar-frosted gumdrops. In a case at one end of the counter were squares of thick white paper covered with rows of small pink, also white, 'peppermint buttons,' small sticks, two inches in length, of chewing gum in waxed paper, a white, tasteless, crystalline substance resembling paraffine. What longing eyes I frequently cast at the small scalloped cakes of maple sugar, prohibitive as regards cost. They sold for a nickel, am I was always inordinately fond of maple sugar, but the price was prohibitive. I seldom possessed more than a penny to spend in those days, and not always that. Father raised a large family, money was never plentiful, and we relished the plain, cheap candies usually sold in those days more than many children of the present day do the finest and most expensive cream chocolates, to many of whom in this extravagant age a dollar is not valued more highly than was a penny by us in years gone by. And 'Candy Secrets!' I don't believe you know what they are like. I've not seen any for years. They were small, square pieces of taffy-like candy, wrapped in squares of gilt or silver paper, inclosing a small strip of paper containing a couple of sentimental lines or jingle. Later came 'French Secrets.' They consisted of a small oblong piece of candy about an inch in length, wrapped in tissue paper of different colors, having fringed ends, twisted together at either end. These also inclosed a tiny strip of paper containing a line or two. Small, white candy hearts contained the words in pink letters, 'Little Sweetheart,' 'I Love You,' 'Name the Day,' etc. These were invariably distributed among the young folks at small parties and created no end of merriment."

"Mary, old as I am, I still remember the delight I experienced when a little, rosy-cheeked urchin surreptitiously passed me around the corner of my desk at the old 'Cross Roads School' a 'Secret,' with the words, 'Do you love me?' My grandmother always kept a supply of hoarhound and peppermint lozenges in her knitting basket to give us children should we complain of hoarseness. My, but 'twas astonishing to hear us all cough until grandmother's supply of mints was exhausted. I think. Mary, I must have had a 'sweet tooth' when a child, as my recollections seem to be principally about the candy kept in my grandfather's store. I suppose in those early days of my childhood candy appealed to me more than anything else, as never having had a surfeit of sweets, candy to me was a rare treat. I remember, Mary, when a little child, my thrifty mother, wishing to encourage me to learn to knit my own stockings, she, when winding the skein of German yarn into a ball, occasionally wound a penny in with the yarn. I was allowed to spend the penny only after I had knitted the yarn and the penny had fallen from the ball. What untold wealth that penny represented! And planning how to spend it was greater pleasure still. Many a pair of long old-fashioned, dark blue and red-striped stockings, were finished more quickly than otherwise would have been done without the promised reward. I became proficient in knitting at an early age," continued Aunt Sarah; "a truly feminine occupation, and as I one time heard a wise old physician remark, 'Soothing to the nerves,' which I know to be true, having knitted many a worry into the heel of a sock. I learned at an early age the value of money, and once having acquired the saving habit, it is not possible to be wasteful in later life."

CHAPTER XXV.

AN ELBADRITCHEL HUNT.

Fritz Schmidt, like many another Bucks County boy, had frequently heard the rural tale of a mythical bird called the "Elbadritchel," supposed to be abroad, particularly on cold, dark, stormy nights, when the wind whistled and blew perfect gales around exposed corners of houses and barns. 'Twas a common saying among "Pennsylvania Germans," at such times, "'Tis a fine night to catch 'Elbadritchels.'"

CATCHING ELBADRITCHELS

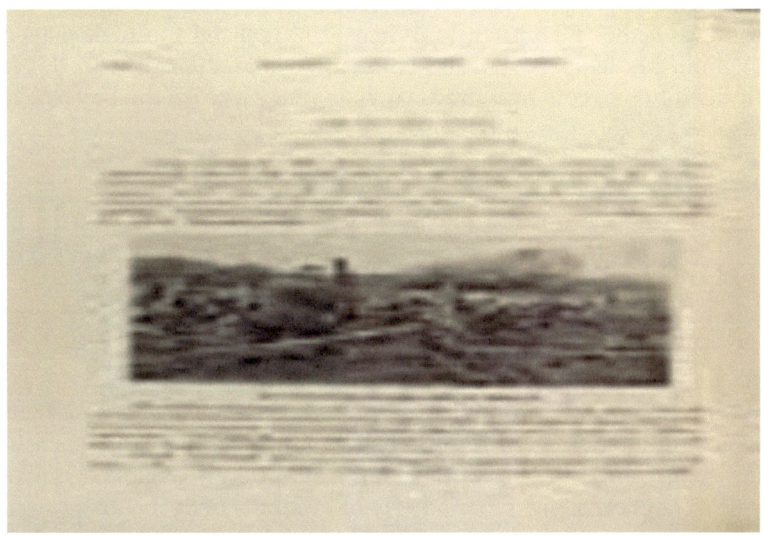

For the information of those who may not even have heard of this remarkable creature, it is described as being a cross between a swallow, a goose and a lyre bird. Have you ever seen an "Elbadritchel?" No one has to my certain knowledge, so I cannot vouch for the truth of this description of it.

Fritz Schmidt had never taught to question the truth of the tale. So, when one cold, stormy night several boys from neighboring farms drove up to the Schmidt homestead and asked Fritz to join them in a hunt for "Elbadritchels," he unhesitatingly agreed to make one of the number, unaware that he had been selected as the victim of a practical joke, and, as usual, was one of the jolliest of the crowd. They drove through a blinding downpour of rain and dismounted on reaching a lonely hill about three miles distant. They gave Fritz a bag to hold. It was fashioned of burlap and barrel hoops, inside of which they placed a lighted candle, and Fritz was instructed how to hold it in order to attract the "Elbadritchel." They also gave him a club with which to strike the bird when it should appear.

The boys scampered off in different directions, ostensibly to chase up the birds, but in reality they clambered into the waiting wagon and were rapidly driven home, leaving Fritz alone awaiting the coming of the "Elbadritchel." When Fritz realized the trick played on him, his feelings may be better imagined than described. He trudged home, cold and tired, vowing vengeance on the boys, fully resolved to get even with them.

CHAPTER XXVI

THE OLD SHANGHAI ROOSTER.

Much of Aunt Sarah's spare time was devoted to her chickens, which fully repaid her for the care given them. She was not particular about fancy stock, but had quite a variety—White Leghorns, Brown Leghorns, big, fat, motherly old Brahma hens that had raised a brood of as many as thirty-five little chicks at one time, a few snow-white, large Plymouth Rocks and some gray Barred one. The *latter* she *liked* particularly because she said they were much, more talkative than any of the others; they certainly did appear to chatter to her when she fed them. She gave them clean, comfortable quarters, warm bran mash on cold winter mornings, alternating with cracked corn and "scratch feed" composed of a mixture of cracked corn, wheat and buckwheat, scattered over a litter of dried leaves on the floor of the chicken house, so they were obliged to work hard for their food.

Old Egg Basket

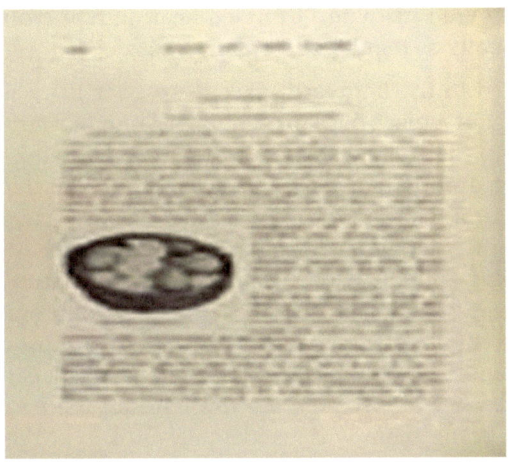

A plentiful supply of fresh water was always at hand, as well as cracked oyster shell. She also fed the chickens all scraps from the table, cutting all meat scraps fine with an old pair of scissors hung conveniently in the kitchen.

She was very successful with the little chicks hatched out when she "set" a hen and the yield of eggs from her hens was usually greater and the eggs larger in size than those of any of her neighbors. This I attribute to her excellent care of them, generous diet, but principally to the fact of the elimination of all the roosters among the flock during the season between the "first of May and December first," with one exception. "Brigham," an immensely large, old, red Shanghai rooster, a most pompous and dignified old chap. A special pet of Aunt Sarah's, she having raised him from a valuable "setting" of eggs given her, and as the egg from which "Brigham," as he was called, emerged, was the only one of the lot which proved fertile, he was valued accordingly and given a longer lease of life than the other roosters, and was usually either confined or allowed to roam outside the chicken yard during the summer months; in the winter, being a swift runner, he usually gobbled up two shares of food before the hens arrived. That accounted for his great size. The old rooster was also noted for his loud crowing.

One day in early Spring, John Landis came into the house hurriedly, saying, "Sarah, your old Shanghai rooster is sick."

"Yes," answered his wife, "I missed hearing him crow this morning; he is usually as regular as an alarm clock."

She hurried to the barnyard, picked up poor Brigham, wrapped him carefully in a piece of blanket and laid him in a small shed. The next morning she was awakened by the lusty crowing of Brigham, who was apparently as well as ever. The next day the same thing happened. Aunt Sarah found him, as she supposed, in a dying condition, and the following morning he was fully recovered. It was quite puzzling until one day John Landis came into the kitchen laughing heartily and said, "Sarah, I am sorry to inform you of the intemperate habits of your pet, Brigham. He is a most disreputable old fellow, and has a liking for liquor. He has been eating some of the brandied cherries which were thrown into the barnyard when the jug containing them was accidentally broken at house cleaning time.

"Well, Sarah, old Brigham was not sick at all—only 'ingloriously' drunk." In the fall of the same year Aunt Sarah spied Brigham one day on top of one of the cider barrels in the shed busily engaged eating the pummace which issued from the bung-hole of the barrel. John Landis, on hearing of Brigham's last escapade, decided, as the rooster was large as an ordinary-turkey, to serve him roasted at Mary's wedding.

Fritz Schmidt remarked one day in the presence of Sibylla: "Chickens must possess some little intelligence; they know enough to go to bed early. Yes, and without an 'alarm clock,' too, Sibylla, eh?"

She walked away without a word to Fritz. The alarm clock was a sore subject with her, and one about which she had nothing to say. Sibylla had never quite forgiven Fritz for the prank played on her. He, happening to hear John Landis tell Sibylla a certain hour he thought a proper time for Jake Crouthamel to take his departure Sunday evenings, Fritz conceived the brilliant (?) idea of setting the alarm clock to "go off" quite early in the evening. He placed the clock at the head of the stairs, and in the midst of an interesting conversation between the lovers the alarm sounded with a loud, whizzing noise, which naturally made quick-tempered Sibylla very angry. She said on seeing Fritz the next morning: "It was not necessary to set the 'waker' to go off, as I know enough to send 'Chake' home when it's time."

Fritz, happening to tell the story to the editor of a small German Mennonite paper, edited in a near-by town, it was printed in that paper in German, which caused Sibylla, on hearing it, to be still more angry at the Professor's son.

CHAPTER XXVII.

"A POTATO PRETZEL."

In the early part of September Mary's Aunt suggested she try to win the prize offered at the Farmers' Picnic in a near-by town for the best "Raised Potato Cake." Aunt Sarah's rye bread invariably captured first prize, and she proposed sending both bread and cake with Sibylla and Jake, who never missed picnic or fair within a radius of one hundred miles.

"POTATO PRETZEL"

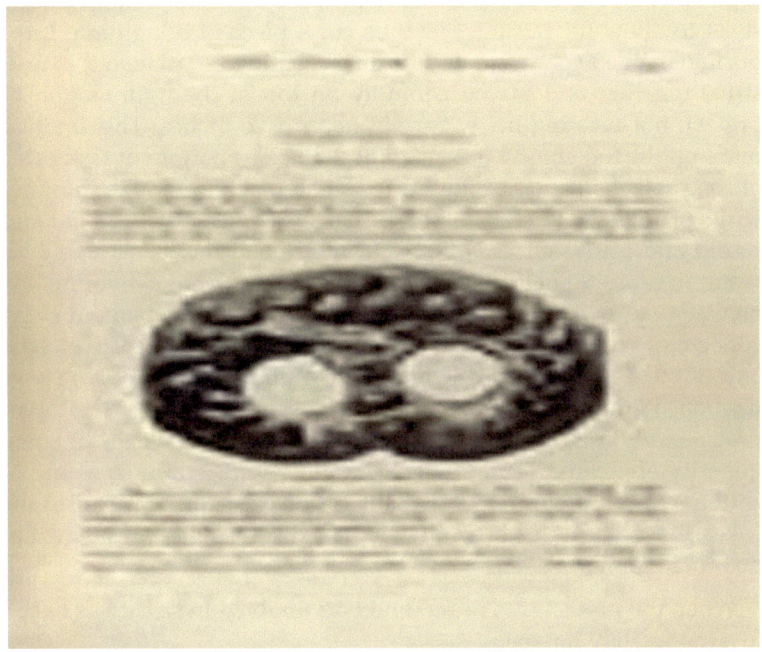

Mary set a sponge the evening of the day preceding that of the picnic, using recipe for "Perfection Potato Cake," which Aunt Sarah considered her best recipe for raised cakes, as 'twas one used by her mother for many years.

On the day of the picnic, Mary arose at five o'clock, and while her Aunt was busily engaged setting sponge for her loaf of rye bread, Mary kneaded down the "potato cake" sponge, set to rise the previous evening, now rounded over top of bowl and light as a feather.

She filled a couple of pans with buns, molded from the dough, and set them to rise. She then, under her Aunt's direction, fashioned the "Pretzel" as follows: She placed a piece of the raised dough on a large, well-floured bake board, rolled it over and over with both hands until a long, narrow roll or strip was formed about the width of two fingers in thickness and placed this strip carefully on the baking sheet, which was similar to the one on which Aunt Sarah baked rye bread; shaped the dough to form a figure eight (8) or pretzel, allowing about two inches of space on either side of baking sheet to allow for raising. She then cut a piece of dough into three portions, rolled each as thick as a finger, braided or plaited the three strips together and placed carefully on top of the figure eight, or pretzel, not meeting by a space of about two inches. This braided piece on the top should not be quite as thick as bottom or first piece of the pretzel. She then rolled three small pieces of dough into tiny strips or rolls the size of small lead pencils, wound them round and round and round into small scrolls, moistened the lower side with water to cause them to adhere, and placed them on the dividing line between the two halves of the figure eight. She placed an old china coffee cup without a handle, buttered on outside, in centre of each half of the figure eight, which kept the pretzel from spreading over the pan. With a small, new paint brush she brushed over the top of Pretzel and Buns, a mixture, consisting of one yolk of egg, an equal quantity of cream or milk (which should be lukewarm so as not to chill the raised dough) and one tablespoon of sugar. This causes the cakes, etc., to be a rich brown when baked, a result to be obtained in no other manner.

When the pretzel was raised and had doubled in size 'twas baked in a moderately hot oven.

Mary's surprise and delight may easily be imagined when Sibylla, on her return from the picnic, handed her the prize she had won, a two-pound box of chocolates, remarking, "Mary, you and Aunt Sarah both got a prize—her's is in the box what Jake's got."

The box on being opened by Aunt Sarah contained a very pretty, silver-plated soup ladle, the prize offered for the best loaf of rye bread.

"Aunt Sarah," inquired Mary one day, "do you think it pays a housekeeper to bake her own bread?"

LOAF OF RYE BREAD

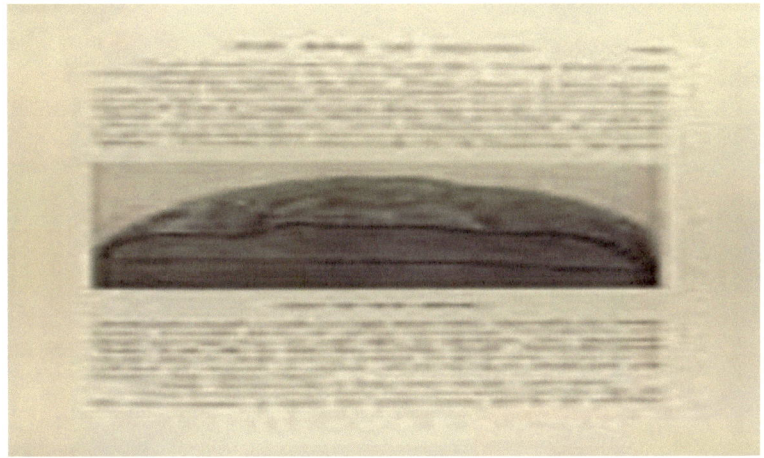

"Certainly, it pays, my dear. From a barrel of flour may be baked three hundred or more one-pound loaves of bread; should you pay five cents a loaf, the bread which may be made from one barrel of flour if bought from a bake shop would cost you fifteen dollars. Now, you add to the cost of a barrel of flour a couple of dollars for yeast, salt, etc., which altogether would not possibly be more than ten dollars, and you see the housewife has saved five dollars. It is true it is extra work for the housewife, but good, wholesome bread is such an important item, especially in a large family, I should advise the thrifty housekeeper to bake her own bread and bake less pie and cake, or eliminate less important duties, to be able to find time to bake bread. From the bread sponge may be made such a number of good, plain cakes by the addition of currants or raisins, which are more wholesome and cheaper than richer cakes."

"I think what you say is true, Aunt Sarah," said Mary.

"Frau Schmidt always bakes her own bread, and she tells me she sets a sponge or batter for white bread, and by the addition of Graham flour, cornmeal or oatmeal, always has a variety on her table with a small expenditure of time and money."

A "BROD CORVEL" OR BREAD BASKET

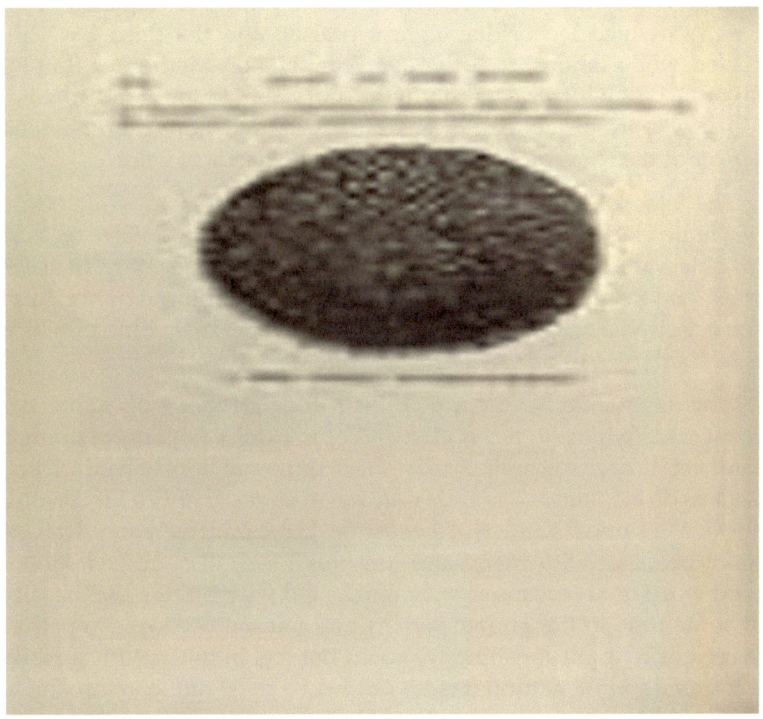

CHAPTER XXVIII.

FAITHFUL SERVICE.

The home-making instinct was so strongly developed in Mary that her share in the labor of cooking and baking became a pleasure. Occasionally she had failures—what inexperienced cook has not?—yet they served only to spur her on to fresh efforts. She had several small scars on her wrist caused by her arm coming in contact with the hot oven when baking. She laughingly explained: "One bar on my arm represents that delicious 'Brod Torte' which Frau Schmidt taught me to bake; the other one I acquired when removing the sponge cake from the oven which Uncle John said 'equaled Aunt Sarah's' (which I consider highest praise), and the third bar I received when taking from the oven the 'Lemon Meringue,' Ralph's favorite pie, which he pronounced 'fine, almost too good to eat.'" Mary was as proud of her scars as a young, non-commissioned officer of the chevron on his sleeve, won by deeds of valor.

The lessons Mary learned that summer on the farm while filling her hope chest and preparing her mind for wifehood were of inestimable value to her in later years. She learned not only to bake, brew and keep house, but from constant association with her Aunt she acquired a self-poise, a calm, serene manner, the value of which is beyond price in this swift, restless age.

One day, while having a little heart-to-heart talk with Mary, her Aunt said: "My dear, never allow an opportunity to pass for doing a kind act. If ever so small, it may cheer some sad, lonely heart. Don't wait to do *big things*. The time may never come. If only a kind word, speak it at once. Kind words cost so little, and we should all be more prodigal with them; and to a tired, sad, discouraged soul, a kind word or act means so very much; and who is there that has not at some time in life known sorrow and felt the need of sympathy? Were our lives all sunshine we could not feel in touch with sorrowing friends. How natural it is for our hearts to go out in sympathy to the one who says 'I have suffered.' Give to your friend the warm hand-clasp and cheery greeting' which cost us nothing in the giving. 'Tis the little lifts which help us over stones in our pathway through life. We think our cross the heaviest when, did we but know the weight of others, we'd not willingly exchange; and remember Mary, 'there are no crown-bearers in Heaven that were not cross-bearers below.' Have you ever read the poem, 'The Changed Cross?' No? Well, I will give it to you to copy in your book of recipes. Should you ever, in future years, feel your cross too heavy to bear, read the poem. How many brave, cheery little women greet us with a smile as they pass. But little do we or any one realize that instead of a song in their hearts the smiles on their lips conceal troubles the world does not suspect, seeking to forget their own sorrows while doing kindly acts for others. They are the real heroes whom the world does not reward with medals for bravery, 'To stand with a smile upon your face against a stake from which you cannot get away, that, no doubt, is heroic; but the true glory is not resignation to the inevitable. To stand unchained, with perfect liberty to go away, held only by the higher claims of duty, and let the fire creep up to the heart, that is heroism.' Ah! how many good women have lived faithful to duty when 'twould have been far easier to have died!"

"FAITHFUL OVER A FEW THINGS."

Matt. xxv: 23.

It may seem to you but a trifle, which you have been called
to do;
Just some humble household labor, away from the public
view,
But the question is, are you faithful, and striving to do your
best,
As in sight of the Blessed Master, while leaving to Him the
rest?
It may be but a little corner, which you have been asked to
fill;
What matters it, if you are in it, doing the Master's will?
Doing it well and faithfully, and doing it with your might;
Not for the praise it may bring you, but because the thing is
right.
In the sight of man you may never win anything like success;
And the laurel crown of the victor may never your temples
press;
If only you have God's approval, 'twill not matter what else
you miss,
His blessing is Heaven beginning, His reward will be perfect
bliss.
Be faithful in every service, obedient to every call;
Ever ready to do His bidding, whether in great things or
small;
You may seem to accomplish little, you may win the praise
of none;

But be sure you will win His favor, and the Master's great "Well Done."
And when at His blessed coming, you stand at His judgment seat;
He'll remember your faithful service and His smile will be Oh! so sweet!
He will bid you a loving welcome, He'll make you to reign for aye,
Over great things and o'er many, with Him, through eternal day.

"THE CHANGED CROSS."

It was a time of sadness, and my heart,
Although it knew and loved the better part,
Felt wearied with the conflict and the strife,
And all the needful discipline of life.
And while I thought on these as given to me,
My trial tests of faith and love to be,
It seemed as if I never could be sure
That faithful to the end I should endure.
And thus, no longer trusting to His might,
Who says, "We walk by faith and not by sight";
Doubting and almost yielding to despair,
The thought arose — My cross I cannot bear.
Far heavier its weight must surely be
Than those of others which I daily see;
Oh! if I might another burden choose,
Methinks I should not fear my crown to lose.
A solemn silence reigned on all around,
E'en nature's voices uttered not a sound;
The evening shadows seemed of peace to tell,
And sleep upon my weary spirit fell.
A moment's pause and then a heavenly light
Beamed full upon my wondering, raptured sight;
Angels on silvery wings seemed everywhere,
And angels' music filled the balmy air.

Then One more fair than all the rest to see—
One to whom all the others bowed the knee—
Came gently to me as I trembling lay,
And, "Follow Me!" He said, "I am the Way."
Then speaking thus, He led me far above,
And there, beneath a canopy of love,
Crosses of divers shapes and sizes were seen,
Larger and smaller than my own had been.
And one there was, most beauteous to behold,
A little one, with jewels set in gold;
Ah! this methought, I can with comfort wear,
For it will be an easy one to bear.
And so, the little cross I quickly took,
But all at once, my frame beneath it shook;
The sparkling jewels fair were they to see,
But far too heavy was their weight for me.
"This may not be," I cried, and looked again
To see if there was any here could ease my pain;
But one by one I passed them slowly by,
Till on a lovely one I cast my eye.
Fair flowers around its sculptured form entwined,
And grace and beauty seemed in it combined;
Wondering, I gazed and still I wondered more,
To think so many should have passed it o'er.
But Oh! that form so beautiful to see,
Soon made its hidden sorrows known to me;
Thorns lay beneath those flowers and colors fair;
Sorrowing, I said. "This cross I may not bear."
And so it was with each and all around,
Not one to suit my need could there be found;
Weeping, I laid each heavy burden down,
As my guide gently said: "No cross, no crown."
At length to him I raised my saddened heart,
He knew its sorrows, bid its doubts depart;
"Be not afraid," He said, "but trust in Me,
My perfect love shall now be shown to thee."
And then with lightened eyes and willing feet,
Again I turned my earthly cross to meet;

With forward footsteps, turning not aside
For fear some hidden evil might betide.
And there, in the prepared, appointed way,
Listening to hear, and ready to obey,
A cross I quickly found of plainest form,
With only words of love inscribed thereon.
With thankfulness, I raised it from the rest,
And joyfully acknowledged it the best;
The only one of all the many there
That I could feel was good for me to bear.
And while I thus my chosen one confessed,
I saw a heavenly brightness on it rest;
And as I bent my burden to sustain,
I recognized my own old cross again.
But, oh! how different did it seem to be!
Now I had learned its preciousness to see;
No longer could I unbelievingly say:
"Perhaps another is a better way."
Oh, no! henceforth my own desire shall be
That He who knows me best should choose for me,
And so whate'er His love sees good to send,
I'll trust its best, because He knows the end.
And when that happy time shall come
Of endless peace and rest,
We shall look back upon our path
And say: "It was the best."

CHAPTER XXIX.

MARY, RALPH, JAKE AND SIBYLLA VISIT THE ALLENTOWN FAIR.

Late in September Jake and Sibylla drove to the Allentown Fair. It was "Big Thursday" of Fair week. They started quite early, long before Ralph Jackson, who had come from the city the day previous, to take Mary to the Fair, had arisen.

SECOND CHURCH BUILDING

Sheltered Liberty Bell, 1777-78. Photographed from the print of an old wood cut used in a German newspaper in the year 1840

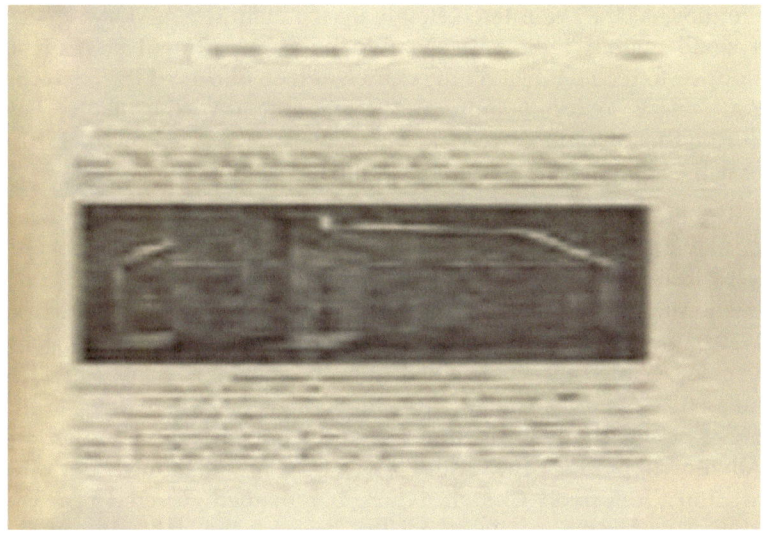

Mary, while appreciating Sibylla's good qualities, never failed to be amused at her broad "Pennsylvania German" dialect.

The morning of the "Fair," Mary arose earlier than usual to allow Sibylla and Jake to get an early start, as it was quite a distance from the farm to the Fair grounds. As they were about to drive away, Sibylla, alighting from the carriage, said, "I forgot my 'Schnup-ftuch.'" Returning with it in her hand, she called, as she climbed into Jake's buggy, "Gut-by, Mary, it looks fer rain."

"Yes" said Jake, "I think it gives rain before we get back yet. The cornfodder in the barn this morning was damp like it had water on it."

And said Mary, "The fragrance of the flowers was particularly noticeable early this morning." Jake, as it happened, was no false prophet. It did rain before evening.

Later in the day, Mary and Ralph drove to a near-by town, leaving horse and carriage at the hotel until their return in the evening, and boarded a train for Allentown. On arriving there, they decided to walk up Hamilton Street, and later take a car out to the Fair grounds. As they sauntered slowly up the main street, Mary noticed a small church built between two large department stores and stopped to read a tablet on the church, which informed the passerby that "this is to commemorate the concealment of the Liberty Bell during the Revolutionary War. This tablet was erected by the Liberty Bell Chapter of the Daughters of the Revolution."

The First Zion's Reformed Church was founded in 1762. In front of the Church a rough block of granite, erected to the memory of John Jacob Mickley, contained the following inscription: "In commemoration of the saving of the Liberty Bell from the British in 1777. Under cover of darkness and with his farm team, he, John Mickley, hauled the Liberty Bell from Independence Hall, Philadelphia, through the British lines, to Bethlehem, where the wagon broke down. The Bell was transferred to another wagon, brought to Allentown, placed beneath the floor of the *Second* Church building of Zion's Reformed Church, where it remained secreted nearly a year. This *tablet* was placed by the order of the Assembly of the Commonwealth of Pennsylvania, June 2nd, 1907, under the auspices of the Pennsylvania Daughters of the Revolution."

This was all very interesting to a girl who had been born and reared in Philadelphia; one who in earliest childhood had been taught to love and venerate the "old Bell."

Ralph was quite as interested in reading about the old Bell as was Mary, and said; "Did you know that the City of Philadelphia purchased the State House property, which included the Bell, in 1818, in consideration of the sum of seventy thousand dollars? No building is ever to be erected on the ground inside the wall on the south side of the State House, but it is to remain a public green and walk forever?"

Liberty Bell Tablet

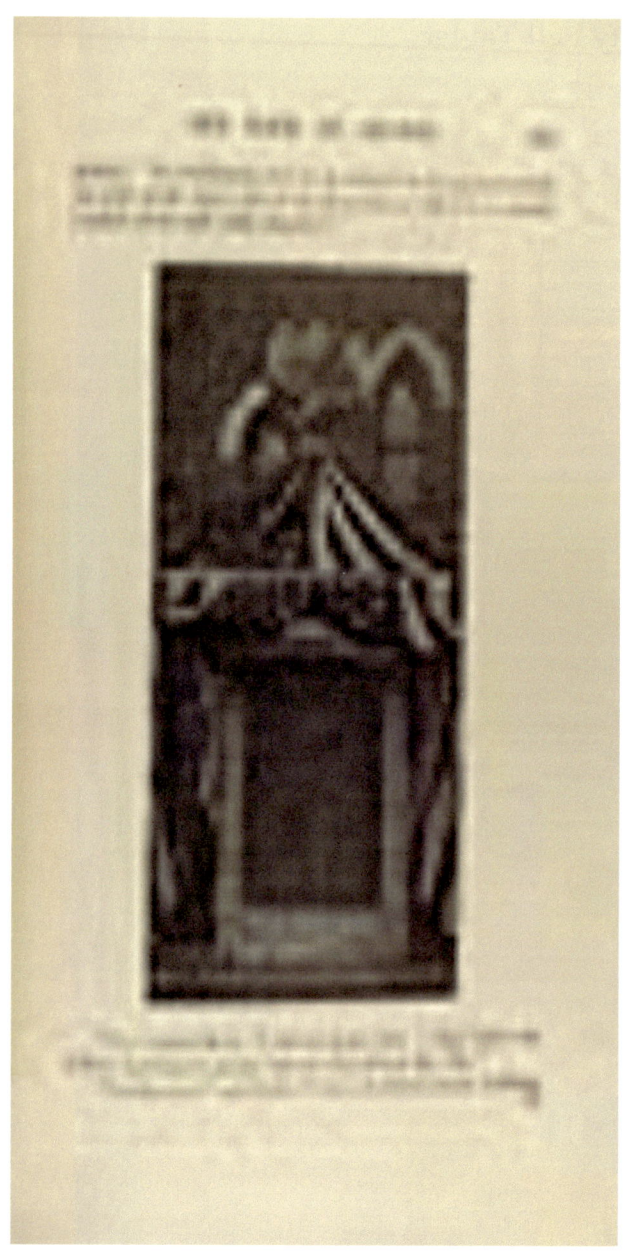

"No," replied Mary, "I did not know that. I don't think we will see anything of greater interest than this at the Fair."

"I understand," said Ralph, "this is the third church building built on this site, where the original church stood in which the Bell was secreted."

Mary, possessing a fair share of the curiosity usually attributed to the "female of the species," on noticing the church door standing ajar, asked Ralph to step inside with her, thinking to find the caretaker within; but no one was visible. A deep silence reigned in the cool, dim interior of the House of God.

One could almost feel the silence, 'twas so impressive. Slowly they walked up the wide church aisle and stood before the quaint baptismal font. A stray sunbeam glancing through one of the beautiful, variously-colored memorial windows, lighted up the pictured saint-like faces over the chancel, making them appear as if imbued with life. Mary softly whispered to Ralph, as if loath to profane the sacredness of the place by loud talking, "I seem to hear a voice saying, 'The Lord is in His holy temple.'" Quietly retracing their steps, they, without meeting any one, emerged into the bright sunlight and were soon in the midst of the turmoil and traffic incident to the principal business street of a city.

The young folks boarded a trolley and in a short time reached the Fair grounds, which offered many attractions to Ralph as well as Mary. The latter was interested in the fine display of needlework, fruits, flowers and vegetables of unusual size. Aunt Sarah's bread won a prize. A blue ribbon attached to Frau Schmidt's highly-prized, old-fashioned, patchwork quilt, showed it to be a winner. Ralph, being interested in the pens of fancy chickens, prize cattle, etc., Mary reluctantly left the woman's department of fancy work, and other interesting things, and accompanied him. On their way to the outlying cattle sheds they noticed two lovers sitting on a bench. Upon a second glance they were convinced that it was Jake and Sibylla. Jake, beaming with happiness, said, "Sibylla vos side by me yet?" They were busily engaged eating a lunch consisting of rolls with hot "weiners" between the two halves, or, as Jake called them, "Doggies," munching pretzels and peanuts between sips of strong coffee, both supremely happy. A yearly visit to the Allentown Fair on "Big Thursday," was *the event* in their dull, prosaic lives.

DURHAM CAVE

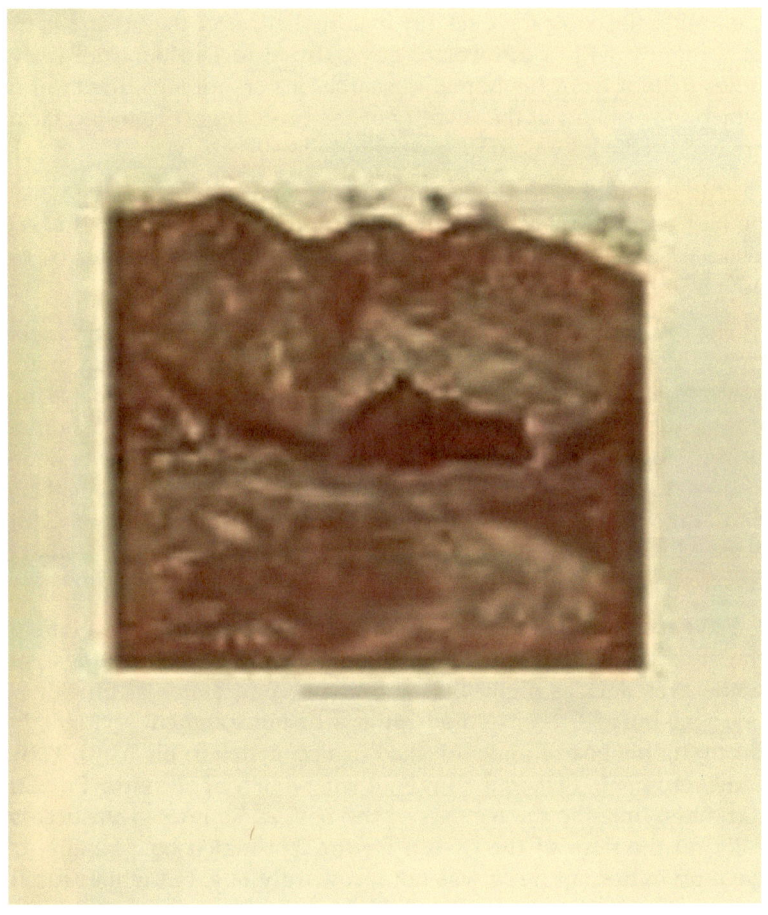

CHAPTER XXX.

FRITZ SCHMIDT EXPLORES DURHAM CAVE.

It appeared to be nothing new for Fritz Schmidt to get into trouble; rather the contrary. One day in early Fall, after the first frost, he, in company with a number of boys, drove to Durham, not many miles distant from his home, in search of persimmons, the crop of which, on account of the severity of the preceding winter, old farmers had predicted would be exceedingly heavy.

Fritz did not tell the boys of his intention to explore a cave which he had been told was in the neighborhood, thinking it would be a good joke to explore the cave first, then tell the boys later of his adventure.

The old gentleman from whom Fritz gained his information relative to the cave aroused the boy's curiosity by saying, "Very many years ago, a skeleton was found in Durham cave and one of the bones, on examination, proved to be the thigh bone of a human being. How he came there, or the manner of his death, was never known." A large room in the cave is known as "Queen Esther's Drawing Room," where, tradition has it "Queen Esther," or Catharine Montour, which was her rightful name, at one time inhabited this cave with some of her Indian followers.

Fritz accidentally stumbled upon the mouth of the cave. None of the other boys being in sight, Fritz quickly descended into the cave, which was dark as night. By lighting a second match as quickly as one was burned, he explored quite a distance, when, accidentally dropping his box of matches, the burning match in his hand, at the same moment, flickered faintly, then went out, leaving Fritz in darkness. Imagine the feelings of the boy, as he groped unsuccessfully on the floor of the cavern for the lost match box. Finally, he gave up in despair. Fritz was not a cowardly boy, but while searching for the matches, he, without thinking, had turned around several times, lost his bearings and knew not in which direction to go to reach the opening of the cave. He heard strange noises which he imagined were bats flopping their wings. There appeared to be something uncanny about the place, and Fritz devoutly wished himself out in the sunshine, when a quotation he had frequently heard his father use came into his mind: "More things are wrought by prayer than this world dreams of." So Fritz knelt down and prayed as he had been taught to pray at his mother's knee, but more earnestly than he had ever prayed before in his life, that God would help him find his way out of the cave, believing that his prayer would be answered. And who shall say it was not answered? For, stumbling onward in the darkness, not knowing if he were coming toward the cave's entrance or going in the opposite direction, he eventually hailed with joy a faint streak of light which he followed, and it soon brought him to the mouth of the cave. He was surprised on joining his companions to find they had not been alarmed at his absence. He had been in the cave only thirty minutes, but to him it had seemed hours. Fritz says to this day he has a horror of Durham Cave or "The Devil's Hole," as it was formerly called.

THE WOODLAND STREAM

CHAPTER XXXI.

MARY'S MARRIAGE.

His vacation ended, after a busy season at the farm, Ralph Jackson returned to his work in the city, strong and robust. He had acquired the coat of tan which Mary's Uncle had predicted. Physically strong as the "Cave Man" of old, he felt capable of moving mountains, and as was natural, he being only a human man, longed for the mate he felt God had intended should one day be his, as men have done since our first gardener, Adam, and will continue to do until the end of time.

When visiting the farm, an event which occurred about every two weeks, Ralph constantly importuned Mary to name an early day for their marriage.

Mary, with a young girl's impulsiveness, had given her heart unreservedly into the keeping of Ralph Jackson, her first sweetheart. Mary was not naturally cold or unresponsive, neither was she lacking in passion. She had had a healthy girlhood, and a wholesome home life. She had been taught the conventional ideals of the marriage relations that have kept the race strong throughout the centuries. Mary possessed great strength of character and fine moral courage. Frequently, not wishing to show her real feeling for the young man; too well poised to be carried off into the wrong channel, defended and excused by many over-sentimental and light-headed novelists of the day, she sometimes appeared almost indifferent to the impetuous youth with warm, red blood leaping in his veins, who desired so ardently to possess her.

Mary's Aunt had taught her the sanctity of parenthood, also that women are not always the weaker sex. There are times when they must show their superiority to "mere man" in being the stronger of the two, mentally if not physically, and Ralph Jackson knew when he called Mary "wife" she would endow him with all the wealth of her pure womanhood, sacredly kept for the clean-souled young man, whose devotion she finally rewarded by promising to marry him the second week in October.

Sibylla Linsabigler, a good but ignorant girl, accustomed to hearing her elder brothers speak slightingly regarding the sanctity of love and marriage, was greatly attached to Mary, whom she admired exceedingly, and looked up to almost as a superior being. She unconsciously imitated many of Mary's ways and mannerisms, and sought to adopt her higher ideals of life and standard of morals.

One Sunday, as Jake Crouthamel was spending the evening with Sibylla, as was his usual custom, he attempted some slight familiarity, which annoyed Sibylla greatly. Jake, noticing the young girl's displeasure at his action, remarked, "I think me Sibylla, you are stuck up yet" (a grave fault in the Bucks County farm hand's opinion).

"No, Chake," Sibylla replied, "I ain't, but Mary, she say a man gives a girl more respect what keeps herself to herself before she is married, and I lofe you Chake and want that you respect me if we marry."

Fritz and Elizabeth Schmidt, on hearing the news of Mary's approaching marriage, promptly begged the privilege of decorating the old farm house parlor for the expected ceremony. They scoured the surrounding woods and countryside for decorations; along old stone fences and among shrubbery by the roadside they gathered large branches of Bitter Sweet. Its racemes of orange-colored fruit, which later in the season becomes beautiful, when the orange gives place to a brilliant red, the outer covering of the berry turns back upon the stems, forming one of the prettiest pictures imaginable in late Autumn. They also gathered branches of feathery wild clematis, which, after the petals had fallen, resembled nothing so much as a cluster of apple seeds, each seed tipped with what appeared like a tiny osprey feather. From the woods near the farm they gathered quantities of trailing ground pine and rainbow-tinted leaves from the numerous brilliant scarlet and yellow maples, which appeared brighter in contrast to the sober-hued trees of shellbark, oak and chestnut.

POLLY SCHMIDT.

The wedding gifts sent to Mary were odd, useful and numerous. The Campfire Girls, to whom she became endeared, gave her a "Kitchen Shower," consisting of a clothes basket (woven by an old basketmaker from the willows growing not far distant), filled to overflowing with everything imaginable that could possibly be useful to a young housekeeper, from the half dozen neatly-hemmed linen, blue ribbon tied, dish clothes, to really handsome embroidered articles from the girls to whom she had given instructions in embroidery during the past summer.

Sibylla's wedding present to Mary was the work of her own strong, willing hands, and was as odd and original as useful. 'Twas a "door mat" made from corn husks, braided into a rope, then sewed round and round and formed into an oval mat. Mary laughingly told Sibylla she thought when 'twas placed on her kitchen doorstep she'd ask every one to please step over it, as it was too pretty to be trod on, which greatly pleased the young girl, who had spent many hours of loving thought and labor on the simple, inexpensive gift.

Mary received from Professor Schmidt a small but excellent copy of one of the world's most famous pictures, "The Night Watch," painted by Rembrandt, in 1642.

"My dear," said the old Professor, "I saw what *was said to be* the original of this painting, the property of Queen Wilhelmina of Holland, at the St. Louis Exposition in 1904. It was in a small, separate building. The size of the picture was fifteen feet by twenty feet. It is the largest and best known of Rembrandt's works. It acquired the wrong title of 'Night Watch' in a period when, owing to the numerous coats of varnish and the effect of smoke and dust, it had gotten so dark in appearance that only the most lucid parts could be discerned. Nowadays, nobody doubts that the light falling from the left on the boisterous company is that of the sun. The musketeers are remarching out of the high archway of their hall, crossing the street in front of it, and going up a bridge. The architecture of the building is a product of Rembrandt's imagination. The steps, also, which we see the men descending, were put there simply to make those at the back show out above those in the front ranks. The march out was to be above all a portrait group. Sixteen persons had each paid their contributions, a hundred guilders on the average, to have their likenesses transmitted to posterity, and every one of them was therefore to be fully visible."

"It is certainly a wonderful picture," said Mary, "and while I have seen few pictures painted by old masters, I think, even with my limited knowledge of art, I cannot fail to appreciate this excellent copy, and I thank you heartily. Professor, and shall always be reminded of you when I look at this copy of a great work."

Mary would not go empty-handed to Ralph at her marriage. Her "hope chest" in the attic was full to overflowing, and quite unique in itself, as it consisted of an old, in fact ancient, wooden dough-tray used in times past by Aunt Sarah's grandmother. Beside it stood a sewing table, consisting of three discarded broom handles supporting a cheese-box cover, with wooden cheese-box underneath for holding Mary's sewing; stained brown and cretonne lined. Mary valued it as the result of the combined labor of herself and Ralph Jackson. A roll of new, home-made rag carpet, patchwork quilts and "New Colonial" rugs, jars of fruit, dried sweet corn, home-made soap, crocks of apple butter, jellies, jams and canned vegetables all bore evidence of Mary's busy Summer at the farm.

The day of Mary's marriage, the twelfth of October, dawned clear and bright, sunshine warm as a day in June. In the centre of the gayly-decorated old farm house parlor, wearing a simple, little, inexpensive dress of soft, creamy muslin, we find Mary standing beside Ralph, who is looking supremely satisfied and happy, although a trifle pale and nervous, listening to the solemn words of the minister. Ralph's "I will" sounded clearly and distinctly through the long room. Mary, with a sweet, serious, faraway look in her blue eyes, repeated slowly after the minister, "I promise to love, honor and"—then a long pause. She glanced shyly up at the young man by her side as if to make sure he was worth it, then in a low, clear tone, added, "obey."

Ralph Jackson certainly deserved the appellation "Cave Man" given him by Fritz Schmidt. He was considerably more than six feet in height, with broad, square shoulders, good features, a clear brain and a sound body. He had never used intoxicants of any description. He sometimes appeared quite boyish in his ways, for on account of his matured look and great size he was frequently judged to be older than he really was.

Aunt Sarah had provided a bounteous repast for the few friends assembled, and while looking after the comfort of her guests tears dimmed the kindly, gray eyes at the thought of parting from Mary.

Small Polly Schmidt, as flower girl at the wedding, was so excited she scarcely knew if she should laugh or cry, and finally compromised by giving Mary what she called a "bear hug," much to Mary's amusement. Fritz gravely said: "Allow me to congratulate you, Mr. Jackson," and turning to Mary, "I wish you a beautiful and happy life, Mrs. Jackson." Mary blushed becomingly on hearing her new name for the first time.

Bidding farewell to friends, Mary and Ralph, accompanied by her Uncle, were driven by "Chake" to the depot in a near-by town, where they boarded the train for the little, newly-furnished home in the suburbs of Philadelphia, the deed of which was Mary's wedding gift from her Uncle, in appreciation of her faithful service on the farm during the summer and for the brightness she had brought into his life and the lives of those with whom she had come in contact, as every one at the farm had felt the captivating charm and winning sweetness of the young girl.

As the train came in sight, the old gentleman, in a voice husky with emotion, bade the young couple, just starting the journey of life together, an affectionate farewell, and repeated solemnly, almost as a benediction, "Es Salamu Aleikum."

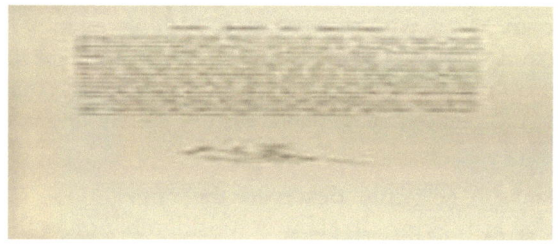

MARY'S COLLECTION OF RECIPES

❖❖❖❖❖❖❖❖❖❖❖❖❖❖

SMALL ECONOMIES, "LEFT-OVERS" OR "IVERICH BLEIBST" AS AUNT SARAH CALLED THEM.

"The young housewife," said Aunt Sarah to Mary, in a little talk on small economies in the household, "should never throw away pieces of hard cheese. Grate them and keep in a cool, dry place until wanted, then spread lightly over the top of a dish of macaroni, before baking; or sprinkle over small pieces of dough remaining after baking pies, roll thin, cut in narrow strips like straws, and bake light brown in a hot oven, as 'Cheese Straws.'"

Wash and dry celery tips in oven, and when not wished for soup they may be used later for seasoning. The undesirable outer leaves of a head of lettuce, if fresh and green, may be used if cut fine with scissors, and a German salad dressing added. The heart of lettuce should, after washing carefully, be placed in a piece of damp cheese cloth and put on ice until wanted, then served at table "au natural," with olive oil and vinegar or mayonnaise dressing to suit individual taste. Should you have a large quantity of celery, trim and carefully wash the roots, cut them fine and add to soup as flavoring. Almost all vegetables may be, when well cooked, finely mashed, strained, and when added to stock, form a nourishing soup by the addition of previously-cooked rice or barley. Add small pieces of meat, well-washed bones cut from steaks or roasts, to the stock pot. Small pieces of ham or bacon (left-overs), also bacon or ham *gravy* not thickened with flour may be used occasionally, when making German salad dressing for dandelion, endive, lettuce or water cress, instead of frying fresh pieces of bacon.

AN OLD FASHIONED BUCKS COUNTY BAKE OVEN

It is a great convenience, also economical, to keep a good salad dressing on hand, and when the white of an egg is used, the yolk remaining may he added at once to the salad dressing (previously prepared). Mix thoroughly, cook a minute and stand away in a cool place. Young housekeepers will be surprised at the many vegetables, frequently left-overs, from which appetizing salads may be made by the addition of a couple tablespoonfuls of mayonnaise, besides nut meats, lettuce, watercress, celery and fruit, all of which may be used to advantage. A good potato salad is one of the cheapest and most easily prepared salads. A German dressing for dandelions, lettuce or potatoes may be prepared in a few minutes by adding a couple of tablespoonfuls of salad dressing (which the forehanded housewife will always keep on hand) to a little hot ham or bacon gravy. Stirring it while hot over the salad and serving at once.

A cup of mashed potatoes, left over from dinner, covered and set aside in a cool place, may be used the next day, with either milk or potato water, to set a sponge for "Dutch Cake," or cinnamon buns with equally good results as if they had been freshly boiled (if the potatoes be heated luke-warm and mashed through a sieve); besides the various other ways in which cold boiled potatoes may be used.

Fruit juices or a couple tablespoonfuls of tart jelly or preserved fruit may be added to mincemeat with advantage. Housewives should make an effort to give their family good, plain, nourishing, wholesome food. The health of the family depends so largely on the quality of food consumed. When not having time, strength or inclination to bake cake, pies or puddings, have instead good, sweet, home-made bread and fruit; if nothing else, serve stewed fruit or apple sauce. Omit meat occasionally from the bill of fare and serve instead a dish of macaroni and cheese and fruit instead of other dessert. Serve a large, rich, creamy rice pudding for the children's lunch. When eggs are cheap and plentiful make simple custards, old-fashioned cornmeal puddings, tapioca, bread puddings and gelatine with fruits. These are all good, wholesome, and not expensive, and in Summer may be prepared in the cool of the early morning with small outlay of time, labor or money. Plan your housework well the day before and have everything in readiness. The pudding may be placed in the oven and baked white preparing breakfast, economizing coal and the time required for other household duties.

Every wife and mother who does her own housework and cooking these days (and their number is legion) knows the satisfaction one experiences, especially in hot weather, in having dinner and luncheon planned and partly prepared early in the morning before leaving the kitchen to perform other household tasks.

Another small economy of Aunt Sarah's was the utilizing of cold mashed potatoes in an appetizing manner. The mashed potatoes remaining from a former meal were put through a small fruit press or ricer to make them light and flaky. To one heaped cup of mashed potatoes (measured before pressing them through fruit press) she added ¾ cup of soft, stale bread crumbs, ¼ cup of flour sifted with ¼ teaspoonful of baking powder. Mix in lightly with a fork yolk of one egg, then the stiffly beaten white, seasoned with salt and a little minced onion or parsley, or both. With well-floured hands she molded the mixture into balls the size of a shelled walnut, dropped into rapidly boiling water and cooked them uncovered from 15 to 20 minutes, then skimmed them from the water and browned in a pan with a little butter and served on platter with meat, a pot roast or beef preferred. From the above quantity of potatoes was made five potato balls.

THE MANY USES OF STALE BREAD

Never waste stale bread, as it may be used to advantage in many ways. The young housewife will be surprised at the many good, wholesome and appetizing dishes which may be made from stale bread, with the addition of eggs and milk.

Take a half dozen slices of stale bread of equal size and place in a hot oven a few minutes to become crisped on the outside so they may be quickly toasted over a hot fire, a delicate brown. Butter them and for breakfast serve with a poached egg on each slice.

A plate of hot, crisp, nicely-browned and buttered toast is always a welcome addition to the breakfast table.

Serve creamed asparagus tips on slices of toast for luncheon.

The economical housewife carefully inspects the contents of her bread box and refrigerator every morning before planning her meals for the day, and is particular to use scraps of bread and left-over meat and vegetables as quickly as possible. Especially is this necessary in hot weather. Never use any food unless perfectly sweet and fresh. If otherwise, it is unfit for use.

Loaves of bread which have become stale can be freshened if wrapped in a damp cloth for a few minutes, then remove and place in a hot oven until heated through.

For a change, toast slices of stale bread quite crisp and serve a plate of hot, plain toast at table, to be eaten broken in small pieces in individual bowls of cold milk. Still another way is to put the stiffly-beaten white of an egg on the centre of a hot, buttered slice of toast, carefully drop the yolk in the centre of the beaten white and place in hot oven a few minutes to cook. Serve with a bit of butter on top, season with pepper and salt. Serve at once.

Another way to use stale bread is to toast slices of bread, spread with butter, pour over 1 cup of hot milk, in which has been beaten 1 egg and a pinch of salt. Serve in a deep dish. Or a cup of hot milk may be poured over crisply-toasted slices of buttered bread, without the addition of an egg.

"BROD GRUMMELLA"

In a bowl containing 1 cup of soft bread crumbs pour 1 cup of sweet milk, then add the slightly-beaten yolks of three eggs, a little pepper and salt, then the stiffly-beaten whites of the three eggs. Place in a fry-pan a tablespoonful of butter and 1 of lard or drippings; when quite hot pour the omelette carefully in the pan. When it begins to "set" loosen around the edges and from the bottom with a knife. When cooked turn one side over on the other half, loosen entirely from the pan, then slide carefully on a hot platter and serve at once. Garnish with parsley.

CROUTONS AND CRUMBS

Still another way is to make croutons. Cut stale bread into small pieces, size of dice, brown in hot oven and serve with soup instead of serving crackers. Small pieces of bread that cannot be used otherwise should be spread over a large pan, placed in a moderate oven and dried until crisp. They may then be easily rolled fine with a rolling-pin or run through the food chopper and then sifted, put in a jar, stood in a dry place until wanted, but not in an air-tight jar. Tie a piece of cheese-cloth over the top of jar. These crumbs may be used for crumbing eggplant, oysters, veal cutlets or croquettes. All should be dipped in beaten white of eggs and then in the crumbs, seasoned with salt and pepper, then floated in a pan of hot fat composed of ⅔ lard and ⅓ suet. All except veal cutlets. They should be crumbed, not floated in deep fat, but fried slowly in a couple tablespoonfuls of butter and lard.

Also fry fish in a pan of hot fat. Shad is particularly fine, prepared in this manner (when not baked). Cut in small pieces, which when breaded are floated in hot fat. If the fat is the right temperature when the fish is put in, it absorbs less fat than when fried in a small quantity of lard and butter.

"ZWEIBACH"

Cut wheat bread in slices not too thin. Place in a warm, not hot, oven, and allow it to remain until thoroughly dry and crisp. Place in a toaster or a wire broiler over a hot fire and toast a golden brown and allow it to remain in the oven until toasted. Keep in cool place until used. Zweibach is considered more wholesome than fresh bread.

"GERMAN" EGG BREAD

Cut stale bread into slices about ¾ inch thick. Cut slices in half, and soak for a few minutes, turning frequently, in the following mixtures: 1 pint of sweet milk, 3 eggs, 1 teaspoonful flour mixed smooth with a little of the cold milk and a pinch of salt. Fry half dozen slices of thinly-sliced bacon in a pan. Put bacon, when fried, in oven to keep hot. Dip the slices of soaked bread in fine, dried bread crumbs and fry quickly in the bacon fat (to which has been added one tablespoon of butter) to a golden brown. Serve at once on the same platter with the bacon, or instead of using bacon fat, fry the crumbed bread in sweet drippings, or a tablespoonful each of lard and butter. This is an appetizing and wholesome breakfast or luncheon dish, served with a tart jelly, either currant or grape.

CREAMED TOAST

Partly fill a large tureen with slices of crisply-browned and buttered toast. (Slices of bread which have become dry and hard may be used for this dish.) When ready to serve, not before, pour over the toasted slices 1 quart of hot milk to which 1 teaspoonful of flour or cornstarch has been added, after being mixed smoothly with a little cold milk or water and cooked a few minutes until thick as cream. Add also a pinch of salt.

If milk is not plentiful, prepare one pint of milk and dip each slice of toasted bread quickly in a bowl of hot water; place in a deep dish and quickly pour over the hot milk, to which a tablespoonful of butter has been added, and serve at once.

BREAD AND ROLLS

Bread, called the "Staff of Life," on account of its nutritive value, should head the list of foods for human consumption. Bread making should stand first in the "Science of Cooking," as there is no one food upon which the comfort, health and well-being of the average family so largely depends as upon good bread. There is absolutely no reason why the housewife of the present day should not have good, sweet, wholesome, home-made bread, if good yeast, good flour and common-sense are used. The milk or water used to mix with flour for making bread sponge should be lukewarm. If too hot, the loaves will be full of holes and coarse grained. If too cold the bread, chilled, will not rise as it should have done had the liquid used been the right temperature. Good bread may be made by using milk, potato water or whey (drained from thick sour milk), and good bread may be made by simply using lukewarm water. I prefer a mixture of milk and water to set sponge. Milk makes a fine-grained, white bread, but it soon dries out and becomes stale. Bread rises more slowly when milk is used. When mashed potatoes are used, the bread keeps moist a longer time. Should you wish extra fine, white, delicate bread, add one cup of sweet cream to the liquid when setting sponge. When milk is used the dough is slower in rising, but makes a creamy-looking and fine-flavored bread. When one Fleischman yeast cake is used in any recipe the ordinary half-ounce cake of compressed yeast is intended, twenty-eight cakes in a pound. These are usually kept in a large refrigerator in a temperature of 44 degrees and should not be kept longer in the home than three days in Summer or six days in Winter, and should always be kept in a cool place until used, if the cook would have success when using.

Use the best hard, Spring wheat flour obtainable for baking bread, or any sponge raised with yeast, as this flour contains a greater quantity of gluten and makes bread of high nutritive value.

Winter wheat maybe used for cake-making and for baking pastry with excellent results, although costing less than Spring wheat.

Always sift flour before using, when setting sponge for bread. When mixing sponge use one quart liquid to about three pounds of flour. "Aunt Sarah" always cut several gashes with a sharp knife on top of loaves when ready to be placed in oven. She also made several cuts across the top of loaves with a hot knife when set to rise to allow gas to escape. If an impression made on a loaf of bread with the finger remains, the bread is light. If the dent disappears, then the loaf is not light enough to be placed in the oven; give it more time to rise. An experienced cook, noted for the excellence and size of her loaves of bread, said she always inverted a pail over the pan containing loaves of bread when set to rise, and allowed the bread to remain covered after being placed in the oven. Loaves will rise to a greater height if this is done. Remove the covering to allow loaves to brown a short time before taking them from the oven. "Aunt Sarah" frequently placed four loaves in her large roasting pan, covered the pan, when set to rise, and allowed the cover to remain until loaves were nearly baked. She brushed the top and sides of loaves with melted butter when set to rise to allow of their being broken apart easily. A more crusty loaf is secured by placing each loaf singly in medium-sized bread tins.

Aunt Sarah considered Fleischman's compressed yeast the best commercial yeast in use, both quick and reliable, but thought better bread was never made than that made by her mother, as she had been taught to make it in years past, by the old-fashioned and slower "sponge method." She was invariably successful in making sweet, wholesome bread in that manner. She used home-made potato yeast or "cornmeal yeast cakes," under different names, always with good results.

Good bread may be made either by the old-fashioned "sponge" method or "straight." Sponge method consists of a batter mixed from liquid yeast (usually home-made potato yeast is used) and a small part of the flour required for making the bread. This batter was usually set to rise at night and mixed up in the centre of a quantity of flour, in an old-fashioned wooden dough tray. The following morning enough flour was kneaded in to form a dough, and when well-raised and light, this dough was formed into loaves and placed in pans for the final rising. The more easily and more quickly made "straight" dough, when using Fleischman's compressed yeast, is mixed in the morning and all the ingredients necessary are added at one time. It is then set to rise and, when the dough has doubled in bulk, it is kneaded down and when risen to once and half its size, shaped into loaves, placed in pans to rise and, when risen to top of pans, bake.

Better bread may be made from flour not freshly milled. Flour should be kept in a dry place; it improves with moderate age. Stand flour in a warm place to dry out several hours before using if you would have good bread.

When baking bread the heat of the oven should not be *too great* at *first*, or the outside of the bread will harden too quickly and inside the loaves will not be thoroughly baked before the crust is thick and dark. The temperature of the oven and time required for baking depend upon the size of the loaves, yet the bread should be placed in rather a quick oven, one in which the loaves should brown in about fifteen minutes, when the heat may be reduced, finishing the baking more slowly.

Small biscuits and rolls can stand a much hotter oven and quicker baking than large loaves, which must be heated slowly, and baked a longer time. A one-pound loaf should bake about one hour. On being taken from the oven, bread should be placed on a sieve, so that the air can circulate about it until it is thoroughly cooled. In the *Farmers' Bulletin*, we read: "The lightness and sweetness depend as much on the way bread is made as on the materials used." The greatest care should be used in preparing and baking the dough and in cooking and keeping the finished bread. Though good house-keepers agree that light, well-raised bread can readily be made, with reasonable care and attention, heavy, badly-raised bread is unfortunately very common. Such bread is not palatable and is generally considered to be unwholesome, and probably more indigestion has been caused by it than by any other badly-cooked food. As compared with most meats and vegetables, bread has practically no waste and is very completely digested, but it is usually too poor in proteins to be fittingly used as the sole article of diet, but when eaten with due quantities of other food, it is invaluable and well deserves its title of "Staff of Life."

When the housewife "sets" bread sponge to rise over night, she should mix the sponge or dough quite late, and early in the morning mold it at once into shapely-looking loaves (should the sponge have had the necessary amount of flour added the night before for making a stiff dough).

Being aware of the great nutritive value of raisins and dried currants, Aunt Sarah frequently added a cup of either one or the other, well-floured, to the dough when shaping into loaves for the final rising.

Aunt Sarah frequently used a mixture of butter and lard when baking on account of its being more economical, and for the reason that a lesser quantity of lard may be used; the shortening qualities being greater than that of butter. The taste of lard was never detected in her bread or cakes, they being noted for their excellence, as the lard she used was home-rendered, almost as sweet as dairy butter, free from taste or odor of pork. She always beat lard to a cream when using it for baking cakes, and salted it well before using, and I do not think the small quantity used could be objected to on hygienic principles.

I have read "bread baking" is done once every three or four weeks, no oftener, in some of the farm houses of Central Europe, and yet stale bread is there unknown. Their method of keeping bread fresh is to sprinkle flour into a large sack and into this pack the loaves, taking care to have the top crusts of bread touch each other. If they have to lie bottom to bottom, sprinkle flour between them. Swing the sack in a dry place. It must swing and there must be plenty of flour between the loaves. It sounds more odd than reasonable, I confess.

"BUCKS COUNTY" HEARTH-BAKED RYE BREAD (AS MADE BY AUNT SARAH)

- 1 quart sweet milk (scalded and cooled).
- 1 tablespoonful lard or butter.
- 2 table spoonsful sugar.
- ½ tablespoonful salt.
- 1 cup wheat flour.
- 3 quarts rye flour (this includes the one cup of wheat flour).
- 1 Fleischman yeast cake or 1 cup of potato yeast.

"BUCKS COUNTY" RYE BREAD

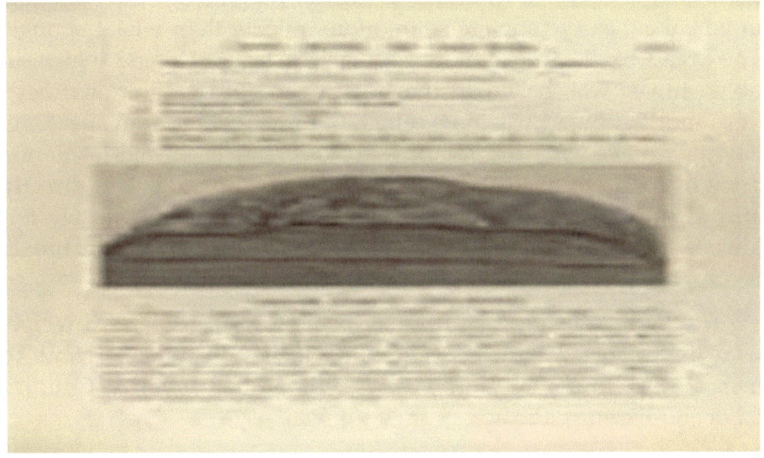

Pour 1 quart of luke-warm milk in a bowl holding 7 quarts. Add butter, sugar and salt, 1½ quarts rye flour and 1 cup of yeast, or one Fleischman's yeast cake, dissolved in a little lukewarm water. Beat thoroughly, cover with cloth, and set in a warm place to rise about three hours, or until it almost reaches the top of bowl. When light, stir in the remaining 1½ quarts of rye flour, in which one cup of wheat is included; turn out on a well-floured bake board and knead about twenty minutes. Shape dough into one high, round loaf, sprinkle flour *liberally* over top and sides of loaf, and place carefully into the clean bowl on top of a *well-floured* cloth. Cover and set to rise about one hour, when it should be light and risen to top of bowl. Turn the bowl containing the loaf carefully upside down on the centre of a hot sheet iron taken from the hot oven and placed on top of range. A tablespoonful of flour should have been sifted over the sheet iron before turning the loaf out on it. Remove cloth from dough carefully after it has been turned from bowl and place the sheet iron containing loaf *immediately in the hot oven*, as it will then rise at once and not spread. Bake at least sixty minutes. Bread is seldom baked long enough to be wholesome, especially graham and rye bread. When baked and still hot, brush the top of loaf with butter and wash the bottom of loaf well with a cloth wrung out of cold water, to soften the lower, hard-baked crust. Wrap in a damp cloth and stand aside to cool where the air will circulate around it. Always set rye bread to rise early in the morning of the same day it is to be baked, as rye sponge sours more quickly than wheat sponge. The bread baked from this recipe has the taste of bread which, in olden times, was baked in the brick ovens of our grandmother's day, and that bread was unexcelled. I know of what I am speaking, having watched my grandmother bake bread in an old-fashioned brick oven, and have eaten hearth-baked rye bread, baked directly on the bottom of the oven, and know, if this recipe be closely followed, the young housewife will have sweet, wholesome bread. Some Germans use Kumel or Caraway seed in rye bread.

Aunt Sarah's loaves of rye bread, baked from the above recipe, were invariably 3½ inches high, 14½ inches in diameter and 46 inches in circumference and always won a blue ribbon at Country Fairs and Farmers' Picnics.

In the oven of Aunt Sarah's range was always to be found a piece of sheet iron 17 inches in length by 16 inches in width. The three edges of the sheet iron turned down all around to a depth of half an inch, the two opposite corners being cut off about a half inch, to allow of its being turned down. It is a great convenience for young housewives to possess two of these sheet-iron tins, or "baking sheets," when baking small cakes or cookies, as being raised slightly from the bottom of the oven, cakes are less liable to scorch and bake more evenly. One sheet may be filled while baking another sheetful of cakes. In this manner a large number of cakes may be baked in a short time. This baking sheet was turned the opposite way, upside down, when baking a loaf of rye bread on it, and when the loaf of bread was partly baked the extra baking sheet was slipped under the bottom of the one containing the loaf, in case the oven was quite hot, to prevent the bottom of the bread scorching. Wheat bread may be baked in the same manner as rye bread, substituting wheat flour for rye. These baking sheets may be made by any tinsmith, and young housewives, I know, would not part with them, once they realize how invaluable they are for baking small cakes on them easily and quickly.

"FRAU SCHMIDTS" GOOD WHITE BREAD (SPONGE METHOD)

To one quart of potato water, drained from potatoes which were boiled for mid-day dinner, she added about ½ cup of finely-mashed hot potatoes and stood aside. About four o'clock in the afternoon she placed one pint of lukewarm potato water and mashed potatoes in a bowl with ¼ cup of granulated sugar and ½ a dissolved Fleischman's yeast cake, beat all well together, covered with a cloth and stood in a warm place until light and foamy. About nine o'clock in the evening she added the reserved pint of (lukewarm) potato water and ½ tablespoonful of salt to the yeast sponge, with enough warmed, well-dried flour to stiffen, and kneaded it until dough was fine-grained. She also cut through the dough frequently with a sharp knife. When the dough was elastic and would not adhere to molding-board or hands, she placed it in a bowl, brushed melted lard or butter over top to prevent a crust forming, covered warmly with a cloth and allowed it to stand until morning. Frau Schmidt always rose particularly early on bake day, for fear the sponge might fall or become sour, if allowed to stand too long. She molded the dough into four small loaves, placed it in pans to rise until it doubled its original bulk. When light she baked it one hour. Bread made according to these directions was fine-grained, sweet and wholesome. She always cut several gashes across top of loaf with a sharp knife when loaves were set to rise, to allow gas to escape.

EXCELLENT "GRAHAM BREAD"

At 6.30 A.M. place in a quart measure ½ cup of sweet cream and 3½ cups of milk, after being scalded (1 quart all together). When lukewarm, add 1 Fleischman yeast cake, dissolved in a little of the luke warm milk, 3 tablespoonfuls sugar and 1 tablespoonful salt. Add 3 cups each of white bread flour and 3 cups of graham flour (in all 6 cups or 1½ quarts of flour). Mix well together and stand in a warm place, closely covered, a couple of hours, until well-risen. Then stir sponge down and add about 2½ cups each of graham and of white flour. (Sponge for graham bread should not be quite as stiff as a sponge prepared from white flour.) Set to rise again for an hour, or longer; when light, stir down sponge and turn on to a well-floured board. Knead well, divide into four portions, mold into four small, shapely loaves, brush with soft butter, place in well-greased pans, set to rise, and in about one hour they should be ready to put in a moderately-hot oven. Bake about fifty minutes. Graham bread should be particularly well-baked. Brush loaves, when baked, with butter, which makes a crisp crust with a nutty flavor.

Should cream not be available, one quart of scalded milk, containing one tablespoonful of butter, may be used with good results. If cream be used with the milk, no shortening is required in the bread. Bread is considered more wholesome when no shortening is used in its preparation.

GRAHAM BREAD (AN OLD RECIPE)

- 2 cups sour milk
- 2 cups sweet milk or water.
- 1 teaspoon soda (Salaratus)
- Graham flour.
- ½ cup molasses.
- 1 tablespoonful melted butter.
- Pinch of salt.

Stiffen about as thick as ordinary molasses cake. Bake at once.

"MARY'S" RECIPE FOR WHEAT BREAD

- 1 cup sweet milk (scalded).
- 1 cup cold water.
- 1 cake Fleischman's yeast (dissolved in a small quantity of luke-warm water).

- 1½ teaspoonfuls sugar.
- 1 rounded teaspoonful salt.
- 1 tablespoonful butter.
- Flour, about 1½ quarts.

This makes good bread and, as bread is apt to chill if set over night in a cold kitchen, or sour if allowed to stand over night in summer, set this sponge early in the morning. Stiffen with flour and knead about 25 minutes; place the dough in a covered bowl in a warm place to rise about two hours and when well-risen and light, knead and stand one hour. Then mold into shapely loaves, place in pans, brush tops of loaves with melted butter, and when doubled in bulk, in about 45 minutes put in an oven which is so hot you can hold your hand in only while you count thirty, or if a little flour browns in the oven in about six minutes, it is hot enough for bread. The oven should be hot enough to brown the bread slightly five minutes after being put in. Medium-sized loaves of bread require from ¾ of an hour to one hour to bake. When bread is sufficiently baked it can be told by turning the loaf over and rapping with the knuckles on the bottom of the loaf. If it sounds hollow, it is thoroughly baked, and should be taken from the oven. Stand loaves up on end against some object, where the air can circulate around them, and brush a little butter over the top to soften the crust. An authority on the chemistry of foods cautious housewives against cooling loaves of bread too rapidly after taking from the oven, and I should like to add a word of caution against eating fresh breads of any kind. Bread should be baked at least twelve hours before being eaten. The sponge for this bread was set at 6 o'clock in the morning; bread was baked at 10.30.

From 1 pint of liquid, 1 cake of yeast and about 1½ quarts of flour were made two loaves of bread. More yeast is required to raise a sponge containing sugar, eggs and shortening than is required to raise bread sponge containing only liquid, flour and yeast.

"FRAU SCHMIDTS" EASILY-MADE GRAHAM BREAD

Should you care to have a couple of loaves of graham bread instead of all-wheat, take a generous cup of the above sponge before it is stiffened beyond a thick batter, and add one tablespoonful of brown sugar or molasses, stiffen with graham flour (not quite as stiff as when making wheat bread), rub butter or lard on top of dough, cover and set in a warm place to rise. When light, mold into one small loaf (never make graham bread into large loaves), place in oblong pan, cover, let stand until light, about 1½ hours, when it should have doubled in size; put in oven and bake thoroughly. When the loaf is taken from the oven, brush butter over the top. This keeps the crust moist.

If a wholesome loaf of "Corn Bread" is wished, use fine, yellow, granulated cornmeal to stiffen the sponge instead of graham flour; do not make dough too stiff.

WHOLE-WHEAT BREAD

- 1 pint boiling water.
- 1 pint sweet milk.
- ½ Fleischman's yeast cake dissolved in luke-warm water.
- ½ tablespoon salt.
- Flour.

When the milk and water are lukewarm add the yeast cake and salt. Then add enough whole wheat flour to make a thin batter. Let stand in a warm place three or four hours. Then stir in as much wheat flour (whole wheat) as can be stirred in well with a large spoon, and pour into well-greased pans. Let rise to double its bulk; then bake from three-fourths to one hour, according to the size of the loaves. This quantity makes three loaves.

NUT BREAD

- 3 cups graham flour.
- 1 cup wheat flour.
- 4 teaspoons baking powder.
- 1 cup chopped English walnuts.
- 1 cup sugar.
- 1 small teaspoon "Mapleine" flavoring (if liked).
- ½ cup milk.
- Pinch salt.
- ½ cup floured raisins (seeded).

Put in a good-sized bread pan and bake on hour in a moderate oven. Strange as it may seem, this bread is lighter and better if allowed to stand a half hour before being placed in the oven to bake.

FRAU SCHMIDTS "QUICK BREAD"

The Professor's wife seldom used any liquid except water to set a sponge for bread. She seldom used any shortening. She taught Mary to make bread by the following process, which she considered superior to any other. From the directions given, housewives may think more time devoted to the making of a couple of loaves of bread than necessary; also, that too great a quantity of yeast was used; but the bread made by "Frau Schmidt" was excellent, quickly raised and baked.

The whole process consumed only about four hours' time, and how could time be more profitably spent than in baking sweet, crusty loaves of bread, even in these strenuous days when the efficient housekeeper plans to conserve strength, time and labor?

First, two Fleischman's compressed yeast cakes were placed in a bowl and dissolved with 4 tablespoonfuls of luke-warm water; she then added 1 cup of lukewarm water, ½ tablespoonful of sugar and ½ teaspoonful of salt and stirred all well together. The bowl containing this yeast foam was allowed to stand in a warm place, closely covered, one hour.

At the end of that time the yeast mixture should be light and foamy. It was then poured into the centre of a bowl containing about 4½ cups of *warmed* flour, mixing the foamy yeast with a *portion* of the flour to make a soft sponge, leaving a wall of flour around the inside edge of bowl, as our grandmothers used to do in olden times when they mixed a sponge for bread of liquid flour and yeast, in one end of the old-fashioned wooden "dough tray," using a wooden stick or small paddle for stirring together the mixture.

The bowl containing the sponge was placed in a warm place to rise. In about 15 or 20 minutes ½ cup of lukewarm water was added to the sponge, stirring in all the outside wall of flour until a dough, the proper consistency for bread, was formed. The dough was turned out on the molding board and given a couple of quick, deft turns with the hands for several minutes, then placed in the bowl and again set to rise in a warm place, free from draughts, for 25 or 30 minutes. When light, with hands slightly greased with butter, she kneaded the dough a short time, until smooth and elastic, divided the dough into two portions, placed each loaf in warmed, well-greased bread pans and stood in a warm place about ¼ hour. Then turned the contents of bread pans onto bake-board, one at a time. Cut each loaf into three portions, rolled each piece into long, narrow strips with the palms of the hands. Pinched ends of the three strips together and braided or plaited them into a braid almost the length of bread pan. Placed each braided loaf in a bread pan and set to raise as before. When well-raised, brush the top of loaves with melted butter. Bake about three-quarters of an hour in a moderately-hot oven. An old-fashioned way of testing the heat of the oven was to hold the hand in the oven while counting thirty. Should one be unable to bear the heat of oven a longer time, then the temperature was correct for baking bread. Should one be able to allow the hand to remain in the oven a longer time, the heat of the oven should be increased.

As a result of carefully following these minute directions, even an inexperienced housewife should have sweet, wholesome bread.

Frau Schmidt insisted that rolling portions of dough separately before combining in a loaf, as for braided loaves, caused the bread to have a finer texture than if just shaped into round loaves.

AN "OATMEAL LOAF"

For a loaf of oatmeal bread, place 1 cup of crushed oats, or common oatmeal, in a bowl, pour over ½ cup of hot milk. When luke warm, add 1 cup of sponge, or batter, reserved from that raised over night for making loaves of white bread; 1 teaspoonful butter, 1 teaspoonful sugar and ½ teaspoonful salt, and about 2 scant cups of white flour. Knead a few minutes, set to rise in a warm place, closely covered, about one hour or until doubled in bulk. Then knead down and form into a shapely loaf, place in a pan, brush melted butter over lop (this improves crust), and when raised, doubled in bulk (in about one hour), place in a moderately hot oven and bake from 40 to 45 minutes. Raisins may be added to this loaf, if liked. Mary preferred this oatmeal loaf to graham bread.

The sponge or batter from which this oatmeal-loaf was made had been prepared in the following manner:

To 1½ cups of luke-warm potato water was added 1 teaspoonful of sugar, 1 cake of yeast; when dissolved, add 1½ cups of white bread flour. Beat all together well, stand closely-covered in a warm place until the following morning. From one cup of this sponge was made one oatmeal loaf, and to the other cup of sponge white flour was added for a loaf of white bread or rolls.

AUNT SARAH'S WHITE BREAD (SPONGE METHOD)

Prepare the following "Yeast Sponge" at noon, the day preceding that on which you bake bread: Place in a bowl (after the mid-day meal) 1 quart of potato water (containing no salt), in which potatoes were boiled; also two medium-sized, finely-mashed potatoes, 1 tablespoonful of sugar and, when luke warm, add 1 cup of good home-made or baker's yeast. Mix all well together; then divide this mixture and pour each half into each of two 1-quart glass fruit jars. Place covers tightly on jars and shake each jar well, to mix yeast and potato-water thoroughly. Stand yeast in a warm place near the kitchen range over night. Jars should be *covered only* with a napkin. The sponge should become light and foamy. In the morning use this freshly-prepared yeast to set sponge for bread.

When preparing to set bread, place in a large bowl 1 pint of potato water, 1 tablespoonful of sugar, 1 pint of the yeast sponge, ½ teaspoonful of salt, and use about 3 pounds of sifted flour, well-dried and warmed. Knead from 15 to 20 minutes, until a stiff dough is formed. The dough should be fine-grained and elastic and not stick to bake board. Place dough in the bowl to rise; this should lake about four hours. When well-risen and light knead down and set to rise again, about 1½ hours. When light, mold into three large, shapely loaves; place in pans and allow to stand one hour. When loaves have doubled in bulk, are very light and show signs of cracking, invert a pan over top of loaves (if that was not done when loaves were put in pans), and place in a rather hot oven to bake. Brush melted butter over loaves of bread when set to rise, it will cause bread to have a crisp crust when baked. The old-fashioned way of testing the heat of an oven was to hold the hand in the oven, if possible, while one counted thirty.

The pint of yeast remaining in jar may be kept in a cool place one week, and may be used during this time in making fresh "yeast foam." This should always be prepared the day before baking bread. Always prepare double the quantity of "yeast foam." Use half to set bread, and reserve half for next baking. Bread baked from this recipe has frequently taken first prize at County Fairs and Farmers' Picnics.

When baking bread, the oven should be quite hot when bread is first placed therein, when the bread should rise about an inch; then the heat of the oven should he lessened and in a half hour a brown crust should begin forming; and during the latter part of the hour (the time required for baking an ordinary-sized loaf) the heat of the oven should be less, causing the bread to bake slowly. Should the heat of the oven not be great enough, when the loaves are placed within for baking, then poor bread would be the result. This method of making bread will insure most satisfactory results, although more troublesome than ordinary methods.

RECIPE FOR "PULLED BREAD"

Take a Vienna loaf of bread, twelve-hours old, cut away all the crust with a clean-cut knife, then break away gently (with your fingers only) small finger-lengths of the bread, place in a moderate oven and brown a golden brown, and it is ready to serve. 'Tis said six loaves will be required for one pound of this pulled bread. 'Tis easily prepared in the home, but quite costly, when purchased. Many people prefer "pulled bread" to fresh bread, as it is more wholesome.

AUNT SARAH'S "HUTZEL BROD"

- 2 pounds dried pears.
- 2 pounds dried prunes.
- 2 quarts juice of fruit and water.
- 1 pound dried currants.
- 1 pound seeded raisins.
- 1 pound blanched and shredded almonds.
- 1 pound chopped English walnut meats.
- 1½ ounces finely-shredded citron.
- 1½ ounces orange peel.
- ½ ounce chopped figs.
- 1 ounce ground cinnamon.
- ¼ ounce ground cloves.
- 2½ ounces anise seed.
- 6 pounds flour (warmed and sifted).
- 2 cakes compressed yeast.
- 1½ cups sugar.
- 1 large tablespoon butter.
- 1 tablespoon salt.
- 4 tablespoons brandy or sherry.

The whole recipe will make 12 loaves of bread.

This delicious German bread was usually made by "Aunt Sarah" one week before Christmas. It may be kept two weeks, and at the end of that time still be good. It is rather expensive as regards fruit and nuts, but as no eggs are used, and a very small quantity of butter; and as bread containing fruit is so much more wholesome than rich fruit cake. I think American housewives would do well to bake this German bread occasionally. Mary took one-fourth the quantity of everything called for in the recipe, except yeast. She used ¾ of a cake of Fleischman's yeast and ¼ of each of the other ingredients, and from these baked three loaves of bread. The prunes and pears should be covered with cold water at night and allowed to stand until the following morning, when, after stewing until tender, the juice should be drained from the fruit and water added to the fruit-juice to measure two quarts. Remove pits from prunes, cut pears and prunes in small pieces; stand aside. Clean currants and raisins, blanch and shred almonds, chop walnut meats, citron, orange peel and figs; add cinnamon, cloves and anise seed. Mix together flour and one quart of the fruit juice; add the compressed yeast cakes (dissolved in a little warm water), knead well, set a sponge as for ordinary bread; when raised, add the remaining quart of fruit juice, sugar, butter and salt. A small quantity of brandy or sherry may be added, but if not liked, fruit juice may be substituted.

Add the remaining ingredients, and knead thoroughly. Allow dough to raise from two to three hours and when light form into loaves and allow to stand an hour, when bake. This quantity of dough should be made into twelve small loaves. Should the flour and liquid used be warmed before mixing, the dough will raise more quickly. It simplifies the work if the fruits and nuts be prepared the day before the bread is baked.

AUNT SARAH'S WHITE BREAD AND ROLLS

- 1 quart potato water.
- 1 mashed potato.
- 1 tablespoonful butter or lard.
- 1 tablespoonful sugar.

- 1 Fleischman yeast cake, or 1 cup good yeast.
- ½ tablespoonful salt.
- Flour to stiffen (about three quarts).

At 9 o'clock in the evening put in a large bowl the mashed potato, the quart of luke-warm potato water (water in which potatoes were boiled for dinner), butter or sweet lard, sugar, salt, and mix with flour into a batter, to which add the Fleischman's or any good yeast cake, dissolved in a little luke-warm water. Beat well and stir in flour until quite stiff, turn out on a well-floured bake-board and knead well about 25 minutes, until the dough is smooth, fine-grained and elastic, and does not stick lo the bake-board or hands. Chop a knife through the dough several times; knead and chop again. This makes the bread finer and closer-grained, or, so Aunt Sarah thought. Knead in all the flour necessary when first mixing the bread. When sufficiently kneaded, form into a large, round ball of dough, rub all over with soft lard, or butter, to prevent forming a crust on top and keep from sticking to bowl, and set to rise, closely-covered with a cloth and blanket, in a warm place until morning. In the morning the bread should be very light, doubled in quantity. Take out enough dough for an ordinary loaf, separate this into three parts, roll each piece with the hand on the bake-board into long, narrow pieces. Pinch the three pieces together at one end and braid, or plait, into a narrow loaf. Brush over top with melted butter; set to rise in a warm place in a bread pan, closely-covered, until it doubles in size—or, if preferred, mold into ordinary-shaped loaves, and let rise until doubled in size, when bake in a moderately-hot oven with steady heat.

Frequently, when the "Twist" loaves of bread were quite light and ready to be placed in the oven, Aunt Sarah brushed the tops with yolk of egg, or a little milk, then strewed "Poppy Seeds" thickly over. The poppy seeds give an agreeable flavor to the crust of the bread.

AUNT SARAH'S RAISED ROLLS (FROM BREAD DOUGH)

A portion of the white bread dough may be made into raised rolls. These rolls are excellent without additional shortening, or, in fact, without anything else being added. Mold pieces of the bread dough into balls the size of a walnut; roll each piece flat with the rolling pin, dip in melted butter, fold and place close together in a bake pan. Let rise *very* light, then bake about 15 minutes in a very hot oven. If a teaspoonful of flour browns in about two minutes in the oven, it is the right temperature for rolls.

CLOVER-LEAF ROLLS

Take pieces of the bread dough, the size of a walnut, cut into three pieces, mold with the hand into round balls the size of small marbles; dip each one in melted butter, or butter and lard, and place three of these in each Gem pan. (These pans may be bought six or twelve small pans fastened together, and are much more convenient than when each one must be handled separately when baking). Allow small rolls to become *very light*, bake in a hot oven, and you will find them excellent. Dipping the rolls in melted butter makes them crisp. Serve hot, or place in a hot oven a few minutes until heated through, if served after they have become cold.

"POLISH" RYE BREAD (AS MADE IN BUCKS COUNTY)

This excellent, nutritious bread, is made from the whole-ground grain. Every part of the grain is used in the flour, when ground. To bake this bread, sift together one quart of this "whole-ground" rye flour and two quarts of white-bread flour. Early in the morning of the day on which bread is to be baked, prepare a thick batter, or sponge, consisting of one quart of potato water (or the same quantity of luke-warm, scalded milk, or a mixture of the two); add one tablespoonful of a mixture of lard and butter and two boiled, mashed potatoes. Two tablespoonfuls of sugar, one-half tablespoonful of salt and one Fleischman's compressed yeast cake, dissolved in a small quantity of water; add about five cups of the mixed, sifted flour, beat the batter well, and stand in a warm place, covered, from one and a half to two hours. When well-risen and light, stir in balance of flour gradually, until all except one cup has been added; then turn onto a bake-board and knead well. This sponge should not be quite as stiff as for wheat bread. Turn the dough onto a clean, well-floured cloth in a large bowl, set to rise and bake according to directions for baking "Hearth-baked Rye Bread" or, if preferred, form into loaves, place in bread pans and, when light, bake.

PERFECT BREAKFAST ROLLS

One quart of scalded milk, when lukewarm, add the following: ½ cup of butter and lard (mixed), 1 egg, 1 tablespoonful of sugar, 1 teaspoonful of salt and 1 Fleischman's yeast cake; add flour to form a thick batter; beat all thoroughly. Mix the above at 9.30 P.M., stand in a warm place, closely-covered, over night. The following morning add more flour; dough should not be mixed quite as stiff as for bread. Allow it to raise in a warm place. When well-risen, place on bread board, roll, cut into small biscuits; dip each biscuit in melted butter, fold together, place in pans a distance apart, and when they have doubled in size, bake in a hot oven.

"AN OLD RECIPE" FOR GOOD BREAD

This country cook invariably baked good bread and always used potato-water in preference to any other liquid for setting sponge. She stood aside water, in which potatoes had been boiled for dinner (usually about one quart or less) and added two finely-mashed potatoes. About 3 or 4 o'clock in the afternoon of the day *before* that on which she intended baking bread, she dissolved one cake of yeast (she used the small cornmeal commercial yeast cakes, sold under different names, such as National, Magic, etc.) in a half-cup of luke-warm water, added ½ teaspoon of salt and sufficient warmed, well-dried flour to make a thin batter. She placed all in a bowl and stood it in a warm place, closely-covered, until about 9 o'clock in the evening, when she added this sponge, which should be light and foamy, to the potato water, which should be lukewarm. She also added 1 tablespoon of salt and enough flour to make a rather thick batter. Heat thoroughly and allow this sponge to stand, well-covered, in a warm place until morning, when add 1 tablespoon sugar, 1 tablespoon butter or lard and warmed flour enough to make a stiff dough. Turn out on the bread board and knead for about twenty minutes, until the dough does not stick to the hands. Place stiffened dough into howl; allow it to rise until bulk is doubled. Mold into loaves, adding as little extra flour as possible. Cut several gashes on top of loaves, brush with melted butter, place in bread pans, and when loaves have doubled in bulk, place in moderately hot oven and bake about one hour.

STEAMED BROWN BREAD

Place in a bowl ¾ cup graham flour and ½ cup of yellow, granulated cornmeal. Sift into this ¾ cup of white flour, 1 teaspoonful of baking powder and ½ teaspoonful of salt. Mix all ingredients together to form a batter by adding 1 cup of sour milk, in which has been dissolved ¾ teaspoonful of soda. Then add 2 tablespoonfuls of molasses. Pour into a well-greased quart can (the tin cans in which coffee is frequently sold will answer nicely), cover closely, place in a kettle of boiling water, steam about three hours; stand in oven a short time after being steamed. Cut in slices and serve as bread, or, by the addition of raisins or currants, and a little grated nutmeg or other flavoring, a very appetizing and wholesome pudding may be served hot, with sugar and cream or any pudding sauce preferred.

A WHOLESOME BREAD (MADE FROM BRAN)

Place in a bowl 4 cups of clean bran and 2 cups of white flour, sifted with 2 teaspoonfuls of baking powder, 1 teaspoonful of salt, 1 tablespoonful of melted butter. Mix into a soft batter with 2 cups of sweet milk; add ½ cup of molasses. Fill two layer cake pans and bake in a hot oven about 25 minutes. This is so easily and quickly made. The young housewife may mix, when commencing to prepare lunch, and when the meal is ready to serve the bread will be baked, and it is an excellent laxative.

FRAU SCHMIDT'S "HUTZEL BROD"

- 1 quart dried pears.
- 1 pint of pear juice.
- 1 Fleischman's yeast cake.
- 1 scant cup brown sugar.
- 2 eggs.
- ¼ teaspoonful soda.
- 1 pound of soaked raisins.
- ¾ cup of a mixture of lard and butter.
- 1 teaspoonful of fennel seed.
- Pinch of salt.
- 2 teaspoonfuls of ground cinnamon.
- Flour to stiffen, as for ordinary bread.

Cover one quart of dried pears with cold water and cook slowly about 20 minutes until they have cooked tender, but not soft (the night before the day on which the bread is to be baked).

Then drain the juice from stewed pears, which should measure 1 pint; when lukewarm, add 1 yeast cake, dissolved in a small quantity of lukewarm water, and about 3 cups of flour and a pinch of salt. Stand, closely-covered, in a warm place over night to raise.

The following morning, add ¼ teaspoonful of baking soda, dissolved in a little warm water, to counteract any acidity of batter. Cream together sugar, butter and lard, add eggs one at a time, men the well-floured, diced pears, also raisins, cinnamon and fennel seed, and enough flour to stiffen as for ordinary bread. Knead well, let rise; it will require some time, as the fruit retards the raising process. When light, turn onto a bake-board, cut into four portions, mold into four shapely loaves, place in pans, brush with melted butter and when quite light, place in a moderate oven and bake one hour. This bread will keep well several weeks, if kept in a tin cake box.

This recipe is much simpler than Aunt Sarah's recipe for making "Hutzel Brod," but bread made from this recipe is excellent.

"AUNT SARAH'S" QUICKLY-MADE BROWN BREAD

- 2 cups of buttermilk, or thick, sour milk.
- ½ cup of sugar.
- ¼ cup of molasses.
- 1 tablespoonful of melted butter.
- 1 egg.
- 1 teaspoonful of soda.
- ¾ teaspoonful of salt.
- 3½ cups of graham flour.
- ½ cup of white flour, sifted with ¾ teaspoonful of baking powder.

The egg was placed in a bowl, and not beaten separately; sugar and butter were creamed together, before being added; then mix in salt and molasses, and gradually add buttermilk, in which the soda had been dissolved; then add white and graham flour, ¾ cups of raisins may be added, if liked. Bake in a bread pan in a moderately hot oven.

"STIRRED" OATMEAL BREAD

Early in the morning 1 cup of oatmeal porridge, left over from that which had been cooked for breakfast, was placed in a bowl and added gradually 2 cups of scalded, luke-warm milk, 1 tablespoon of a mixture of lard and butter, ¼ cup New Orleans molasses and one Fleischman's yeast cake, dissolved in a little of the milk; stir in about 3 cups of bread flour and stand in a warm place about 1¼ hours to rise; then add 3½ cups more of bread flour and 1 teaspoonful of salt.

Stir well with a spoon, and pour into three small bread tins; let rise, when well-risen, bake about ¾ of an hour in a moderately hot oven. This is a delicious and wholesome bread and no kneading is necessary. 1½ cups of the cooked oatmeal might be used, then use less white bread flour when mixing.

NUT AND RAISIN BREAD

- 2 cups buttermilk, or sour milk.
- ½ cup brown sugar,
- 2 cups graham flour.
- 1 cup wheat flour.
- 1 teaspoonful of soda, dissolved in a little of the milk.
- 1 teaspoonful of baking powder, sifted with the wheat flour.

Mix all together, add one cup of seeded raisins, ¼ cup of ground peanuts and ¼ cup chopped walnut meats. Bake in an ordinary bread pan.

"SAFFRON" RAISIN BREAD

For this old-fashioned, "country" bread, set a sponge in the evening, consisting of 1 cup of luke-warm water, 1 Fleischman's compressed yeast cake and 2 tablespoonfuls of saffron water, obtained by steeping ½ tablespoonful of dried saffron flowers in a small quantity of boiling water a short time. Use about 2 cups of flour to stiffen the sponge. Cover bowl containing sponge and stand in a warm place until morning, when add the following: ¾ cup of soft A sugar, ¼ cup lard and ⅛ cup of butter (beaten to a cream); then add one egg. Beat again and add this mixture to the well-risen sponge. Add also ¾ cup of seeded raisins and about 1¾ cups of flour.

The dough should be almost as stiff as ordinary bread dough. Set to rise about one hour. Then divide the dough and mold into two shapely loaves. Place in oblong bread pans. Let rise about 1½ hours. Brush melted butter over top of loaves and bake in a moderately hot oven, as one would bake ordinary bread.

This bread is a rich, golden yellow, with a distinctive, rather bitter, saffron flavor, well-liked by some people; saffron is not unwholesome.

"Speaking of saffron bread," said John Landis, to his niece, Mary, "I am reminded of the lines I was taught when quite a small boy:"

> "Wer will gute kuchen haben, der muss sieben sachen haben;
> Eier, butter un schmalz, milch, zucker un mehl;
> Un saffron mach die kuchen gehl."

"Of course, Mary, you do not understand what that means. I will translate it for you. 'Who would have good cakes, he must have seven things—eggs, butter and lard, milk, sugar and flour, and saffron makes the cakes yellow.'"

RAISED ROLLS

- 2 quarts of sifted flour.
- 1 pint of boiled milk (lukewarm).
- 1 tablespoon sugar.
- ½ cup butter and lard, mixed.
- ¾ cake compressed yeast, or ¾ cup yeast.
- 1 teaspoon salt.

At 5 o'clock P.M. set sponge with half or three-fourths of the flour and all the other ingredients.

About 9 o'clock in the evening, knead well, adding the balance of the flour. Cover and let stand in a warm place until morning. In the morning, roll out about ¾ of an inch thick, cut into small rolls, place in baking pans far enough apart so they will not touch, and when raised quite light, bake.

Or, take the same ingredients as above (with one exception; take one whole cake of compressed yeast), dissolved in half a cup of luke-warm water, and flour enough to make a thin batter. Do this at 8.30 in the morning and let rise until 1 o'clock; then knead enough flour in to make a soft dough, as soft as can be handled. Stand in a warm place until 4.30, roll out quite thin; cut with small, round cake-cutter and fold over like a pocketbook, putting a small piece of butter the size of a pea between the folds; set in a warm place until 5.30, or until very light; then bake a delicate brown in a hot oven. If made quite small, 70 rolls may be made from this dough.

To cause rolls of any kind to have a rich, brown glaze, when baked, before placing the pan containing them in the oven, brush over the top of each roll the following mixture, composed of—yolk of 1 egg, 1 tablespoon of milk, and 1 teaspoon of sugar.

"GRANDMOTHER'S" FINE RAISED BISCUITS

- 1 quart scalded milk (lukewarm).
- ¾ cup of butter, or a mixture of butter and lard.
- ½ cup of sugar.
- 1 teaspoonful of salt.
- 2 Fleischman's yeast cakes.
- Whites of 2 eggs.
- Flour.

Quite early in the morning dissolve the two yeast cakes in a little of the milk; add these, with one-half the quantity of sugar and salt in the recipe, to the remainder of the quart of milk; add also 4 cups of flour to form the yeast foam. Beat well and stand in a warm place, closely-covered, one hour, until light and foamy.

Beat the sugar remaining and the butter to a cream; add to the yeast foam about 7 to 8 cups of flour, and the stiffly-beaten whites of the two eggs.

Turn out on a well-floured bread board and knead about five minutes. Place in a bowl and let rise again (about one hour or longer) until double in bulk, when roll out about one inch in thickness. Cut small biscuits with a ½ pound Royal Baking Powder can.

Brush tops of biscuits with a mixture consisting of yolk of one egg, a teaspoonful of sugar and a little milk; this causes the biscuits to have a rich brown crust when baked.

Place biscuits on pans a short distance apart, let rise until doubled in bulk; bake in a rather quick oven.

From this recipe was usually made 55 biscuits. One-half of this recipe would be sufficient for a small family.

Mary's Aunt taught her the possibilities of what she called a "Dutch" sponge — prepared from one Fleischman's yeast cake. And the variety a capable housewife may give her family, with the expenditure of a small amount of time and thought.

About 9 o'clock in the evening Mary's Aunt placed in a bowl 2 cups of potato water (drained from potatoes boiled for dinner). In this she dissolved one Fleischman's yeast cake, stirred into this about 3 cups of well-warmed flour, beat thoroughly for about ten minutes. Allowed this to stand closely covered in a warm place over night. On the following morning she added to the foamy sponge 1½ cups lukewarm, scalded milk, in which had been dissolved 1 tablespoonful of a mixture of butter and lard, 2 generous tablespoonfuls of sugar and 1 teaspoonful of salt. About 6¼ cups of well-dried and warmed flour; she stirred in a part of the flour, then added the balance. Kneaded well a short time, then set to raise closely covered in a warm place 2½ to 3 hours.

When dough was light it was kneaded down in bowl and allowed to stand about one hour, and when well risen she placed 2 cups of light bread sponge in a bowl, and stood aside in warm place; this later formed the basis of a "Farmers' Pound Cake," the recipe for which may be found among recipes for "Raised Cakes."

From the balance of dough, or sponge, after being cut into 3 portions, she molded from the one portion 12 small turn-over rolls, which were brushed with melted butter, folded together and placed on tins a distance apart and when *very* light baked in a quick oven.

From another portion of the sponge was made a twist or braided loaf.

And to the remaining portion of dough was added ½ cup of currants or raisins, and this was called a "Currant" or "Raisin Loaf," which she served for dinner the following day.

The rolls were placed in the oven of the range a few minutes before breakfast and served hot, broken apart and eaten with maple syrup or honey and the delicious "Farmers' Pound Cake" was served for supper.

Aunt Sarah baked these on ironing day. The kitchen being unusually warm, as a result of the extra heat required in the range for heating flatirons, caused the dough to rise more quickly than otherwise would have been the case.

STIRRED BREAD

Frau Schmidt thought bread more easily digested and wholesome if ingredients of a loaf be stirred together instead of kneaded. This is the method she taught Mary. She poured into a bowl 3 cups of lukewarm water, added 1 cake of Fleischman's yeast, dissolved in a little of the water; sifted in gradually about 8½ cups of flour, added 1 tablespoonful of sugar, ½ teaspoonful of salt, mixed all well together with a spoon until a stiff dough was formed, which she molded into two shapely loaves, handling as little as possible; placed in bread pans, allowed to stand several hours to raise, and when light baked. Mary said, "This bread may be more wholesome than old-fashioned bread, which has been kneaded, but I prefer Aunt Sarah's bread, well-kneaded, fine-grained and sweet," but, she continued, "I will make an exception in favor of Aunt Sarah's 'Stirred Oatmeal' bread, which, I think, fine."

POTATO BISCUITS

At 6 o'clock in the morning place in a bowl 1 cup of finely-mashed (boiled) potatoes (the cup of left-over mashed potatoes may be used as a matter of economy). Add 1 cup of potato water (the water drained from boiled potatoes), in which ½ cake of Fleischman's yeast had been dissolved, add 1 cup of flour and 1 teaspoon of sugar. Stand in a warm place to raise, from 1 to 1¼ hours. At the expiration of that time add to the foamy sponge 1 large tablespoonful of butter or lard, 1 egg and ½ teaspoonful of salt, beaten together before adding. Add about 2 cups of flour, beat thoroughly and allow to raise another hour; then roll out the dough about 1 inch in thickness and cut into small biscuits, dip each one in melted butter and place on pans, a short distance apart, stand about one hour to raise, when bake in a rather hot oven. These Potato Biscuits are particularly nice when freshly baked, and resemble somewhat biscuits made from baking powder. From this recipe was made two dozen biscuits.

AUNT SARAH'S POTATO YEAST

- 9 medium-sized potatoes.
- 5 tablespoons sugar.
- 2 tablespoons salt.
- 1 quart water.

Grate the raw potatoes quickly, so they will not discolor, pour over the grated potato the quart of boiling water, add salt and sugar, cook several minutes until the consistency of boiled starch, let cool, and when lukewarm add 1 cup of good yeast. Stir all together in a crock, cover and let stand in a warm place three or four hours, when it is foamy and rises to top of crock, stir down several times, then fill glass fruit jars, cover and stand away in a cool place until needed.

This yeast will keep about ten days. Use one cup to about three pounds of flour, or one quart of liquid, when setting sponge for bread. Save one cup of this yeast to start fresh yeast with.

PERFECTION POTATO CAKES

- 1 cup of boiled mashed potatoes.
- 1 cup sweet milk.
- 1 cup water in which 1 Fleischman yeast cake was dissolved.
- 2 cups soft A sugar.
- ½ cup butter and lard mixed.
- 2 eggs.
- A little salt.
- About 7 cups of flour.

Cream the sugar, butter and eggs together. Add mashed potatoes, milk and cup of water containing yeast, alternately with the flour, until about 7 cups of flour have been used, making a dough as stiff as can be stirred with a spoon. Stand, covered, in a warm place by the range until morning. These should be set to rise about nine o'clock in the evening. The following morning take pieces of the dough, on a well-floured bake board; roll about one inch thick, to fit in pie tins, place in pie tins to raise; when doubled in bulk spread with melted butter and sprinkle sugar thickly over top and bake in a moderately hot oven until lightly browned on top. This quantity of dough makes six cakes.

Instead of brushing the cakes with above mixture, place in a bowl ½ cup of soft A sugar, ½ cup flour, a tiny pinch of salt and baking powder each and 2 tablespoonfuls of butter (not melted), mix all together as crumbly as possible, then the crumbs were sprinkled thickly over tops of cakes, which had been brushed with a mixture of milk and sugar. Place cakes in oven when raised; bake 20 minutes.

This recipe was given Mary by an old "Bucks County" cook, noted for the excellence of her raised cakes.

MARY'S RECIPE FOR CINNAMON BUNS

Early in the morning mix a sponge or batter consisting of ½ cup of potato water (water drained from boiled potatoes) and ½ cup of lukewarm, scalded milk, one Fleischman's compressed yeast cake, dissolved in the ½ cup of lukewarm potato water, 1 teaspoonful sugar, pinch of salt and about 1½ cupfuls of warmed flour. Stand this sponge in a warm place, closely covered, about ¾ of an hour, to raise. At the end of that time add to the light, well-risen sponge, the following: 3 tablespoonfuls of a mixture of lard and butter, and ⅓ cup of soft A sugar, creamed together. Add one large egg. Beat well. Lastly, add about 2 cupfuls of flour. Mix all together thoroughly, and let raise again about 1½ hours. Divide the well-risen sponge into four portions. Roll each piece with rolling-pin into lengthwise pieces about ½ inch thick and spread with one tablespoonful of melted butter, scant 2 tablespoonfuls of brown sugar, dust over this a small quantity of cinnamon, and 1 tablespoonful of dried currants. Shape into a long, narrow roll with the hands, on a well-floured bake-board. Cut each roll into five pieces. Pinch one end of each piece together and place each bun, cut side down, a short distance apart, in an iron pan which has been well greased, having brushed a little melted butter and a sprinkling of sugar over pan. Allow these to rise in a warm place as before, about 1½ hours, until quite light, as having the extra sugar, butter and currants added retards their rising as quickly as would plain biscuits.

Bake 20 to 25 minutes in a moderate oven.

From this quantity of material was made 20 cinnamon buns.

"KLEINA KAFFE KUCHEN" (LITTLE COFFEE CAKES)

- Scant ½ cup lard and butter.
- 2 cups sifted flour.
- 2 whole eggs and the yolks of 2 more.
- 3 tablespoons sugar.
- ¼ cup cream.
- ¼ milk.
- 1 Fleischman's yeast cake.

- ⅛ teaspoon salt.

The yeast cake was dissolved in the ¼ cup lukewarm milk, a couple tablespoons of flour were added and mixed into a batter, and stood in a warm place to rise. The butter and sugar were stirred to a cream, salt was added, the eggs were beaten in, one at a time, next was added the sponge containing the yeast, the lukewarm cream, and the sifted flour. Grease slightly warmed Gem pans, sift a little flour over them, fill two-thirds full with the soft dough, set in a warm place to rise to tops of pans, and when quite light bake in a medium hot oven about 25 minutes. The oven should be hot enough to allow them to rise quickly. Put something underneath the pans in the oven to prevent bottom of cakes from burning. These may be set about 8 o'clock in the morning if cakes are wished for lunch at noon. These are not cheap, as this quantity makes only 12 cakes, but they are light as puffballs. The Professor's wife served them when she gave a "Kaffee Klatch." She doubled the recipe, baked the cakes in the morning, and placed them in the oven to heat through before serving. The cakes should be broken apart, not cut. The cakes made from this recipe are particularly fine.

GROSSMUTTER'S POTATO CAKES

- 1 cup hot mashed potatoes.
- 1½ cups sugar.
- 1 scant cup butter and lard.
- 1 cup home-made yeast or 1 yeast cake dissolved in 1 cup
- lukewarm water.
- 3 eggs.
- Flour.

At 5 o'clock in the afternoon set to rise the following: One cup of sugar and one cup of hot mashed potatoes; when lukewarm add one cup of flour and one cup of yeast; beat all together, stand in a warm place to rise and at 9 o'clock in the evening cream together 1 cup of a mixture of lard and butter, 1 cup of sugar, 3 eggs and pinch of salt; add the sponge and beat well. Stir as stiff as you can stir it with a large spoon, cover, set in a warm place to rise until morning, when roll out some of the dough into cakes about one inch thick, put in pie tins to rise, and when light, make half a dozen deep impressions on top of each cake with the forefinger, spread with melted butter and strew light-brown sugar thickly over top, or mix together 1 cup sugar, butter size of an egg, 2 tablespoons flour, 2 tablespoons boiling water, beat well and spread the mixture on cakes just before placing in oven. Bake the cakes about 20 minutes in a moderate oven. This is a very old recipe used by Aunt Sarah's grandmother, and similar to the well-known German cakes called "Schwing Felders."

AUNT SARAH'S "BREAD DOUGH" CAKE

- 1 cup bread dough.
- 1 egg.
- ½ cup soft A sugar.
- 1 tablespoon lard or butter.
- ¼ teaspoon soda.

When her bread dough was raised and ready to put in the pans she placed a cupful of it in a bowl and added the egg, sugar, butter, soda (dissolved in a little hot water); some dried raisins or currants, and just enough flour so it might be handled easily. Put in a small agate pan four inches deep, let rise until light, dust pulverized sugar over top and bake about 25 or 30 minutes in a moderate oven.

Double the materials called for, using 2 *cups* of well-risen bread dough or sponge, and you will have a good-sized cake.

GOOD, CHEAP DUTCH CAKES

To a bowl containing 1 cup of scalded milk, add 1 tablespoonful of lard and 1 cup of sugar. When lukewarm add 1 yeast cake (Fleischman's), dissolved in 1 cup of lukewarm water, and about 5 cups of good flour. Set to rise at night about nine o'clock, the next morning roll out pieces about one and a half inches thick, to fit in medium-sized pie tins. Set in a warm place to rise. When light, brush top with melted butter and strew sugar thickly over and bake from 15 to 20 minutes in a moderately hot oven. These cakes are *inexpensive* and *good*; *no eggs* or *butter* being used.

RECIPE FOR "LIGHT CAKES" (GIVEN MARY BY A FARMER'S WIFE)

In the evening mix a sponge consisting of ½ cup of mashed potatoes, ½ cup sugar, 1 cup of yeast or 1 cake of Fleischman's yeast dissolved in a cup of lukewarm water; ½ cup of a mixture of butter and lard and a pinch of salt and flour to thicken until batter is quite thick. Stand in a warm place, closely covered, until morning, when add 2 eggs and ½ cup of sugar and flour to stiffen as thick as sponge can be stirred with a spoon. Set to rise; when light roll out one inch thick, place in pie tins, brush tops with melted butter and brown sugar, set to rise, and, when well risen, bake.

BUTTER "SCHIMMEL"

Place in a mixing bowl 2 cups of warm, mashed potatoes and add ¾ of a cup of shortening (a mixture of lard and butter), (or use Aunt Sarah's substitute for butter); one cup of A sugar and 1 teaspoonful salt.

Beat all to a cream. When lukewarm, add 2 eggs and either 1 yeast cake dissolved in 1 cup of lukewarm water, or 1 cup of potato yeast; use about 2 cups of flour to make a thin batter. Set to raise over night or early in the morning. When well risen add about 4 cups of flour. Make about as stiff a dough as can be stirred well with a mixing spoon. Place soft dough on a bake-board; roll out into a sheet about one-half inch thick; cut into squares about the size of a common soda cracker; bring each of the four corners together in the centre like an envelope; pinch together; place a small piece of butter (about one-eighth teaspoonful) on the top where the four corners join. Stand in a warm place to rise. When well risen and light place in the oven. When baked, take from oven, and while hot dip all sides in melted butter and dust granulated or pulverized sugar over top. These are not as much trouble to prepare as one would suppose from the directions for making. The same dough may be cut in doughnuts with a tin cutter and fried in hot fat after raising, or the dough may be molded into small, round biscuits if preferred, and baked in oven.

"BUCKS COUNTY" DOUGHNUTS

About nine o'clock in the evening a batter was mixed composed of the following:

- 1 cup milk.
- 1 cup hot water.
- 1 teaspoonful of sugar.
- 1 cup yeast (or one cake of Fleischman's yeast dissolved in
- one cup of lukewarm water).
- 1 pinch of salt.
- 3½ cups of flour.

Stand in a warm place until morning. Then add ½ cup of butter and 1½ cups of soft A sugar, creamed together, and from 3 to 4 cups of flour. The dough should be as stiff as can be stirred with a spoon. Set to rise in a warm place; when light and spongy, roll out on a well-floured bake-board and cut into round cakes with a hole in the centre. Let rise again, and when well risen fry a golden brown in deep fat and sift over pulverized sugar. This recipe will make 45 doughnuts. These are good and economical, as no eggs are used in this recipe.

EXTRA FINE "QUAKER BONNET" BISCUITS

For these quaint-looking, delicious biscuits, a sponge was prepared consisting of:

- 1 pink milk.
- 3 eggs.
- ½ cup mixture butter and lard.
- 1 yeast cake (Fleischman's).
- About 7 cups flour.

Set to rise early in the morning. When well risen (in about 3 hours), roll dough into a sheet about ¼ inch in thickness, cut with a half-pound baking powder can into small, round biscuits, brush top of each one with melted butter (use a new, clean paint brush for this purpose), place another biscuit on top of each one of these, and when raised very light and ready for oven brush top of each biscuit with a mixture consisting of half of one yolk of egg (which had been reserved from the ones used in baking), mixed with a little milk. Biscuits should have been placed on a baking sheet some distance apart, let rise about one hour until quite light, then placed in a quick but not *too hot* an oven until baked a golden brown on top.

Mary gave these the name of "Quaker Bonnet" Biscuits, as the top biscuit did not raise quite as much as the one underneath and greatly resembled the crown of a Quaker bonnet.

From this quantity of dough was made three dozen biscuits. These are not cheap, but extra fine.

BUCKS COUNTY CINNAMON "KUCHEN"

Explicit directions for the making of these excellent raised cakes was given Mary by an old, experienced Pennsylvania German cook. They were prepared from the following recipe: Early in the morning 1 pint of milk was scalded. When lukewarm, add 3½ cups of flour and 1 cake of Fleischman's compressed yeast (which had been dissolved in 1 tablespoonful of lukewarm water). Beat the mixture well. Cover and stand in a warm place to rise. When well risen, which should be in about 2 hours, add the following mixture, composed of ¾ cup of sugar and ½ cup of butter, creamed together; ½ teaspoonful of salt; 1 egg was beaten into the mixture, and about 2 cups of flour were added, enough to make a dough as stiff as can be stirred with a spoon. Dough should not be as stiff as for bread. Let stand about 1 hour. When well risen and light, divide into four portions. Roll out each piece of dough to thickness of one inch. Place cakes in medium-sized pie tins and allow them to stand about one hour. When well risen, doubled in bulk, make half dozen deep impressions on top of each cake with the forefinger. Brush top of each cake with ½ tablespoonful of melted butter. Sprinkle over 2 tablespoonfuls of soft A sugar and sift over a little pulverized cinnamon, if liked, just before placing cakes in oven. Bake cakes from 20 to 25 minutes in a moderately hot oven. From this dough may be made four cakes.

Excellent biscuits may also be made from this same dough, by simply moulding it into small biscuits and place in a pan some distance apart. Let rise and brush tops of biscuits with a mixture composed of a part of an egg yolk, a tablespoonful of milk and ½ teaspoonful sugar. This causes the biscuits to have a rich, brown color when baked.

The sponge from which these cakes or biscuits were made was mixed and set to rise at 6 o'clock in the morning, and the baking was finished at 11 o'clock. Sponge should be set to rise in a warm room. If these directions are carefully followed the housewife will invariably have good results. Always use hard Spring wheat for bread or biscuits, raised with yeast; and Winter wheat, which costs less, will answer for making cake and pastry. In cold weather always warm flour before baking, when yeast is used for baking raised cakes. Soft A sugar or a very light brown is to be preferred to granulated.

MORAVIAN SUGAR CAKES

At 5 o'clock P.M. set a sponge or batter, consisting of 1 cup of mashed potatoes, 2 cups of sugar, 1 cup of sweet milk, scalded and cooled, ½ cake of yeast, dissolved in 1 cup of lukewarm water, 2 eggs ¾ cup of a mixture of lard and butter, add 3 cups of flour, beat well, stand in a warm place to raise; at 9 o'clock add about 6 cups of flour. Stand until morning in a warm place, near the range. The following morning turn out on a floured bake-board, roll out cakes one inch thick, place in pie tins, when ready for the oven; punch half a dozen small holes in the top of cakes, in which place small bits of butter. Sprinkle sugar over liberally and cinnamon if liked. Bake in a moderate oven.

MARY'S POTATO CAKES

- 1 cup freshly-boiled mashed potatoes.
- 1 cup scalded sweet milk.
- 1 cup sugar.
- Flour about 6 cups.
- 1 cake Fleischman's yeast.
- 2 eggs.
- ½ cup butter and lard mixed.
- ½ cup potato water.

At 7 o'clock in the morning Mary mixed a sponge consisting of a cup of mashed potatoes, 1 cup scalded milk, ½ cup sugar, 1½ cups of flour and the cake of Fleischman's yeast, dissolved in half a cup of lukewarm potato water. This was set to rise in a warm place near the range for several hours until light. Then she creamed together ½ cup of sugar, 2 eggs and ½ cup of butter and lard, or use instead the "Substitute for Butter." Added the creamed sugar, butter and eggs to the well-risen sponge and about 4½ cups of flour. Sift a couple of tablespoons of flour over top of sponge, and set to rise again about 1½ hours. When light, take cut pieces of the sponge on a well-floured bread-board, knead for a minute or two, then roll out with a rolling-pin inlo pieces about one inch thick, place in well-greased small pie tins, over which a dust of flour has been sifted, set to rise about 1½ hours. When light and ready for oven brush top with milk, strew crumbs over or brush with melted butter and strew sugar over top; after punching half dozen holes in top of each cake, bake in a moderately hot oven from 20 to 25 minutes until a rich brown, when cakes should be baked. Five potato cakes may be made from this sponge, or four cakes and one pan of biscuits if preferred. Use soft "A" sugar rather than granulated for these cakes, and old potatoes are superior to new. Or when these same cakes were raised, ready to be placed in the oven, Mary frequently brushed the tops of cakes with melted butter, strewing over the following: 1 cup of flour mixed with ½ cup of sugar and yolk of 1 egg, and a few drops of vanilla. This mixture rubbed through a coarse sieve and scattered over cakes Mary called "Streusel Kuchen."

GERMAN RAISIN CAKE (RAISED WITH YEAST)

Place in a bowl 1 cup of milk, scalded and cooled until lukewarm; add 1 tablespoonful of sugar and dissolve one cake of yeast in the milk. Mix in 1 cup of flour and stand in a warm place to raise ¾ of an hour. Then cream together in a separate bowl ½ cup soft "A" sugar, ½ cup of butter or "butter substitute," add 1 egg and a pinch of salt; stir in 1¼ cups of flour, ½ cup of well-floured raisins, and ½ teaspoonful of vanilla flavoring. Add the yeast mixture and allow it to raise about 2 hours longer. At the expiration of that time turn the well-risen sponge out on a floured bake-board. After giving the dough several deft turns on the board with the hand, place in a well-greased fruit cake pan, which has been dusted with flour. Stand pan containing cake in a warm place, let rise until very light, probably 1¼ hours, when brush the top of cake with a small quantity of a mixture of milk and sugar. Sift pulverized sugar thickly over top. Place the cake in a moderately hot oven, so the cake may finish rising before commencing to brown on the top. Bake about 35 minutes.

"KAFFEE KRANTZ" (COFFEE WREATH)

- 1 cup sugar.
- ¾ cup butter and lard.
- 4 eggs.
- 1 pint milk.
- 1 Fleischman's yeast cake.
- 4 cups flour.

Cream together the sugar, butter, lard and eggs, add the milk, which has been scalded and allowed to cool; flour, and yeast cake, dissolved in a half cup of lukewarm water; beat well. Set this sponge to rise in a warm place, near the range, as early as possible in the morning. This will take about 1½ hours to rise. When the sponge is light add about 3 cups more of flour. The dough, when stiff as can be stirred with a spoon, will be right. Take about 2 cups of this sponge out on a well-floured bake-board, divide in three pieces, and braid and form into a wreath or "Krantz," or they may be made out into flat cakes and baked in pie tins after they have been raised and are light. Sprinkle sugar thickly over top after brushing with milk containing a little sugar, before placing in oven. These should rise in about 1½ hours. Place in a moderately hot oven and bake from 20 to 25 minutes. This recipe Frau Schmidt translated from the German language for Mary's especial benefit.

This coffee wreath is particularly fine if small pieces of crushed rock candy be sprinkled liberally over the top and blanched almonds stuck a couple of inches apart over the top just before placing the cake in the oven, after the cakes had been brushed with a mixture of milk and sugar.

"MONDEL KRANTZ" OR ALMOND CAKE (AS MADE BY FRAU SCHMIDT)

- 1 pint sweet milk.
- ¾ cup sugar.

- 3 eggs.
- 1 yeast cake or 1 cup yeast.
- ⅓ cup butter.
- 2 tablespoons rock candy.
- 1 orange.
- 2 tablespoons chopped almonds.
- Flour.

Set to rise early in the morning. To the scalded milk, when luke-warm, add the yeast and flour enough to make a batter, cover, set to rise until light, near the range, which will take several hours. Then add the sugar, butter and eggs beaten to a cream, grated rind and juice of orange, a couple tablespoons finely-chopped almonds, and add enough flour to make a soft dough, as stiff as can be stirred with a spoon; set to rise again, and when light, divide the dough in two portions, from which you form two wreaths. Roll half the dough in three long strips on the floured bake-board with the hands, then braid them together. Place a large coffee cup or bowl inverted on the centre of a large, round or oval, well-greased pan, lay the wreath around the bowl. The bowl in the centre of the pan prevents the dough from running together and forming a cake. Brush the top of the wreath with a little milk, containing teaspoon of sugar, over the top of the wreath, stick blanched, well-dried almonds, and strew thickly with crushed rock candy or very coarse sugar.

Let rise until light, then bake. This makes two quite large wreaths.

The Professor's wife told Mary when she gave her this recipe, this almond wreath was always served at the breakfast table on Christmas morning at the home of her parents in Germany, and was always baked by her mother, who gave her this recipe, and it was found on the breakfast table of Frau Schmidt Christmas morning as regularly as was made "Fast Nacht Kuchen" by Aunt Sarah every year on "Shrove Tuesday," the day before the beginning of the Lenten season.

THE PROFESSOR'S WIFE'S RECIPE FOR "DUTCH CAKES"

- 2 tablespoons of butter or lard.
- 2 eggs.
- 1 cup "Soft A" sugar.
- ½ yeast cake.
- 1 pint milk.
- ½ teaspoonful of salt.
- Flour.

She scalded the milk, added butter and eggs, well beaten, when the milk was lukewarm, then added yeast, dissolved in a little lukewarm water, sugar, salt and flour to make a thin batter. Beat all together five minutes, stood the batter, closely covered, in a warm place, over night. In the morning, added flour to make a soft dough, kneaded lightly for ten minutes, placed in bowl and set to rise again. When light, she rolled out dough one inch in thickness, placed in pie tins, and when raised a second time spread over the cakes the following mixture before placing in oven: 1 cup sugar, 2 tablespoonfuls of flour, 2 tablespoonfuls of boiling water and butter size of an egg, beaten well together. Bake 20 minutes.

"FARMERS' POUND CAKE" (AS AUNT SARAH CALLED THIS)

Place in a bowl 2 cups of light, well-raised bread sponge (when all flour necessary had been added and loaves were shaped ready to be placed in bread pan for final rising). Cream together ¾ cup of a mixture of lard and butter, add 2 eggs, first yolks then stiffly beaten whites, also add 1½ cups soft A sugar. Add to the 2 cups of bread sponge in bowl and beat well until fully incorporated with the dough, then add ½ cup of lukewarm milk, in which had been dissolved ½ teaspoonful of salaratus.

Beat all together until mixture is smooth and creamy, then add 2 cups of bread flour and ½ teaspoon of lemon flavoring. Beat well and add 1½ cups of either currants or raisins, dusted with flour. Pour mixture into an agate pudding dish (one holding 3 quarts, about 2½ inches in depth and 30 inches in circumference). Stand in a warm place 3 to 4 hours to raise; when raised to top of pan place in a moderately hot oven and bake about 40 minutes, when, taken from oven, dust with pulverized sugar thickly over top of cake.

This cake should be large as an old-fashioned fruit cake, will keep moist some time in a tin cake box, but is best when freshly baked.

GERMAN "COFFEE BREAD"

- ⅓ cup sugar
- ⅓ cup butter
- 1 cup hot milk
- 1 yeast cake
- 2 eggs
- 2½ cups flour.

As Aunt Sarah taught Mary to bake this, it was fine. She creamed together in a bowl the sugar and butter, poured the hot milk over this, and when lukewarm, added the compressed yeast cake, dissolved in ¼ cup of lukewarm water. She then added two small, well-beaten eggs, about 2½ cups flour, or enough to make a stiff *batter*, and ½ teaspoonful salt. Beat thoroughly, cover and set to rise in a warm place about 1½ hours or until doubled in bulk. This was set to rise quite early in the morning. When light, beat thoroughly and with a spoon spread evenly on top of well-greased, deep pie tins, which have been sprinkled with a little flour.

Spread the crumbs given below over the top of cakes, cover and let rise 15 minutes and bake a rich brown in moderate oven.

For the crumbs, mix together in a bowl 1 heaped cup of fine, soft, stale bread crumbs, 2½ tablespoonfuls light brown sugar, ¾ of a teaspoonful cinnamon, pinch of salt, ¼ cup of blanched and chopped almonds, and 2 tablespoonfuls of soft butter. This sponge or dough should be unusually soft when mixed, as the crumbs sink into the dough and thicken it. Add only the quantity of flour called for in recipe.

"FAST NACHT KUCHEN" (DOUGHNUTS)

- 3 tablespoons honey.
- ¾ quart milk.
- 2 quarts flour.
- 1 yeast cake.
- ½ cup butter.
- 2 eggs.

Without fail, every year on Shrove Tuesday, or "Fast Nacht," the day before the beginning of Lent, these cakes were made. Quite early in the morning, or the night before, the following sponge was set to rise: The lukewarm, scalded milk, mixed into a smooth batter with 1 quart of flour; add 1 Fleischman's yeast cake, dissolved in a very little water. Beat well together, set in a warm place to rise over night, or several hours, and when light, add the following, which has been creamed together: eggs, butter and lard, a little flour and the honey. Beat well, and then add the balance of the flour, reserving a small quantity to flour the board later. Set to rise again, and when quite light roll out on a well-floured board, cut into circles with a doughnut cutter, cut holes in the centre of cakes, let rise, and then fry in deep fat; dust with pulverised sugar and cinnamon, if liked. These are regular German doughnuts, and are never very sweet. If liked sweeter, a little sugar may be added. From this batter Mary made 18 "fried cakes," or "Fast Nacht Kuchen," as the Germans call them. She also made from the same dough one dozen cinnamon buns and two Dutch cakes. The dough not being very sweet, she sprinkled rivels composed of sugar, flour and butter, generously over the top of the "Dutch cakes." The dough for doughnuts, or fried cakes, should always have a little more flour added than dough for "Dutch cakes" or buns; baked in the oven. If *too soft*, they will absorb fat while frying.

""KAFFEE KUCHEN" (COFFEE CAKE)

- 2 cups milk.
- 1 heaped cup soft A sugar.
- ½ cup butter and lard.
- 1 egg.
- 1 Fleischman's yeast cake.
- Flour.

These German Coffee Cakes should be set to rise either early in the morning or the night before being baked. Scald 2 cups sweet milk and set aside to cool. Cream together in a bowl 1 heaped cup of A sugar, ½ cup butter and lard and the yolk of egg. Add this to the lukewarm milk alternately with 6½ cups flour and the yeast cake dissolved in ⅓ cup lukewarm water. Beat all together, and, lastly, add the stiffly-beaten white of egg. Cover and set in a warm place to rise over night, or, if set to rise in the morning, stand about 2½ hours until light. Put an extra cup of flour on the bake-board, take out large spoonfuls of the dough, mix in just enough flour to roll out into flat cakes, spread on well-greased pie tins, stand in a warm place until light, about 1¼ hours. When the cakes are ready for the oven, brush melted butter over the top, strew thickly with brown sugar, or spread rivels over top, composed of ½ cup sugar, ½ cup flour and 2 tablespoonfuls of butter, crumbled together. Strew these over the cakes just before placing them in the oven of range.

"STREUSEL KUCHEN"

For these German-raised cakes, take ½ cup mashed potatoes and ½ cup of potato water, ½ cup lard and butter mixed, creamed with ½ cup sugar. Mix with these ingredients about 3½ cups of flour and 1 cup of yeast. Set this sponge to rise at night in a warm place, well covered. The next morning add to the light, well-risen sponge, 2 eggs, ½ cup sugar and about 1½ cups flour. Let stand in a warm place until light. Then roll out pieces size of a plate, one inch thick; place on well-greased pie tins, let rise, and when light and ready for the oven brush over tops with melted butter and strew over the tops of cakes the following: Mix 1 cup of flour, ½ cup of sugar and yolk of 1 egg. Flavor with a few drops of vanilla (or use vanilla sugar, which is made by placing several vanilla beans in a jar of sugar a short time, which flavors sugar). Rub this mixture of flour, sugar and yolk of egg through a coarse sieve and strew over tops of cakes.

Or, this same recipe may be used by taking, instead of 1 cup of yeast, one Fleischman yeast cake, dissolved in 1 cup of lukewarm water. Instead of sponge being set to rise the night before the day on which the cakes are to be baked, the sponge might be set early in the morning of the same day on which they are to be baked — exactly in the same manner as if sponge was set the night before; when light, add eggs, sugar and balance of flour to sponge, and proceed as before.

MUFFINS, BISCUITS, GRIDDLE CAKES AND WAFFLES

Use 1 scant cup of liquid to 1 good cup of flour, usually, for "Griddle Cake" batter. Use baking powder with sweet milk, 1 heaping teaspoonful of Royal baking powder is equivalent to 1 teaspoonful of cream of tartar and ½ teaspoonful of salaratus (baking soda) combined. Use either baking powder or salaratus and cream of tartar combined, when using sweet milk. Use 1 teaspoonful of baking soda to 1 pint of sour milk. Allow a larger quantity of baking powder when no eggs are used. Have all materials cold when using baking powder. When milk is only slightly sour, use a lesser quantity of soda and a small quantity of baking powder.

SALLY LUNN (AS AUNT SARAH MADE IT)

As "Aunt Sarah" made this, it required 1 cup of sweet milk, 2 eggs, 1 tablespoonful of butter, 3 tablespoonfuls of sugar, flour to make a stiff batter, about 2¾ cups (almost three cups) of flour sifted with 3 scant teaspoonfuls of baking powder. Served immediately when taken from the oven, this is an excellent substitute for bread for lunch.

AUNT SARAH'S RECIPE FOR "JOHNNY CAKE"

One and one half cups of sour milk, ⅓ cup of shortening, a mixture of lard and butter, 1½ tablespoonfuls of sugar, 2 cups of yellow cornmeal, 1 cup of white bread flour, 1 egg, 1 teaspoon of soda, dissolved in a little hot water, a little salt. Mix all together, add the stiffly-beaten white of egg last. Pour batter in an oblong bread tin, bake about 45 minutes in a quick oven. Granulated corn meal was used for this cake.

MARY'S BREAKFAST MUFFINS

- 3 cups sifted flour.
- 1 teaspoon salt.
- 1 teaspoon sugar.
- 1 tablespoon butter and lard.
- ¼ cake Fleischman's yeast.
- 2 eggs.
- 2 cups boiled milk.

Place the flour, salt, sugar, butter, lard and yeast cake, dissolved in water, in a bowl and mix well; then add the eggs and milk, which should be lukewarm. Set to rise in a warm place over night. In the morning do not stir at all, but carefully place tablespoonfuls of the light dough into warm, well-greased Gem pans, let stand a short time, until quite light, then bake in a hot oven 15 to 20 minutes and serve hot for breakfast. These should be light and flakey if made according to directions.

RICE MUFFINS

- 1 cup cold boiled rice.
- Yolk of egg and white beaten separately.
- 1 teaspoon sugar.
- ½ teaspoon salt.
- 1 cup sweet milk.
- 2 cups flour.
- 2 teaspoonfuls baking powder.

Put the rice, yolk of egg, sugar and salt in a bowl and beat together; then add 1 teacup sweet milk alternately with the flour, in which has been sifted the baking powder. Add the stiffly-beaten white of egg; bake in muffin pans in hot oven. This makes about fifteen muffins.

INDIAN PONE

Beat together, in the following order, 2 eggs, 1 tablespoonful of white sugar, 1½ cups of sweet milk, 1 teaspoonful of salt; to which add 1 cup of granulated yellow corn meal and 2 cups of white flour, sifted, with 3 scant teaspoonfuls of Royal baking powder. Lastly, add 1 tablespoonful of melted (not hot) butter. Pour batter in bread pan and bake in a hot oven 25 to 30 minutes. Serve hot. Do not cut with a knife when serving, but break in pieces. When the stock of bread is low this quickly-prepared corn bread or "pone" is a very good substitute for bread, and was frequently baked by Mary at the farm. Mary's Aunt taught her to make a very appetizing pudding from the left-over pieces of corn bread, which, when crumbled, filled 1 cup heaping full; over this was poured 2 cups of sweet milk; this was allowed to stand until soft; when add 1 large egg (beaten separately), a generous tablespoonful of sugar, a couple of tablespoonfuls of raisins, a pinch of salt; mix well, pour into a small agate pudding pan, grate nutmeg over the top, and bake in a moderate oven 1 hour or a less time. Serve with sugar and cream.

"PFANNKUCHEN" (PANCAKES)

Four eggs, whites and yolks were beaten separately, 2 tablespoonfuls of milk, were added; 1 teaspoonful of chopped parsley; mix lightly together, add salt to season. Place 2 tablespoonfuls of butter in a fry pan. When butter has melted, pour mixture carefully into pan. When cooked, sprinkle over a small quantity of finely minced parsley. Roll like a "jelly roll." Place on a hot platter and serve at once, cut in slices.

"EXTRA FINE" BAKING POWDER BISCUITS

One quart of flour was measured; after being sifted, was placed in a flour sifter, with 4 heaping teaspoonfuls of Royal baking powder and 1 teaspoonful of salt. Sift flour and baking powder into a bowl, cut through this mixture 1 tablespoonful of butter and lard each, and mix into a soft dough, with about 1 cup of sweet milk. 1 egg should have been added to the milk before mixing it with the flour. Reserve a small quantity of the yolk of egg, and thin with a little milk. Brush this over the top of biscuits before baking.

Turn the biscuit dough onto a floured bake-board. Pat out about one inch thick. Cut into rounds with small tin cake cutter. Place a small bit of butter on each biscuit and fold together. Place a short distance apart on baking tins and bake in a quick oven.

"FLANNEL" CAKES, MADE FROM SOUR MILK

One pint of sour milk, 2 eggs (beaten separately), a little salt, 1 large teaspoon of melted butter, 1 teaspoonful of molasses, 1 good teaspoon of soda, sifted with enough flour to make a smooth batter. Beat hard and then add the 2 yolks and the stiffly-beaten whites of eggs. Bake small cakes on a hot, well-greased griddle. Serve with honey or maple syrup.

"FLANNEL" CAKES WITH BAKING POWDER

Sift together in a bowl 1 pint of flour, 1 teaspoon of salt, 2 teaspoons of Royal baking powder, mixed to a smooth batter, with about 1 pint of sweet milk. Add two yolks of eggs, 1 tablespoon of melted butter. Lastly, add the 2 stiffly-beaten whites of eggs. 1 teaspoon of baking molasses added makes them brown quickly. Bake on a hot griddle, well greased.

FRAU SCHMIDT'S RECIPE FOR WAFFLES

One pint of sour milk, 1 quart of sweet milk, 1 teaspoon salt, 1 tablespoon butter, whites of three eggs and yolks of two and 1 teaspoon of baking soda, and flour to make a rather thin batter. Beat the two yolks of the eggs until light and creamy, then add ½ teaspoon of baking powder, little flour, then the sour milk with soda dissolved in it, stirring all the time. Then add 1 tablespoon of melted or softened butter, then the sweet milk; beat well; and lastly, add the stiffly-beaten whites of the three eggs. Bake in hot waffle iron.

"CRUMB" CORN CAKES

One pint of stale bread crumbs (not fine, dried crumbs), covered with 1 pint of sour milk. Let stand over night. In the morning add 1 tablespoon of butter, yolks of 2 eggs and a little salt, ½ teaspoon of salaratus (good measure), ¾ cup of granulated corn meal, to which add a couple of tablespoons of bread flour, enough to fill up the cup. Stir all well together, add the 2 stiffly-beaten whites of eggs and drop with a tablespoon on a hot, greased griddle. Make the cakes small, as they do not turn quite as easily as do buckwheat cakes. This makes about two dozen cakes. These are good.

"GRANDMOTHER'S" RECIPE FOR BUTTERMILK WAFFLES

Mix to a smooth batter, 4 cups of sour buttermilk, 5 cups of flour, and add 1 tablespoon of melted butter, 1 teaspoon salt, 1 tablespoon of molasses. Add the well-beaten yolks of 3 eggs, 1½ teaspoonfuls of baking soda, dissolved in a little hot water. Lastly, add the stiffly-beaten whites of 3 eggs. Place about 3 tablespoonfuls of the batter on hot, well-greased waffle irons. If buttermilk cannot be procured, sour milk may be used with good results, providing the milk is quite sour. From this quantity of batter may be made twelve waffles. Serve with maple syrup or honey.

BREAD GRIDDLE CAKES

To 1 pint of sour milk add about 3 slices of stale bread and allow the bread to soak in this mixture over night. In the morning beat up smoothly with 1 egg yolk, 1 teaspoonful of soda, a pinch of salt and enough cornmeal and white flour, in equal quantities, to make a moderately thin batter. Lastly, add the stiffly-beaten white of egg, bake on a hot griddle. Cakes should be small in size, as when baked cakes are less readily turned than other batter cakes. These cakes are economical and good.

NEVER FAIL "FLANNEL" CAKES

- 2 cups thick sour milk (quite sour).
- 2 tablespoonfuls sweet milk.
- 1 egg.
- ½ teaspoonful salt.
- 2 cups flour.
- 1 teaspoonful baking soda (good measure).

Pour the milk in a bowl, add yolk of egg. Sift together flour, baking soda and salt, four times. Beat all well together. Then add the stiffly-beaten white of egg, and bake at once on a hot griddle, using about two tablespoonfuls of the batter for a cake. Serve with butter and maple syrup or a substitute.

This recipe, given Mary by an old, reliable cook, was unfailing as to results, if recipe be closely followed. The cakes should be three-fourths of an inch thick, light as a feather, and inside, fine, like bread, not "doughy," as cakes baked from richer batters frequently are.

From this recipe was made eighteen cakes.

WAFFLES MADE FROM SWEET MILK AND BAKING POWDER

Sift together 1 quart of flour, 2 teaspoonfuls of baking powder and ½ teaspoonful of salt. Mix into a batter, a little thicker than for griddle cakes, with sweet milk; add yolks of 3 eggs, 3 tablespoonfuls of melted butter; lastly, stir in lightly the 3 stiffly-beaten whites of eggs. Bake on a hot, well-greased waffle iron and serve with maple syrup.

"BUCKS COUNTY" BUCKWHEAT CAKES

About 12 o'clock noon dissolve 1 cake of yeast (the small, round or square cornmeal cakes) in 1 pint of lukewarm water. Add to this 1 tablespoonful wheat flour, 1 tablespoonful yellow cornmeal, and enough good buckwheat flour to make a thin batter. Set in a warm place near the range to rise. About 6 or 7 o'clock in the evening add this sponge to 1 quart and 1 pint of lukewarm potato water (water drained from boiled potatoes), 1 tablespoonful of mashed potatoes added improves the cakes; add salt. They need considerable. Stir in enough buckwheat flour to make quite a stiff batter, beat hard and set to rise, covered, in a warm place over night. The next morning add 1 teaspoonful saleratus, dissolved in a little hot water; 1 table-spoonful of baking molasses and a little warm milk, to thin the batter; or water will answer. The batter should be thin enough to pour. Let stand a short time, then bake on a hot griddle. Half this quantity will be enough for a small family. Then use only ½ teaspoonful saleratus. Bake golden brown on hot griddle. Serve with honey or maple syrup. If this recipe for buckwheat cakes is followed, you should have good cakes, but much of their excellence depends on the flour. Buy a small quantity of flour and try it before investing in a large quantity, as you cannot make good cakes from a poor brand of flour.

DELICIOUS CORN CAKES

One cup of sweet milk heated to boiling point; stir in 2 heaping tablespoonfuls yellow, granulated cornmeal; add a tablespoonful of butter or lard and salt to taste. As soon as the mixture has cooled, stir in 1 tablespoonful of wheat flour. If the batter should be too thick, stir in enough cold, sweet milk to make it run easily from the spoon. Add 1 heaping teaspoonful of Royal baking powder. Drop spoonfuls on hot, greased griddle, and bake. This quantity makes cakes enough to serve three people, about sixteen small cakes. This is an economical recipe, as no eggs are used.

RICE WAFFLES (AS AUNT SARAH MADE THEM.)

Add 1 tablespoonful of butter and 1 tablespoonful lard to 1 cup of cold, boiled rice; 2 yolks of eggs, the whites beaten separately and added last; 2 cups of flour, 1 teaspoonful salt and 2 teaspoonfuls baking powder, sifted together; 1 teaspoonful of sugar and 1 teaspoonful of molasses, and enough sweet milk to make a thin batter. Bake in hot waffle irons. With these serve either maple syrup or a mixture of sugar and cinnamon.

"GERMAN" EGG-PANCAKES (NOT CHEAP)

These truly delicious pancakes were always baked by "Aunt Sarah" when eggs were most plentiful. For them she used, 1 cup flour, 5 fresh eggs, ½ cup milk.

The yolks of 5 eggs were broken into a bowl and lightly beaten. Then milk and flour were added gradually to form a smooth batter. Lastly, the stiffly-beaten whites of eggs were added. Large spoonfuls were dropped on a hot, well-greased griddle, forming small cakes, which were served as soon as baked. These cakes require no baking powder. Their lightness depends entirely on the stiffly-beaten whites of eggs.

"FRAU SCHMIDT'S" GRIDDLE CAKE RECIPE

The Professor's wife gave Mary this cheap and good recipe for griddle cakes: 1 pint of quite sour, thick milk; beat into this thoroughly 1 even teaspoon of baking soda, ½ teaspoon each of salt and sugar and 2 cups of flour, to which had been added 1 tablespoon of granulated cornmeal and 1 rounded teaspoon of baking powder before sifting. No eggs were used by the Professor's wife in these cakes, but Mary always added yolk of 1 egg to the cakes when she baked them.

MARY'S RECIPE FOR "CORN CAKE"

- 1 cup of white flour.
- ½ cup cornmeal (yellow granulated cornmeal).
- 1 cup of sweet milk.
- 2 teaspoonfuls baking powder.
- 1 tablespoonful sugar.
- ½ teaspoonful salt.
- 1 tablespoonful butter.
- 1 tablespoonful lard.
- 1 egg.

Sift together flour, salt and baking powder, sugar, and add ½ cup of granulated, yellow cornmeal. Mix with 1 cup milk, 1 beaten egg, and the 2 tablespoonfuls of butter and lard. Beat thoroughly. Add a tablespoonful more of flour if not as stiff as ordinary cake batter. Pour in well-greased bread tin and bake about 40 minutes in a hot oven.

AUNT SARAH'S DELICIOUS CREAM BISCUITS

Place in a flour sifter 2 cups of flour, 2 teaspoonfuls baking powder, ½ teaspoonful of salt and ½ teaspoonful of sugar. Sift twice; stir together ½ cup of sweet milk and ½ cup of thick, sweet cream. Quickly mix all together, cutting through flour with a knife, until a soft dough is formed, mixing and handling as little as possible. Drop spoonfuls into warmed muffin tins and bake at once in a hot oven. Serve hot.

These are easily and quickly made, no shortening other than cream being used, and if directions are closely followed will be flakey biscuits when baked.

Aunt Sarah was always particular to use pastry flour when using baking powder, in preference to higher-priced "Hard Spring Wheat," which she used only for the making of bread or raised cakes, in which yeast was used.

MARY'S MUFFINS

- 2 cups of flour.
- 3 even teaspoonfuls of baking powder.
- 2 tablespoonfuls of sugar.
- 1 cup of sweet milk.
- 2 eggs.
- 1 tablespoonful of butter.

Sift flour and baking powder in a bowl; add 1 tablespoonful of sugar and a pinch of salt; add the 2 yolks of eggs to the 1 cup of milk, and mix with the flour and baking powder; lastly, add the stiffly-beaten whites of eggs. Place large spoonfuls of the batter in small Gem pans. Bake in a hot oven 20 minutes. These muffins are fine.

CORN MUFFINS (AS MADE BY "FRAU SCHMIDT")

- 2 eggs.
- 1½ tablespoonfuls of sugar.
- 1 cup of granulated yellow cornmeal.
- 1½ cups of sweet milk.
- 2 cups of white flour.
- 3 teaspoonfuls of baking powder.
- 1 tablespoonful melted butter.
- A pinch of salt

Beat together eggs and sugar, add milk and cornmeal and the white flour, sifted, with baking powder and salt; add the 1 table-spoonful of melted butter. Bake 20 minutes in warmed Gem pans, in a hot oven. Mary's Aunt taught her to utilize any left-over muffins by making a very appetizing pudding from them called "Indian Sponge" Pudding, the recipe for which may be found among pudding recipes.

STRAWBERRY SHORTCAKE (AS FRAU SCHMIDT MADE IT)

- 1 pint of flour.
- 3 teaspoonfuls of baking powder.
- 2½ tablespoonfuls of butter or lard.
- 1 egg.

- ½ teaspoon of salt.
- Milk or water.

Sift together flour, baking powder and salt, and cut butter or lard through the flour. Add 1 beaten egg to about 1 cup of sweet milk, and add gradually to the flour, cutting through it with a knife until a soft dough is formed, mixing and handling as little as possible. Divide the dough into two portions, roll out one portion quickly and place on a large pie tin; spread the top of cake with softened (not melted) butter, lay the other cake on top and bake in a quick oven. When baked and still hot, the cakes may be easily separated without cutting; when, place between layers, and, if liked, on top of the cake, crushed, sweetened strawberries. "Frau" Schmidt thought a crushed banana added to the strawberries an improvement. Serve the hot shortcake with sweet cream and sugar.

Or, the recipe for baking a plain (not rich) layer cake might be used instead of the above. When baked and cooled, spread between the layers the following:

To the stiffly-beaten white of 1 egg, add 1 cup of sugar; beat well. Then add 1 cup of crushed strawberries. Beat all together until the consistency of thick cream. Serve cold.

PERFECTION WAFFLES

Sift together 4 cups of flour, 2 teaspoonfuls of baking soda and 1 teaspoonful of salt, four times.

Separate 3 fresh eggs. Place the yolks in an earthenware mixing bowl. Beat well with a spoon. Then add 3½ cups of sour milk or sour buttermilk and ½ cup of sour cream, and 1 teaspoonful of melted butter. Mix a smooth batter with the sifted flour and soda. Lastly, add the stiffly-beaten whites of 3 eggs. Mix the batter quickly and thoroughly. Bake on a hot, well-greased waffle iron and serve at once.

The waffles may be buttered as soon as baked and sugar sifted over, or a saucer containing a mixture of cinnamon and sugar, or a small jug of maple syrup may be served with them. Twelve waffles were made from this recipe.

RECIPE FOR MAKING "BAKING POWDER"

Sift together three times (through a fine sieve) 8 tablespoonfuls of cream of tartar, 4 tablespoonfuls of baking soda (salaratus), 4 table-spoonfuls of flour. Cornstarch may be substituted for flour. This latter ingredient is used to keep the cream of tartar and soda separate and dry, as soda is made from salt and will absorb moisture. This recipe for making a pure baking powder was given Mary by Fran Schmidt, who had used it for years with good results.

FRITTERS, CROQUETTES, DUMPLINGS AND CRULLERS

When cooking any article to be immersed in fat use about this proportion: 2 pounds of sweet lard to 1 of suet, which had been previously tried out. It is cheaper, also more wholesome, to use part suet than to use all lard. Save all pieces of left-over fat, either raw or cooked, from steaks, roasts, bacon or ham. Cut all up into small pieces and place in a pan in the oven until tried out, or put in a double boiler and stand over boiling water until fat is tried out. Strain and stand aside to be used as drippings. To clarify this fat, pour boiling water over, let cook a short time, strain and stand away in a cool place, when a cake of solid fat will form on top, which may be readily removed and used as drippings, or it may be added to the kettle of fat used for deep frying. Always strain fat carefully after frying croquettes, fritters, etc. Should the frying fat become dark add to the can of soap fat the economical housewife is saving. Return the clear-strained fat to the cook pot, cover carefully, stand aside in a cool place, and the strained fat may be used times without number for frying. The housewife will find it very little trouble to fry fritters, croquettes, etc., in deep fat, if the fat is always strained immediately after using, and returned to the cook pot, kept especially for this purpose. Stand on the hot range when required and the fat will heat in a few minutes, and if the fat is the right temperature, food cooked in it should not be at all greasy. When the housewife is planning to fry fritters or croquettes she should, if possible, crumb the articles to be fried several hours before frying, and stand aside to become perfectly cold. When the fat for frying is so hot a blue smoke arises, drop in the fritters or croquettes, one at a time, in order not to chill the fat or plunge a frying basket, containing only a couple of fritters at a time, in the hot fat, as too many placed in the fat at one time lowers the temperature too quickly and causes the fritters to be greasy and soggy. To test the fat before dropping in the fritters, if a small piece of bread is dropped in the fat and browns in about one minute the fat is the right temperature for frying fritters, and fritters fried at the correct temperature should be a rich brown and not at all greasy. When removing fritters from hot fat place on coarse brown paper to absorb any remaining fat. Fritters composed of vegetables, or oysters, should be served on a platter garnished with parsley, and fritters composed of fruit, should have pulverized sugar sifted over them liberally. Should a small piece of bread brown in the fat while you count twenty, fat is the correct temperature for frying croquettes, but is too hot for frying crullers or any food not previously cooked.

KARTOFFLE BALLA (POTATO BALLS)

Boil until tender, 8 medium-sized (not pared) potatoes; when quite cold remove parings and grate them; fry one finely-chopped onion in a little butter until a yellow-brown; add this, also 1 egg, to the potatoes, season with salt and pepper and add flour enough to mold into balls; use only flour enough to hold the mixture together. The chopped onion may be omitted, and instead, brown small, dice-like pieces of bread in a little butter, shape dumplings into balls the size of walnuts, place a teaspoonful of the browned bread crumbs in the centre of each and add also a little chopped parsley. Drop the dumplings in salted boiling water and cook uncovered from 15 to 20 minutes. When dumplings rise to the top they should be cooked sufficiently, when remove from kettle with a skimmer to a platter; cut dumplings in half and strew over them bread crumbs, browned in butter.

"BOOVA SHENKEL"

For this excellent "Pennsylvania German" dish, which I am positive has never before been published, take 2½ pounds of stewing
meat (beef preferred), season with salt and pepper and cook slowly
several hours until tender.

For the filling for the circles of dough, take 12 medium-sized
white potatoes, pared and thinly sliced, steamed until tender; then
add seasoning to taste of salt and pepper, 2 tablespoonfuls of butter,
2 tablespoonfuls of finely-minced parsley and 1 finely-chopped
onion (small); lastly, add 3 eggs, lightly beaten together, to the mixture. Allow this to stand while the pastry is being prepared in the
following manner:

Pastry—Sift into a bowl 2½ cups of flour, 2 teaspoonfuls of baking
powder and ½ teaspoonful of salt, 1 generous teaspoonful of lard
and 1 of butter. Cut through the flour, mix with water into a dough
as for pie crust. Roll thin, cut into about ten circles, and spread some
of the mixture on each circle of dough. Press two opposite edges
together like small, three-cornered turnover pies; drop these on to
the hot meat and broth in the cook pot, closely covered. Cook slowly from 20 to 30 minutes. Before serving the "Boova Shenkel" pour
over the following:

Cut slices of stale bread into dice and brown in a pan containing 1
large tablespoonful of butter and a couple tablespoonfuls of fat
(which had been skimmed from top of broth before "Boova Shenkel"
had been put in cook pot), add about ½ cup of milk to diced,
browned bread; when hot, pour over the "Boova Shenkel" and serve
with the meat on a large platter.

RICE BALLS WITH CHEESE

Place 2 cups of cold, boiled rice, well drained, in a bowl and add
½ cup of grated cheese, a little salt, ¼ cup flour and the stiffly-
beaten white of one egg. Mix all together and mold into balls about

297

the size of a small egg, with a little of the flour; then roll them in fine, dried bread crumbs, and stand away until perfectly cold. When preparing for lunch, beat the yolk of the egg with a little milk, dip the rice balls into this, then into fine, dried bread crumbs, drop in deep fat and fry a golden brown. Drain on brown paper and serve, garnished with parsley.

"KARTOFFLE KLOSE"

One quart of cold, boiled, skinned potatoes, grated. (Boil without paring the day before they are to be used, if possible.) Put into a frying pan 1 tablespoonful of butter, 1 finely-minced onion (small onion), and fry until a light brown. Remove from fire and mix with this: 2 heaped tablespoonfuls flour, 1 tablespoonful of finely-cut parsley, 2 eggs (whites beaten separately), and 2 slices of bread, cut fine. Add grated potatoes and bread crumbs, alternately, mixing together lightly with a fork; add the other ingredients, season well with salt and pepper, form into round balls the size of a walnut and drop into a stew-pan of boiling, salted water, containing a teaspoon of butter. Do not cover the stew-pan while they are cooking. As soon as the dumplings rise to the top, skim one out and cut in half to see if it is cooked through. They should take from 15 to 20 minutes to cook. Skim out of the boiling water on a platter. Cut each dumpling in half, pour over them bread crumbs browned in a pan containing a little lard and butter, and serve. The onion may be omitted and only finely-chopped parsley used, if desired, or use both. Or place the halved dumplings in pan containing a little lard and butter and chopped onion (if the latter is liked), and brown on each side, then serve.

RICE CROQUETTES (AND LEMON SAUCE)

Boil 1 cup of well-washed rice in 6 or 8 cups of rapidly-boiling water, until tender. The rice, when cooked and drained, should fill 3 cups. Prepare a cream sauce of 1 pint of milk, 3 heaping tablespoonfuls of flour and 2 tablespoons of butter and 2 egg yolks. Stir in 3 cups of flaky, cooked rice, while rice is still hot. When the mixture has cooled, mold into small cone shapes with the hands, stand aside until perfectly cold. Dip the croquettes into the whites of eggs, then roll them in fine, dried bread crumbs and fry in deep fat. If a cube of bread browns in the fat in a little longer time than a half minute, the fat is the right temperature. Eighteen croquettes were made from this quantity of rice.

Lemon Sauce—To serve with rice croquettes, cream together ½ cup of sugar, 1 tablespoonful of butter, 1 egg, 2 cups of boiling water was added and all cooked together until the mixture thickened. When cooled slightly add the juice and grated rind of one lemon. Serve in a separate bowl, and pass with the croquettes.

CORN OYSTERS

Slice off tips of kernels from cobs of corn and scrape down corn-pulp from cobb with a knife. To 1 pint of pulp add 2 eggs, 2 heaping tablespoonfuls of flour, ½ teaspoonful of salt and a pinch of cayenne pepper and of black pepper; add the 2 yolks of eggs, then stir in lightly the stiffly-beaten white of eggs and flour. Fry in only enough butter to prevent them sticking to the pan. Drop into pan by spoon-fuls size of an ordinary fried oyster, brown on both sides and serve hot.

BANANA FRITTERS

From one banana was made 4 fritters. The banana was halved, cut lengthwise and then cut cross-wise. The batter will do for all fruits, clams, corn or oysters. Make a sauce of the liquor, mixed with same quantity of milk, with a tablespoon of butter added, chopped parsley and flour to thicken. When making oyster or clam fritters use same rule as for fruit fritters, using clam juice and milk instead of all milk.

For the "fritter batter," sift together 1 pint of flour, 2 teaspoonfuls baking powder and a pinch of salt. Stir slowly into it a pint of milk, then the well-beaten yolks of 3 eggs, and, lastly, the stiffly-beaten whites of eggs. Beat hard for a few minutes and fry at once in smoking hot fat. Orange sections make delicious fritters, or halves of fresh or canned peaches may be used.

Allow the bananas to stand one-quarter hour in a dish containing a small quantity of lemon juice and sugar before putting them in the batter. Lay the slices of bananas or sections of orange in the batter, then take up a tablespoonful of the batter with one slice of banana for each fritter, drop into hot fat one at a time, and fry a golden brown. Sift pulverized sugar over and serve hot. If a small piece of bread browns in one minute in the fat it is the right temperature to fry any previously uncooked food.

PARSNIP FRITTERS

Scrape and boil 5 or 6 parsnips in salted water until tender and drain. If old parsnips, cut out the centre, as it is tough and woody. Mash parsnips fine, add 1 egg yolk (white beaten separately), and added last a little salt, 1 large tablespoonful flour, ¼ teaspoonful baking powder, mold into small cakes, dredge with flour, and fry quickly to a golden brown in a tablespoonful of butter and one of drippings. Serve at once.

AUNT SARAH'S "SCHNITZ AND KNOPF"

This is an old-fashioned "Pennsylvania German" favorite. The end of a ham bone, containing a very little meat, was placed in a large kettle with a small quantity of water, with "Schnitz," or sliced, sweet, dried apples, which had been dried without removing the parings. When the apples were cooked tender in the ham broth; dumplings, composed of the following, were lightly dropped on top of the apples and broth and cooked, closely covered, from 15 to 20 minutes. Do not uncover kettle the first ten minutes. When dumplings have cooked place them with the "Schnitz" on a large platter, and serve at once.

A VERY OLD RECIPE FOR DUMPLINGS, OR "KNOPF"

One and one-half quarts of flour was sifted with 2½ tablespoonfuls of Royal baking powder, 1 teaspoonful of butter was cut through the flour in small bits, 1 egg was beaten and enough milk or water added to the egg to mix the flour into quite a soft dough. Sometimes instead of molding the dough into balls large spoonfuls were placed over the apples. Aunt Sarah had used this recipe for many years. This is a very old recipe, and from it was made a larger quantity than ordinary housekeepers usually require. Half the quantity, about 1½ pints of flour to 1¼ tablespoonfuls of baking powder, mixed according to the directions given in the first part of recipe, would be about the correct proportions for a family of ordinary size.

Aunt Sarah frequently substituted sour cherries and a teaspoonful of butter was added instead of ham and "Schnitz." Dumplings prepared from this recipe may be dropped on stewed chicken and broth and cooked or steamed, make an excellent pot-pie. Should there be more dough mixed than required for dumplings, place a panful in the oven and bake as biscuits. More baking powder is required when dough is steamed or boiled than when baked in the oven.

"KARTOFFLE KUKLEIN" (POTATO FRITTERS OR BOOFERS)

Place in a bowl 2 cups grated, pared, *raw* potatoes; drain off any liquid formed, then add 1 small onion, also grated; large egg or 2 small eggs, salt and pepper, 1 tablespoonful chopped parsley, ¼ teaspoonful baking powder (good measure), and a couple tablespoonfuls of flour to thicken just enough to make the fritters hold together; then drop by spoonfuls in deep, hot fat, and fry a rich brown. The fritters form into odd shapes a trifle larger than a fried oyster, when dropped in the fat.

Should the fritter batter separate when dropped in the fat, add more flour, but if too much flour is added they are not as good as when a lesser quantity is used. Drain the fritters on brown paper and garnish the platter upon which they are served with parsley. Mary's Uncle was very fond of these fritters. He preferred them to fried oysters, and always called them "potato boofers." I would not answer for the wholesomeness of these fritters. In fact, I do not think any fried food particularly wholesome.

ROSETTES, WAFERS AND ROSENKUCHEN (AS MADE BY FRAU SCHMIDT)

Prepare a batter from the following:

- 1 cup of sweet milk.
- 2 eggs.
- Pinch of salt.
- 1 cup of flour, good measure.

Gradually mix the flour with the milk to form a smooth batter, free from lumps. Add yolks, then the slightly-beaten whites of eggs. Fasten the long handle to a wafer iron, shaped like a cup or saucer, and stand it in hot fat, a mixture of ⅔ lard and ⅓ suet, or oil; when heated, remove at once, and dip quickly into the batter, not allowing the batter to come over top of the wafer iron. Then return it to the hot fat, which should cover the wafer iron, and in about 25 or 30 seconds the wafer should be lightly browned, when the wafer may be easily removed from the iron on to a piece of brown paper to absorb any fat which may remain. This amount of batter should make about forty wafers. On these wafers may be served creamed oysters, vegetables, chicken or fruit. When using the wafers as a foundation on which to serve fruit, whipped cream is a dainty adjunct. One teaspoonful of sugar should then be added to the wafer batter. These wafers may be kept several weeks, when by simply placing them in a hot oven a minute before serving they will be almost as good as when freshly cooked. Or the wafers may be served as a fritter by sifting over them pulverized sugar and cinnamon.

"BAIRISCHE DAMPFNUDELN"

These delicious Bavarian steamed dumplings are made in this manner: 1 cake of Fleischman's compressed yeast was dissolved in a cup of lukewarm milk, sift 1 pint of flour into a bowl, add 1 teaspoonful of sugar and 1 teaspoonful of salt. Mix the flour with another cup of lukewarm milk, 1 egg and the dissolved yeast cake and milk (two cups of milk were used altogether). Work all together thoroughly, adding gradually about 1½ cups of flour to form a soft dough. Do not mix it too stiff. Cover the bowl with a cloth; stand in a warm place until it has doubled the original bulk. Flour the bread board and turn out dough and mold into small biscuits or dumplings. Let these rise for half an hour, butter a pudding pan and place dumplings in it, brushing tops with melted butter. Pour milk in the pan around the dumplings to about two-thirds the depth of the dumplings; set pan on inverted pie tin in oven and bake a light brown. Serve with any desired sauce or stewed fruit. Or, after the shaped dough has raised, drop it in a large pot of slightly-salted boiling water, allowing plenty of room for them to swell and puff up, and boil continuously, closely covered, for 20 minutes. This quantity makes about 30 small dumplings. Should you not wish so many, half the quantity might be molded out, placed in a greased pie tin, and when light, which takes half an hour, bake in a moderately hot oven, and you will have light biscuits for lunch.

The thrifty German Hausfraus make fritters of everything imaginable, and sometimes unimaginable. Mary was told one day by a German neighbor how she prepared a fritter she called:

"HELLER BLUTHER KUKLEIN"

She gathered elderberry blossoms, rinsed off the dust, and when free from moisture dipped the blossoms into fritter batter, holding the stem ends, then dropped them into hot fat, and when golden brown, drained a minute on coarse, brown paper before serving, dusted them with powdered sugar; cinnamon may also be dusted over if liked. Mary pronounced them "fine," after tasteing, and said: "They certainly are a novelty." Perhaps something like this suggested the Rosette Iron, as it is somewhat similar.

APYL KUKLEIN (APPLE FRITTERS)

Pare and core 4 large tart apples. Cut each apple into about 4 round slices and allow the sliced apples to lie a couple of hours in a dish containing 2 tablespoonfuls of brandy, mixed with a half teaspoonful of cinnamon and a half teaspoonful of sugar. Drain the sliced apples, then a few at a time should be dropped in the following batter, composed of: 1 cup of flour sifted with ½ teaspoonful of Royal baking powder, ¼ teaspoonful of salt, add the yolks of 2 eggs and 1 cup of milk to form a smooth batter, then add the stiffly-beaten whites of eggs. Fry light brown, in deep, hot fat, and sift over powdered sugar. "Fried Apples" are an appetizing garnish for pork chops; the apples should be cored, *not pared*, but should be sliced, and when cut the slices should resemble round circles, with holes in the centre. Allow the sliced apples to remain a short time in a mixture of cinnamon and brandy, dry on a napkin, and fry in a pan, containing a couple of tablespoonfuls of sweet drippings and butter.

DUMPLINGS MADE FROM "BREAD SPONGE"

Aunt Sarah's raised dumplings from bread sponge were greatly relished at the farm.

When bread sponge, which had been set to rise early in the morning, and all flour necessary for loaves of bread had been added and loaves were being shaped to place in bread tins, Aunt Sarah reserved an amount of sponge sufficient for one loaf of bread, added a little extra salt, shaped them into small balls, size of a lemon, placed them on a well-floured board some distance apart to raise; when light (at 12 o'clock, if the dinner hour was 12.20), she carefully dropped the light balls of dough into a large pot of rapidly boiling, slightly salted water, covered closely, and boiled about 20 minutes, (Do not have more than one layer of the dumplings in cook pot, and do not place too close together; allow room for them to expand.)

Test by tearing one apart with a fork. Serve at once, and serve with a roast, to be eaten with gravy, with butter, or they may be eaten as a dessert, with jelly or maple syrup.

Aunt Sarah frequently added an equal quantity of fine, dried bread crumbs and flour and a little extra salt to a thin batter of bread sponge (before all the flour required for bread had been added), made about as stiff a dough as for ordinary loaves of bread; molded them into balls. When sufficiently raised, boiled them either in water or meat broth in the same manner as she prepared dumplings; made *only* of *flour*.

This is a small economy, using *bread crumbs* in place of *flour*, and these are delicious if prepared according to directions. Remember to have a large quantity of rapidly boiling water in which to cook the dumplings, not to allow water to stop boiling an instant and to keep cook pot closely covered for 20 minutes before removing one, and breaking apart to see if cooked through. These are particularly nice served with stewed apricots.

"LEBER KLOSE" OR LIVER DUMPLINGS

Boil a good-sized soup bone for several hours in plenty of water, to which add salt and pepper to taste and several small pieces of celery and sprigs of parsley to flavor stock. Strain the broth or stock into a good-sized cook pot and set on stove to keep hot.

For the liver dumplings, scrape a half pound of raw beef liver with a knife, until fine and free from all veins, etc. Place the scraped liver in a large bowl, cut three or four good-sized onions into dice, fry a light brown, in a pan containing 1 tablespoonful of lard and butter mixed. Cut into dice ¾ to a whole loaf of bread (about 2 quarts). Beat 2 eggs together, add 1 cup of sweet milk, season well with salt and pepper, and mix all together with 1 large cup of flour.

If not moist enough to form into balls when mixed together, add more milk. Keep the mixture as soft as possible or the dumplings will be heavy. Flour the hands when shaping the balls, which should be the size of a shelled walnut. Stand the pot containing stock on the front of the stove, where it will boil, and when boiling, drop in the dumplings and boil, uncovered, for 15 minutes. When cooked, take the dumplings carefully from the stock on to a large platter, pour the stock over the dumplings and serve.

These are excellent, but a little troublesome to make. One-half this quantity would serve a small family for lunch.

FRAU SCHMIDT'S "OLD RECIPE FOR SCHNITZ AND KNOPF"

Place a cook pot on the range, containing the end piece of a small ham; partly cover with water. This should be done about three hours before serving, changing the water once. Soak sweet, un-pared, sliced, dried apples over night in cold water. In the morning cook the dried apples (or schnitz) in a small quantity of the ham broth, in a separate stew-pan, until tender. Remove ham from broth one-half hour before serving. Sweeten the broth with a small quanti-ty of brown sugar, and when the broth commences to boil add raised dumplings of dough, which had been shaped with the hands into round balls about the size of an ordinary biscuit. Cook 25 minutes. Do not uncover the cook-pot after the dumplings have been dropped into the broth until they have cooked the required length of time. When the dumplings have cooked a sufficient time carefully remove to a warm platter containing the cooked apple schnitz. Thicken the broth remaining with a little flour, to the con-sistency of cream. Pour over the dumplings and serve at once.

Dumplings — At 9.30 in the evening set a sponge consisting of 1 cup of lukewarm milk, 1 tablespoonful sugar, 1 tablespoonful of butter, 1 egg, ¾ of an yeast cake, add flour enough to form a sponge (as stiff as may be stirred with a mixing spoon). Set to raise in a warm place over night. In the morning add more flour to the risen sponge until nearly as stiff as for bread. Form into round dumplings, place on a well-floured bake-board to rise slowly. Twenty-five minutes before serving drop dumplings into the hot broth in a large cook-pot.

There should be only one layer of dumplings, otherwise they will be heavy.

"BROD KNODEL," OR BREAD DUMPLINGS

- 3 cups of stale bread (cut like dice).
- ¾ cup of flour.
- ½ teaspoonful baking powder.
- ¾ cup milk.
- 2 tablespoonfuls butter.
- 1 egg.
- 1 teaspoonful of finely-minced parsley.
- ½ teaspoonful finely-minced onion (if liked).
- Pinch of salt.

Place two cups of diced bread in a bowl and pour over ¾ cup of milk. (Reserve 1 cup of diced bread, which brown in 1 tablespoonful of butter, to be added to the mixture later.) Allow milk and bread to stand 10 or 15 minutes; then add 1 tablespoonful of melted butter, 1 egg, flour and baking powder, and salt; fried, diced bread and parsley, and mix all together. With well-floured hands form the mixture into balls size of a walnut, and drop at once into rapidly boiling salted water and cook 15 minutes. Stew pan should be closely covered. When cooked, remove to platter with perforated skimmer, and serve at once, or drop dumplings into a pan containing 1 tablespoonful of melted butter, and brown on all sides before serving.

"GERMAN" POT PIE

To serve a family of six or seven, place 2 pounds of beef and 4 pork chops, cut in small pieces, in a cook-pot. Season with a little chopped onion, pepper and salt. This should be done about three or four hours before dinner. One hour before serving prepare the dough for pot pie. Pare white potatoes, slice and dry on a napkin, sift 2 cups of flour with 1 teaspoonful of baking-powder, pinch of salt, cut through the sifted flour, 1 level tablespoonful of shortening. Moisten dough with 1 egg and enough milk to make dough stiff enough to handle. (Almost 1 cup of milk, including the egg.) Cut off a small piece of dough, size of a small teacup, roll thin and take up plenty of flour on both sides. Take up all flour possible. Cut this dough into four portions or squares. Have the meat more than covered with water, as water cooks away.

Place a layer of potatoes on meat (well seasoned), then the pared potatoes and small pieces of dough alternately, never allowing pieces of dough to lap; place potatoes between. Roll the last layer out in one piece, size of a pie plate, and cover top layer of potatoes with it. Cover closely and cook three-quarters of an hour from the time it commences to boil. Then turn out carefully on a platter and serve at once.

"ZWETCHEN DAMPFNUDELN" (PRUNE DUMPLINGS)

In the evening a sponge was prepared with yeast for bread. All the flour required to stiffen the dough for loaves of bread being added at this time. The bread sponge was stood in a warm place to rise over night. In the morning, when shaping the dough into loaves, stand aside about one pint of the bread dough. Later in the morning form the pint of dough into small balls or dumplings, place on a well-floured bake board and stand in a warm place until doubled in size. Then drop the dumplings into a cook pot containing stewed prunes, a small quantity of water, a little sugar and lemon peel, if liked. The dried prunes had been soaked over night in cold water, and allowed to simmer on the range in the morning. The prune juice should be hot when the dumplings are added. Cook dumplings one-half hour in a closely covered cook-pot and turn out carefully on to a warmed platter, surrounded by prune juice and prunes.

GREEN CORN FRITTERS

Grate pulp from six cars of corn; with a knife scrape down the pulp into a bowl, add 2 eggs, beaten separately, a couple tablespoonfuls of milk, 1 large tablespoonful of flour, ¼ teaspoonful of baking powder and a pinch of salt. Drop with a spoon on a well-greased griddle. The cakes should be the size of a silver half dollar. Bake brown on either side and serve hot. These should not be fried as quickly as griddle cakes are fried, as the corn might then not be thoroughly cooked.

"MOULDASHA" (PARSLEY PIES)

Mash and season with butter and salt half a dozen boiled white potatoes, add a little grated onion and chopped parsley. Sift together in a bowl 1 cup of flour, 1 teaspoonful baking powder and a little salt. Add a small quantity of milk to one egg if not enough liquid to mix into a soft dough. Roll out like pie crust, handling as little as possible. Cut into small squares, fill with the potato mixture, turn opposite corners over and pinch together all around like small, three-cornered pies. Drop the small triangular pies into boiling, salted water a few minutes, or until they rise to top; then skim out and brown them in a pan containing a tablespoonful each of butter and lard. I have known some Germans who called these "Garden Birds." Stale bread crumbs, browned in butter, may be sprinkled over these pies when served. Serve hot.

These are really pot pie or dumplings with potato filling. Mary's Aunt always called these "Mouldasha." Where she obtained the name or what its meaning is, the writer is unable to say.

INEXPENSIVE DROP CRULLERS

Cream together 1 cup sugar and 1 egg, then add one cup of milk alternately with 2 cups of flour, sifted with 2 teaspoonfuls of baking powder. Add ½ teaspoonful of vanilla and enough flour to make a stiff batter.

Take about ½ a teaspoonful of the batter at a time and drop into boiling hot fat, and brown on both sides; then drain on coarse, brown paper and, when cool, dust with pulverized sugar. These cakes are cheap and good, and as no shortening is used are not rich. Do not make cakes too large, as they then will not cook through readily.

BATTER BAKED WITH GRAVY

The Professor's wife gave Mary this recipe, given her by an Englishwoman. The recipe was liked by her family, being both economical and good. When serving roast beef for dinner, before thickening the gravy, take out about half a cup of liquid from the pan and stand in a cool place until the day following. Reheat the roast remaining from previous day, pour the half cup of liquid in an iron fry pan, and when hot pour the following batter in the pan with the fat and bake in a moderately hot oven about 25 minutes. Or the batter may be poured in pan about 25 minutes before meat has finished roasting.

The batter was composed of 1 cup of flour, sifted with 1 small teaspoonful of baking powder and ½ teaspoonful of salt, mixed smooth with 1 cup of sweet milk. Add 2 well-beaten eggs. When baked cut in small pieces, surround the meat on platter, serve instead of potatoes with roast. The addition of baked dough extends the meat flavor and makes possible the serving of a smaller amount of meat at a meal.

"GERMAN" SOUR CREAM CRULLERS

One cup sugar, 1 cup sour cream, 2 eggs, 2 tablespoonfuls of butter, 1 teaspoonful soda, pinch of salt. About 3½ cups of flour. (Use extra flour to dredge the bake-board when rolling out crullers.) This is a very good recipe for crullers, in which the economical house-wife may use the cup of cream which has turned sour. This necessi-tates using less shortening, which otherwise would be required. Cream together sugar, butter, add yolks of eggs. Dissolve the soda in a small quantity of sour cream. Mix cream alternately with the flour. Add pinch of salt. Add just enough flour to roll out. Cut with small doughnut cutter with hole in centre. Fry in hot fat. Dust with pulverized sugar.

"GRANDMOTHER'S" DOUGHNUTS

Cream together 1 cup sugar and 2 teaspoonfuls butter, ½ a grated nutmeg, and a pinch of salt. Add 2 eggs, beaten without separating yolks from whites, and 1 cup of sweet milk. Then add 4 cups of flour (or 1 quart), prepared as follows: Measure 1 quart of unsifted flour and sift twice with 2 generous teaspoonfuls of baking powder. Use this to thicken the batter sufficiently to roll out and use about 1 extra cup of flour to flour the bake-board. Turn out one-half the quantity of dough on to a half cup of flour on the bake-board. Roll out dough half an inch thick. Cut out with round cutter, with hole in centre, and drop into deep, hot fat. Use ⅔ lard and ⅓ suet for deep frying; it is cheaper and more wholesome than to use all lard. When fat is hot enough to brown a small piece of bread while you count 60, it is the correct temperature for doughnuts. The dough should be as soft as can be handled. When cakes are a rich brown, take from fat, drain well on coarse, brown paper, and when cool dust with pulverized sugar and place in a covered stone jar. Never use fat as hot for frying doughnuts as that used for frying cro-quettes, but should the fat not be hot the doughnuts would be greasy. These doughnuts are excellent if made according to recipe.

FINE "DROP CRULLERS"

Cream together 1½ cups pulverized sugar, 3 eggs, add 1 cup sweet milk, ½ teaspoonful of salt, 3½ cups of flour, sifted after measuring with 2 teaspoonfuls of baking powder. Drop teaspoonfuls of this carefully into boiling fat.

They should resemble small balls when fried. Batter must not be too stiff, but about the consistency of a cup-cake batter.

Boil them in a mixture of cinnamon and sugar when all have been fried.

SOUPS AND CHOWDERS

Stock is the basis of all soups made from meat, and is really the juice of the meat extracted by long and gentle simmering. In making stock for soup always use an agate or porcelain-lined stock pot. Use one quart of cold water to each pound of meat and bone. Use cheap cuts of meat for soup stock. Excellent stock may be made from bones and trimmings of meat and poultry. Wash soup bones and stewing meat quickly in cold water. Never allow a roast or piece of stewing meat to lie for a second in water. Aunt Sarah did not think that wiping meat with a damp cloth was all that was necessary (although many wise and good cooks to the contrary). Place meat and soup bones in a stock pot, pour over the requisite amount of soft, cold water to extract the juice and nutritive quality of the meat; allow it to come to a boil, then stand back on the range, where it will just simmer for 3 or 4 hours. Then add a sliced onion, several sprigs of parsley, small pieces of chopped celery tops, well-scraped roots of celery, and allow to simmer three-quarters of an hour longer. Season well with salt and pepper, 1 level teaspoonful of salt will season 1 quart of soup. Strain through a fine sieve, stand aside, and when cool remove from lop the solid cake of fat which had formed and use for frying after it has been clarified. It is surprising to know the variety of soups made possible by the addition of a small quantity of vegetables or cereals to stock. A couple tablespoonfuls of rice or barley added to well-seasoned stock and you have rice or barley soup. A small quantity of stewed, sweet corn or noodles, frequently "left-overs," finely diced or grated carrots, potatoes, celery or onions, and you have a vegetable soup. Strain the half can of tomatoes, a "left-over" from dinner, add a tablespoonful of butter, a seasoning of salt and pepper, thicken to a creamy consistency with a little cornstarch, add to cup of soup stock, serve with croutons of bread or crackers, and you have an appetizing addition to dinner or lunch.

The possibilities for utilizing left-overs are almost endless. The economically-inclined housewife will be surprised to find how easily she may add to the stock pot by adding left-over undesirable pieces of meat and small quantities of vegetables. One or two spoonfuls of cold left-over oatmeal may also be added to soup with advantage, occasionally. Always remove the cake of fat which forms on top of soup as soon as cooled, as soup will turn sour more quickly if it is allowed to remain. If soup stock be kept several days in summer time, heat it each day to prevent souring.

Pieces of celery, onion, parsley, beans and peas may all be added to soup to make it more palatable. Also fine noodles. The yolk of a hard-boiled egg dropped into the soup kettle and heated through, allowing one for each plate of soup served, is a quick and appetizing addition to a soup of plain broth or consomme.

VEGETABLE SOUP

Slice thinly 3 potatoes, 3 carrots, 3 turnips, the undesirable parts of 2 heads of celery, 2 stalks of parsley and 3 onions. Cook the onions in a little butter until they turn a yellow brown, then add the other ingredients. Season well with salt and black pepper, also a pinch of red pepper. Put all together in a stew-pan, cover with three quarts of water, stand on range and simmer about three hours. Strain soup into stew-pan, place on range, and when hot add Marklose Balls.

MARKLOSE BALLS

Take marrow from uncooked beef soup bones, enough to fill 2 tablespoons, cut fine, add 2 eggs, 1 teaspoonful grated onion to flavor, pepper and salt, stiffen with 1 cup of bread crumbs, shape into balls size of marbles, drop into hot broth and cook uncovered from 15 to 20 minutes.

Aunt Sarah purchased two good-sized soup bones containing considerable meat. After extracting 2 tablespoonfuls of marrow from the uncooked bones, she put the bones in a stew-pan with a couple of quarts of water, a large onion, chopped fine, and a piece of celery, and cooked for several hours, then skimmed off scum which arises on top of broth, removed the soup bones and meat and added a couple of tablespoonfuls of grated carrot, pepper and salt to taste, cooked a short time, and then added the marrow balls, a little chopped parsley and a couple of tablespoonfuls of boiled rice. Two tablespoonfuls of marrow will make about 15 balls, with the addition of crumbs, eggs, etc.

EGG BALLS FOR SOUP

Mash the yolks of 2 hard-boiled eggs fine and smooth with a little soft butter. Beat the white of 1 egg, and add with about 2 table-spoonfuls of flour, salt and pepper. Mix all together. Use a little flour to mold the mixture into balls the size of quite small marbles. Do not make too stiff. Drop these into hot broth or soup and cook about five minutes. This quantity will make 12 small balls.

"SUPPEE SCHWANGEN"

Mary was taught to make these by the Professor's wife. She beat together either 1 or 2 raw eggs, ½ cup flour, 1 tablespoonful butter, a little salt, and just enough milk to thin the mixture enough so it may be dropped by half teaspoonfuls into hot soup stock or broth. Cook these small dumplings about 10 minutes. Serve in soup broth.

CREAM OF OYSTER BOUILLON

Put two dozen oysters through food chopper, cook oyster liquor and oysters together five minutes, heat 1 pint milk and 1 tablespoon flour, mixed smooth with a little cold milk, and 1 tablespoonful butter. Let come to a boil, watching carefully that it does not burn. Pour all together when ready to serve. Serve in bouillon cups with crackers. This recipe was given Mary by a friend in Philadelphia, who thought it unexcelled.

GERMAN NOODLE SOUP

Place about 3 pounds of cheap stewing beef in a cook-pot with sufficient water and cook several hours, until meat is quite tender; season with salt and pepper. About an hour before serving chop fine 3 medium-sized potatoes and 2 onions and cook in broth until tender. Ten or fifteen minutes before serving add noodle.

To prepare noodles, break 2 fresh eggs in a bowl, fill ½ an egg shell with cold water, add the eggs, and mix with flour as stiff as can conveniently be handled. Add a little salt to flour. Divide dough into sheets, roll on bake-board, spread on cloth a short time and let dry, but not until too brittle to roll into long, narrow rolls. Cut this with a sharp knife into thin, thread-like slices, unroll, drop as many as wished into the stew-pan with the meat and cook about 10 or 15 minutes. Place the meat on a platter and serve the remainder in soup plates. The remaining noodles (not cooked) may be unrolled and dried and later cooked in boiling salted water, drained and placed in a dish and browned butter, containing a few soft, browned crumbs, poured over them when served. The very fine noodles are generally served with soup and the broad or medium-sized ones served with brown butter Germans usually serve with a dish of noodles, either stewed, dried prunes, or stewed raisins. Both are palatable and healthful.

CREAM OF CELERY

Cook 1 large stalk of celery, also the root cut up in dice, in 1 pint of water, ½ hour or longer. Mash celery and put through a fine sieve. Add 1 pint of scalded milk, and thicken with a tablespoonful of flour, mixed with a little cold milk. Add 2 tablespoonfuls of butter, pepper and salt, and simmer a few minutes. Just before serving add a cup of whipped cream. Serve with the soup, small "croutons" of bread.

OYSTER STEW

Rinse a stew-pan with cold water, then put in 1 pint of milk and let come to a boil. Heat 15 oysters in a little oyster liquor a few minutes, until the oysters curl up around the edges, then add the oysters to one-half the hot milk, add a large tablespoonful of butter, season well with salt and pepper, and when serving the stew add the half pint of boiling hot milk remaining. This quantity makes two small stews. Serve crackers and pickled cabbage. When possible use a mixture of sweet cream and milk for an oyster stew instead of all milk. An old cook told Mary she always moistened half a teaspoonful of cornstarch and added to the stew just before removing from the range to cause it to have a creamy consistency.

CLAM BROTH

Clam broth may be digested usually by the most delicate stomach. It can be bought in cans, but the young housewife may like to know how to prepare it herself. Strain the juice from one-half dozen clams and save. Remove objectionable parts from clams, cut in small pieces, add ½ pint of cold water and the clam juice, let cook slowly about 10 minutes, strain and season with pepper and salt, a little butter and milk, and serve hot.

TURKEY SOUP

Take broken-lip bones and undesirable pieces of roast turkey, such as neck, wings and left-over pieces of bread filling, put in stew-pot, cover with water, add pieces of celery, sliced onion and parsley, cook several hours, strain, and to the strained liquor add a couple tablespoonfuls of boiled rice, season with salt and pepper and serve. Some of the cold turkey might also be cut in small pieces and added to the soup.

CREAM OF PEA SOUP

Cook quarter peck of green peas until very tender, reserve one-half cup, press the remainder through a sieve with the water in which they were boiled. Season with salt and pepper. Mix 1 table-spoonful of flour, 1 tablespoonful of butter with 1 cup of hot milk. Mix flour smooth with a little cold milk before heating it. Cook all together a few minutes, then add the one cup of peas reserved. If soup is too thick add a small quantity of milk or water.

TOMATO SOUP

One quart of canned tomatoes, 1 tablespoonful sugar, 1 onion, and a sprig of parsley, cut fine, and 1 carrot and 2 cloves. Stew until soft enough to mash through a fine, wire sieve. Place one quart of sweet milk on the stove to boil. Mix 1 large tablespoonful of corn-starch smooth, with a little cold milk, and stir into the hot milk. Add 1 large tablespoonful of butler and ¼ teaspoonful (good measure) of soda. Let cook one minute, until it thickens, add 1 teaspoonful of salt. Do not add the milk to the strained tomatoes until ready to serve. Then serve at once.

FRAU SCHMIDTS CLAM SOUP

Chop 12 clams fine, add enough water to the clam broth to meas-ure one quart, cook all together about 15 minutes; add 3 pints of scalding hot milk, season with 1½ tablespoonfuls butter and salt and pepper to taste. Serve crackers with the soup.

CLAM CHOWDER

Cut ¼ pound of rather "fat" smoked bacon in tiny pieces the size of dice; fry until brown and crisp. Take 25 fresh clams, after having drained a short time in a colander, run through a food chopper and place in ice chest until required. Pour the liquor from the clams into an agate stew-pan; add 6 medium-sized potatoes and 4 medium-sized onions, all thinly sliced; also add the crisp bits of bacon and fat, which had fried out from the bacon, to the clam juice. Cook all together slowly or simmer 3 or 4 hours. Add water to the clam liquor occasionally as required. Ten or fifteen minutes before serving add 1 cup of hot water and the chopped clams (clam juice if too strong is liable to curdle milk). Allow clams to cook in the clam broth 10 to 15 minutes. Boil 2 quarts of sweet milk, and when ready to serve add the hot milk to the chowder, also 1 teaspoonful of chopped parsley. One-half this quantity will serve a small family. Serve crisp crackers and small pickels, and this chowder, served with a dessert, makes an inexpensive, nourishing lunch.

BROWN POTATO CHOWDER

Put a pint of diced, raw potatoes in a stew-pan over the fire, cover with 1 quart of water, to which a pinch of salt has been added. Cook until tender, but not fine, then add water so that the water in the stew-pan will still measure one quart should some have boiled away. Place a small iron fry-part on the range, containing 1 tablespoonful of sweet lard; when melted, it should measure about 2 tablespoonfuls. Then add 4 tablespoonfuls of flour, a pinch of salt and stir constantly, or rather mash the flour constantly with a spoon, being careful not to allow it to scorch, until a rich brown; add this to the diced potatoes and the quart of water in which they were boiled, stir until the consistency of thick cream, or like clam chowder. Should there be a few, small lumps of the browned flour not dissolved in the chowder, they will not detract from the taste of it; in fact, some are very fond of them. Perhaps some folks would prefer this, more like a soup; then add more hot water and thin it, but be careful to add more seasoning, as otherwise it would taste flat and unpalatable. Very few people know the *good flavor* of *browned flour*. It has a flavor peculiarly its own, and does not taste of lard at all. I would never advocate *any* seasoning except butter, but advise economical housewives to try this, being very careful not to scorch the flour and fat while browning.

A mixture of butter and lard may be used in which to brown the flour should there be a prejudice against the use of lard alone.

BEAN CHOWDER

Another palatable, cheap and easily prepared dish is called Bean Chowder. Small soup beans were soaked over night in cold water. Pour off, add fresh water and cook until tender. Then add browned flour (same as prepared for Potato Chowder) and the water in which the beans were cooked. When ready to serve, the beans were added. More water may be added until broth is thin enough for soup, then it would be called "brown bean soup."

BOUILLON

Buy a soup bone, cook with a chopped onion, one stalk of celery and a sprig of parsley until meat falls from bone. Season with salt and pepper. Strain the broth into a bowl and stand aside until perfectly cold. Then remove the cake of fat formed on top of soup and add it to drippings for frying. The broth may be kept several days if poured into a glass jar and set on ice. When wanted to serve, heat 1 pint of broth, add 2 tablespoonfuls of cream to yolks of 2 eggs. Stir well. Pour boiling hot broth over the cream and yolks of eggs and serve at once in bouillon cups. Serve crackers also. Do not cook mixture after cream and yolks of eggs have been added. This is very nourishing.

FARMER'S RICE

One and one-half quarts of milk, poured into a double boiler and placed on the range to heat. One cup of flour was placed in a bowl; into the flour 1 raw egg was dropped and stirred with a knife until mixed, then rubbed between the fingers into fine rivels. It may take a little *more* flour; the rivels should be dry enough to allow of being rubbed fine. When the milk commences to boil drop the rivels in by handfuls, slowly, stirring constantly. Salt to taste. Let cook 15 minutes. Eat while hot, adding a small piece of butter as seasoning. This should be a little thicker than ordinary rice soup.

PHILADELPHIA "PEPPER POT"

This recipe for far-famed "Philadelphia Pepper Pot" was given Mary by a friend living in the Quaker City, a good cook, who vouched for its excellence:

The ingredients consist of the following:

- 1 knuckle of veal.
- 2 pounds of plain tripe.
- 2 pounds of honeycomb tripe.
- 1 large onion,
- 1 bunch of pot-herbs.
- 4 medium-sized potatoes.
- 1 bay leaf—salt and cayenne pepper to season.
- ½ pound of beef suet—and flour for dumplings.

The day before you wish to use the "Pepper Pot" procure 2 pounds of plain tripe and 2 pounds of honeycomb tripe. Wash thoroughly in cold water place in a kettle. Cover with cold water and boil eight hours; then remove tripe from water, and when cold cut into pieces about ¾ of an inch square. The day following get a knuckle of veal, wash and cover with cold water—about three quarts—bring slowly to the simmering point, skimming off the scum which arises, simmer for three hours. Remove the meat from the bones, cut into small pieces, strain broth and return it to the kettle. Add a bay leaf, one large onion, chopped, simmer one hour; then add four medium-sized potatoes, cut like dice, and add to the broth. Wash a bunch of pot-herbs, chop parsley (and add last), rub off the thyme leaves, cut red pepper in half and add all to broth; then add meat and tripe and season with salt; *if liked hot*, use a pinch of cayenne pepper. For the dumplings, take 1 cup of beef suet, chopped fine, 2 cups flour, pinch of salt, mix well together and moisten with enough cold water to allow of their being molded or rolled into tiny dumplings, the size of a small marble. Flour these well to prevent sticking together. When all are prepared drop into soup, simmer a few minutes, add parsley and serve at once.

GERMAN VEGETABLE SOUP

Take 6 potatoes, half the quantity of onions, carrots, turnips, cabbage and a stalk of celery, cut up into dice-shaped pieces, place all in a stew-pan and cover with a couple quarts of hot water. Let cook about two hours, until all the vegetables are tender, then add 1 tablespoonful of butter, a large cup of milk, and about a tablespoonful of flour mixed smooth with a little cold milk, cook a few minutes, add a tablespoonful minced parsley, and serve.

A CHEAP RICE AND TOMATO SOUP

Take one pint of rice water which has been drained from one cupful of rice boiled in 2½ quarts of water 25 minutes (the rice to be used in other ways), and after the rice has drained in a sieve add to the rice water 1 cup stewed, strained tomatoes (measure after being strained), 1 teaspoonful butter, 1 teaspoonful flour mixed with a little cold water, salt, pepper, and 1 tablespoonful of the cooked rice, and you have a palatable soup, as the water in which the rice was boiled is said to be more nutritious than the rice.

FISH, CLAMS AND OYSTER (BONED SHAD)

How many young cooks know how to bone a shad? It is a very simple process, and one becomes quite expert after one or two trials. And it fully repays one for the extra time and trouble taken, in the satisfaction experienced by being able to serve fish without bones. With a sharp knife cut the fish open along the back bone on the outside of the fish, but do not cut through the bone, then carefully cut the fish loose along the back bone on each side, cut the centre bone away with the smaller bones branching out on each side attached. Cut the shad into sizable pieces after being washed in cold water and dried on a cloth to take up all the moisture. Dip pieces of fish into white of egg containing a teaspoonful of water, roll in fine, dried bread crumbs, season with salt and pepper, drop in hot fat, and fry a rich brown. Serve on a platter, surrounded by a border of parsley.

Some small portions of the fish will adhere to the bones, however carefully the fish has been boned. The meat may be picked from the bones after cooking in salt water until tender. Flake the fish, and either make it into small patties or croquettes.

Shad roe should be parboiled first and then dredged with flour on both sides and fried in drippings or a little butter.

CROQUETTES OF COLD, COOKED FISH

Shred or flake cold, cooked fish, which has been carefully picked from bones. To 2 cups of fish add an equal amount of mashed potatoes, a small half cup of cold milk, 1 tablespoonful butter, yolk of 1 egg, lightly beaten, 1 teaspoonful of chopped parsley, season with salt and pepper. Mix all well together, and when cold, form in small croquettes. Dip into white of egg containing 1 tablespoonful of water, roll in fine, dried bread crumbs and fry in hot fat. Shad, salmon, codfish, or any kind of fish may be prepared this way, or prepare same as "Rice Croquettes," substituting-fish for rice.

SHAD ROE

Shad roe should be carefully taken from the fish, allowed to stand in cold water, to which a pinch of salt has been added, for a few minutes, then dropped in boiling water, cooked a short time and drained. Dredge with flour and fry slowly in a couple tablespoonfuls of butter and lard or drippings until a golden brown. Be particular not to serve them rare. Serve garnished with parsley.

Or the shad roe may be parboiled, then broken in small pieces, mixed with a couple of lightly beaten eggs and scrambled in a frypan, containing a couple of tablespoonfuls of butter and sweet drippings. Serve at once. Garnish with parsley or water cress.

SCALLOPED OYSTERS

Take about 50 fresh oysters. Place a layer of oysters in a baking dish alternately with fine, dried crumbs, well seasoned with pepper and salt and bits of butter, until pan is about two-thirds full. Have a thick layer of bread crumbs for the top, dotted with bits of butter. Pour over this half a cup or less of strained oyster liquor and small cup of sweet milk. Place in oven and bake from 40 to 50 minutes.

DEVILED OYSTERS

- 2 dozen oysters.
- 1 cup rich milk.
- 3 tablespoonfuls flour.
- Yolks of 2 raw eggs.
- 1 generous tablespoonful butter.
- 1 tablespoonful finely-minced parsley.

Drain oysters in a colander and chop rather coarsely.

Mix flour smooth with a little cold milk. Place the remainder of the milk in a saucepan on the range. When it commences to boil add the moistened flour and cook until the mixture thickens, stirring constantly to prevent burning, or cook in a double boiler. Add yolks of eggs and butter, ½ teaspoonful salt and ¼ teaspoonful of black pepper and a pinch of cayenne pepper. Then add chopped oysters, stir all together a few minutes until oysters are heated through. Then turn into a bowl and stand aside in a cool place until a short time before they are to be served. (These may be prepared early in the morning and served at six o'clock dinner.) Then fill good-sized, well-scrubbed oyster shells with the mixture, sprinkle the tops liberally with fine-dried, well-seasoned bread crumbs. (Seasoned with salt and pepper.) Place the filled shells on muffin tins to prevent their tipping over; stand in a hot oven about ten minutes, until browned on top, when they should be heated through. Serve at once in the shells. Handle the hot shells with a folded napkin when serving at table. This quantity fills thirteen oyster shells. Serve with the oysters small pickles, pickled cabbage or cranberry sauce as an accompaniment.

PLANKED SHAD

After eating planked shad no one will wish to have it served in any other manner, as no other method of preparing fish equals this. For planked shad, use an oak plank, at least two inches thick, three inches thick is better. Planks for this purpose may be bought at a department store or procured at a planing mill. Place plank in oven several days before using to season it. Always heat the plank in oven about 15 minutes before placing fish on it, then have plank *very hot*. Split a nicely-cleaned shad down the back, place skin side down, on hot plank, brush with butter and sprinkle lightly with pepper and salt. Put plank containing shad on the upper grating of a hot oven of coal range and bake about 45 minutes. Baste frequently with melted butter. The shad should be served on the plank, although not a very sightly object, but it is the proper way to serve it. The flavor of shad, or, in fact, of any other fish, prepared in this manner is superior to that of any other. Fish is less greasy and more wholesome than when fried. Should an oak plank not be obtainable, the shad may be placed in a large roasting pan and baked in oven. Cut gashes across the fish about two inches apart, and place a teaspoonful of butter on each. Bake in oven from 50 to 60 minutes. Serve on a warmed platter, garnished with parsley, and have dinner plates warmed when serving fish on them. Do not wash the plank with soap and water after using, but instead rub it over with sandpaper.

BROILED MACKEREL

When fish has been cleaned, cut off head and scrape dark skin from inside. Soak salt mackerel in cold water over night, skin side up, always. In the morning; drain, wipe dry and place on a greased broiler, turn until cooked on both sides. Take up carefully on a hot platter, pour over a large tablespoonful of melted butter and a little pepper, or lay the mackerel in a pan, put bits of butter on top, and set in a hot oven and bake. Garnish with parsley.

CODFISH BALLS

Soak codfish several hours in cold water. Cook slowly or simmer a short time. Remove from fire, drain, and when cold squeeze out all moisture by placing the flaked fish in a small piece of cheesecloth. To one cup of the flaked codfish add an equal quantity of warm mashed potatoes, yolk of 1 egg, 1 tablespoonful of milk and a little pepper. Roll into small balls with a little flour. Dip in white of egg and bread crumbs, and when quite cold fry in deep fat. Garnish with parsley.

FRIED OYSTERS

Procure fine, large, fresh oysters for frying. Drain in a colander carefully, look over, and discard any pieces of shell. Roll each oyster in fine, dried bread crumbs, well seasoned with salt and pepper, then dip them in a lightly-beaten egg, and then in bread crumbs. Allow them to stand several hours in a cool place before frying. Place a few oysters at one time in a wire frying basket, and immerse in smoking hot fat. Should too great a number of oysters be placed in the fat at one time it would lower the temperature of the fat and cause the oysters to become greasy. Drain the oysters when fried on heavy, brown paper, to absorb any remaining fat, and serve at once.

For all deep frying use two-thirds lard and one-third suet, as suet is considered to be more wholesome and cheaper than lard. Two items to be considered by the frugal housewife.

If fat for deep frying is the right temperature a crust is at once formed, and the oysters do not absorb as great a quantity of fat as when fried in only enough butter and drippings to prevent scorching, as they must then be fried more slowly. Serve pickled cabbage and tomato catsup when serving fried oysters.

PANNED OYSTERS

Aunt Sarah always prepared oysters in this manner to serve roast turkey. At the very last minute, when the dinner was ready to be served, she placed 50 freshly-opened oysters, with their liquor, in a stew-pan over a hot fire. The minute they were heated through and commenced to curl up, she turned them in a hot colander to drain a minute, then turned the oysters into a stew-pan containing two large tablespoonfuls of hot, melted butter, and allowed them to remain in the hot butter one minute, shaking the pan lo prevent scorching, seasoned them with salt and pepper, and turned all into a heated dish and sent to the table at once. These are easily prepared and are more wholesome than fried oysters.

OYSTERS STEAMED IN THE SHELL

Place well-scrubbed shells, containing fresh oysters, in a deep agate pan, which will fit in a kettle containing a small amount of boiling water. Cover very closely until the shells open easily. These may be served in the shell with hot, melted butter, in a side dish, or they may be removed from the shell to a hot bowl and seasoned with hot butter, salt and pepper.

A RECIPE GIVEN MARY FOR "OYSTER COCKTAIL"

To 2 tablespoonfuls of tomato catsup add ½ tablespoonful of grated horseradish, ½ tablespoonful of lemon juice, ½ teaspoonful of tabasco sause, ½ tablespoonful of vinegar, 1 saltspoonful of salt. Stand on ice one hour at least.

To serve—The freshly-opened oysters on half shell were placed on a plate, in the centre of which was placed a tiny glass goblet containing a small quantity of the mixture, into which the oysters were dipped before being eaten.

OYSTER CROQUETTES

Boil 50 oysters five minutes, drain. When cold, cut into small pieces, add ½ cup of bread crumbs and mix all together with a thick cream sauce composed of ½ cup of cream or milk thickened with flour, to which add 1 large tablespoonful of butter; season with salt, a dash of red pepper and 1 teaspoonful of finely-minced parsley. Stand this mixture on ice until quite cold and firm enough to form into small croquettes. Dip in egg and bread crumbs and fry in deep fat until a golden brown.

Serve at once on a platter garnished with sprigs of parsley From these ingredients was made 12 croquettes.

FRAU SCHMIDTS WAY OF SERVING "OYSTER COCK-TAILS"

Place in a bowl 2 tablespoonfuls of tomato catsup, 1 teaspoonful of grated horseradish, 2 tablespoonfuls of very finely cut celery juice and pulp of 2 lemons. Season with salt and pepper. Mix this with oysters which have been cut in small pieces. Serve in halves of lemons, from which the pulp has been carefully removed.

Place on ice a short time before serving. Crisp crackers should be served at the same time this is served.

SALMON LOAF

One can of salmon, from which all bones have been removed, 1 cup of cracker crumbs, ½ cup of milk, 1 tablespoonful of butter, which had been melted; 2 eggs beaten, salt and pepper to season. Mix all together, bake in a buttered pudding dish one-half hour or until browned on top. Serve hot.

CREAMED SALMON

A half cup of canned salmon, a left-over from lunch the preceding day, may be added to double the quantity of cream dressing, and when heated through and served on crisply-toasted slices of stale bread, make a tasty addition to any meal.

Of course, it is not necessary to tell even unexperienced housewives never under any circumstances allow food to stand in tins in which it was canned; do not ever stand food away in tin; use small agateware dishes, in which food, such as small quantities of leftovers, etc, may be reheated. Never use for cooking agate stew-pans, from the inside of which small parties have been chipped, as food cooked in such a vessel might become mixed with small particles of glazing, and such food when eaten would injure the stomach.

OYSTER CANAPES

- 1 cup cream.
- 4 tablespoonfuls of bread crumbs.

- 1 tablespoonful of butter.
- 3 dozen stewing oysters.

Season with paprika, tiny pinch of nutmeg and salt. Boil the cream, add bread crumbs and butter. Chop oysters fine, add seasoning. Serve hot in pattie cups or on toast. Serve small pickles or olives. Good dish for chafing dish.

MEAT

Every young housewife should be taught that simmering is more effective than violent boiling, which converts water into useless steam. Even a tough, undesirable piece of "chuck" or "pot roast" may be made more tender and palatable by long-continued simmering than it would be if put in rapidly boiling water and kept boiling at that rate. Meat may be made more tender also by being marinated; that is, allowing the meat to stand for some time in a mixture of olive oil and vinegar before cooking it. In stewing most meats a good plan is to put a large tablespoonful of finely-minced beef suet in the stew-pan; when fried out, add a little butter, and when sizzling hot add the meat, turn and sear on both sides to retain the juice in the meat, then add a little hot water and let come to a boil; then stand where the meat will just simmer but not slop cooking for several hours. The meat then should be found quite tender. Cheaper cuts of meat, especially, require long, slow cooking or simmering to make them tender, but are equally as nutritious as high-priced meats if properly prepared.

To quote from *The Farmers' Bulletin*: "The number of appetizing dishes which a good cook can make out of the meat 'left over' is almost endless. Undoubtedly more time and skill are required in their preparation than in the simple cooking of the more expensive cuts. The real superiority of a good cook lies not so much in the preparation of expensive or fancy dishes as in the attractive preparation of inexpensive dishes for every day. In the skillful combination of flavors. Some housewives seem to have a prejudice against economizing. If the comfort of the family does not suffer and the meals are kept as varied and appetizing as when they cost more, with little reason for complaint, surely it is not beneath the dignity of any family to avoid useless expenditure, no matter how generous its income. And the intelligent housekeeper should take pride in setting a good table."

This is such an excellent article, and so ably written and true, that I feel it would be to the advantage of every young housewife to read and profit by it.

"SAUERGEBRATENS" OR GERMAN POT ROAST

Buy about three pounds of beef, as for an ordinary pot roast. Place in a large bowl. Boil vinegar (or, if vinegar is too sharp, add a little water, a couple of whole cloves and a little allspice); this should cover the piece of meat. Vinegar should be poured over it hot; let stand a couple of days in a cool place uncovered; turn it over occasionally. When wanted to cook, take from the vinegar and put in a stew-pan containing a little hot fried-out suet or drippings in which has been sliced 2 onions. Let cook, turn occasionally, and when a rich brown, stir in a large tablespoonful of flour, add 1½ cups of hot water, cover and cook slowly for two or three hours, turning frequently. Half an hour before serving add small pared potatoes, and when they have cooked tender, serve meat, gravy and potatoes on a large platter.

The writer knew an old gentleman who had moved to the city from a "Bucks County farm" when a boy, who said that he'd walk five miles any day for a dish of the above as his mother had prepared it in former years.

Mary was surprised at the amount of valuable information to be obtained from the different *Farmers' Bulletins* received at the farm, on all subjects of interest to housewives, and particularly farmers' wives. All books were to be had free for the asking.

The dishes Mary prepared from recipes in the *Farmers Bulletin* on "economical use of meat in the home," were especially liked at the farm, particularly "Stewed Shin of Beef" and "Hungarian Goulash" (a Hungarian dish which has come to be a favorite in the United States).

HUNGARIAN GOULASH

- 2 pounds top round of beef.
- 1 onion.
- A little flour.
- 2 bay leaves.
- 2 ounces salt pork.
- 6 whole cloves.
- 2 cups of tomatoes.
- 6 peppercorns.
- 1 stalk celery.
- 1 blade mace.

Cut the beef into 2-inch pieces and sprinkle with flour. Fry the salt pork until a light brown; add the beef and cook slowly for about thirty-five minutes, stirring occasionally. Cover with water and simmer about two hours. Season with salt and pepper or paprika. From the vegetables and spices a sauce is made as follows: Cook in sufficient water to cover for 20 minutes; then rub through a sieve, and add to some of the stock in which the meat was cooked. Thicken with flour, using 2 tablespoonfuls (moistened with cold water) to each cup of liquid, and season with salt and paprika. Serve the meat on a platter with the sauce poured over it. Potatoes, carrots and green peppers cooked until tender and cut into small pieces or narrow strips are usually sprinkled over the dish when served, and noodles may be arranged in a border upon the platter.

BROILED STEAK

When buying beefsteak for broiling, order the steak cut 1 inch to 1¼ inches thick. Place the steak on a well-greased, hot broiler and broil over a clear, hot fire, turning frequently. It will take about ten minutes to broil a steak 1-inch thick. When steak is broiled place on a hot platter, season with butter, pepper and salt, and serve at once. Serve rare or otherwise, but serve *at once*. Broil-steak unseasoned, as salt extracts juice from meat. Steak, particularly, loses its savoriness if not served *hot*. What to a hungry man is more nutritious and appetizing than a perfectly broiled, rare, juicy, steak, served hot? And not a few young and inexperienced cooks serve thin steaks, frequently overdone or scorched, containing about the same amount of nourishment a piece of leather would possess, through lack of knowledge of knowing just how. Often, unconsciously. I will admit; yet it is an undiluted fact, that very many young housewives are indirectly the cause of their husbands suffering from the prevailing "American complaint," dyspepsia, and its attendant evils. And who that has suffered from it will blame the "grouchy man" who cannot well be otherwise. So, my dear "Mrs. New Wife," be warned in time, and always remember how near to your husband's heart lies his stomach, and to possess the former you should endeavor to keep the latter in good condition by preparing, and serving, nourishing, well-cooked food.

STEWED SHIN OF BEEF

- 4 pounds of shin of beef.
- 1 medium-sized onion.
- 1 whole clove and bay leaf.
- 1 sprig of parsley.
- 1½ tablespoonfuls flour.
- 1½ tablespoonfuls of butler or savory drippings.
- 1 small slice of carrot.
- ½ tablespoonful of salt.
- ½ teaspoonful of pepper.
- 2 quarts boiling water.

Have the butcher cut the bone in several pieces. Put all the ingredients but the flour and butter in a stew-pan and bring to a boil. Set the pan where the liquid will just simmer for six hours, or after boiling for five or ten minutes put all into the fireless cooker for eight or nine hours. With the butter, flour and ¼ cup of the clear soup from which the fat has been removed make a brown sauce. To this add the meat and marrow removed from the bone. Heat and serve. The remainder of the liquid in which the meat has been cooked may be used for soup.

HAMBURG STEAK

Take the tough ends of two sirloin steaks and one tablespoonful of kidney suet, run through a food chopper; season with pepper and salt, form into small cakes, dredge lightly with flour, fry quickly, same manner steak is fried, turning frequently. The kidney fat added prevents the Hamburg steak being dry and tasteless. "A tender, juicy broiled steak, flaky baked potatoes, a good cup of coffee and sweet, light, home-made bread, a simple salad or fruit, served to a hungry husband would often prevent his looking for an affinity," said Aunt Sarah to her niece Mary.

MEAT STEW WITH DUMPLINGS
STEW.

- 5 pounds of a cheap cut of beef.
- 4 cups of potatoes cut into small pieces.
- ⅔ cup each of turnips and carrots cut into ½-inch cubes.
- ½ an onion chopped.
- ¼ cup of flour.
- Season with salt and pepper.

Cut the meat into small pieces, removing the fat. Fry out the fat and brown the meat in it. When well browned, cover with boiling water. Boil for five minutes and then cook in a lower temperature until meat is done. If tender, this will require about three hours on the stove, or five hours in the fireless cooker. Add carrots, onions, turnips and pepper and salt during the last hour of cooking, and the potatoes fifteen minutes before serving. Thicken with the flour diluted with cold water. Serve with dumplings. If this dish is made in the tireless cooker the mixture must be reheated when the vegetables are put in. Such a stew may also be made of mutton. If veal or pork is used the vegetables may be omitted or simply a little onion used. Sometimes for variety the browning of the meat is dispensed with. When white meat, such as chicken, veal or fresh pork is used, the gravy is often made rich with cream or milk thickened with flour.

DUMPLINGS.

- 2 cups of flour.
- 4 teaspoons (level) of baking powder.
- ⅔ cup of milk or a little more if needed.
- ½ teaspoonful of salt.
- 2 teaspoonfuls of butter.

Mix and sift the dry ingredients. Work in butter with the tips of the fingers. Add milk gradually, roll out to thickness of half inch. Cut with biscuit cutter. Place in a buttered steamer over a kettle of hot water and cook from 12 to 15 minutes. If the dumplings are cooked with the stew enough liquid should be removed to allow of their being placed directly upon the meat and vegetables. Sometimes the dough is baked and served as biscuits, over which the stew is poured. If the stew is made with chicken or veal it is termed a fricassee.

This recipe tells of such an economical way of extending the meat flavor that I think every young housewife should know it. Mary copied it from *The Farmers' Bulletin*, an article on the "Economical Use of Meat in the Home." The dumplings, as she prepared them from this recipe, were regular fluff balls, they were so light and flaky. I would add, the cook-pot should be closely covered while cooking or steaming these dumplings, and the cover should not be raised for the first ten minutes.

A lesser quantity of baking powder might be used with equally good results, but these dumplings are certain to be light and flaky. A larger quantity of baking powder should be used when dough is steamed or boiled than if dough is baked, if one expects good results.

EXTENDING THE MEAT FLAVOR

Mary learned, through reading *The Farmers' Bulletin*, different methods of extending the meat flavor through a considerable quantity of material, which would otherwise be lacking in distinctive taste, one way to serve the meat with dumplings, generally in the dish with it; to combine the meat with crusts, as in meat pies or meat rolls, or to serve the meat on toast or biscuits. Borders of rice, hominy or mashed potatoes are examples of the same principles, applied in different ways.

By serving some preparation of flour, rice, hominy or other food, rich in starch, with the meat, we get a dish which in itself approaches nearer to the balanced ration than meat alone, and one in which the meat flavor is extended through a large amount of the material.

The measurements given in the above recipes call for a level spoonful or a level cup, as the case may be.

In many American families meat is eaten two or three times a day. In such cases, the simplest way of reducing the meat bill would be to cut down the amount used, either by serving it less often or by using less at a time. Deficiency of protein need not be feared, when one good meat dish a day is served, especially if such nitrogenous materials as eggs, milk, cheese and beans are used instead. In localities where fish can be obtained fresh and cheap, it might well be more frequently substituted for meat for the sake of variety as well as economy. Ingenious cooks have many ways of "extending the flavor" of meat; that is, of combining a small quantity with other materials to make a large dish as in meat pies, stews and similar dishes.

The foregoing information may be useful to other young, prospective housekeepers who may never have read "the very instructive articles on The Economical Use of Meat in the Home,' in the *Farmers' Bulletin.*"

PREPARING A POT ROAST

When buying a pot roast, "Aunt Sarah" selected a thick, chunky piece of meat, weighing several pounds, and a small piece of beef suet which she cut into small bits, placed pan containing them on hot range, added a small, sliced onion, and when fat was quite hot she added the quickly rinsed piece of meat, and quickly seared it to retain the juice; added 1 cup of hot water, a sprig of parsley, seasoning of salt and pepper; cooked a short time, then allowed it to stand on the range closely covered, where it would simmer gently several hours; turning the meat frequently, adding a small amount of water occasionally, as the broth was absorbed by the meat. An inexperienced cook will be surprised to find how tender, palatable, and equally nutritious, an inexpensive cut of meat may become by slow simmering. When the pot roast has become tender, remove from the broth and place on a *hot platter*; this latter is a small item, but dishes may be quickly heated in a hot oven and meat and vegetables are more appetizing if served hot on warmed plates. "Forgive this digression; I fear the pot roast will cool even on a warmed platter." After removing the meat from the pan add a large tablespoonful of flour, moistened with a small quantity of cold water, to the broth in the pan for gravy; cook until thickened, strain sliced onion and parsley from the broth, add seasoning of salt and pepper, serve on the platter with the meat; the onion added, gives the gravy a fine flavor and causes it to be a dark, rich brown in color.

STUFFED BREAST OF VEAL

Rub the piece of meat with salt, pepper, ginger and minced onion. Prepare a stuffing as for chicken of crumbled, stale bread, etc., or soak pieces of stale bread in cold water. Squeeze dry and season with a little minced onion, parsley, a little melted butter, salt and pepper, and moisten all with one egg. Fill the breast of veal with this stuffing, sew together, place in roasting pan with a small quantity of water, to which a tablespoonful of butter has been added. Roast in a moderately hot oven until well done, basting frequently.

"GEDAMPFTES RINDERBRUST"

Take breast of beef or veal, without fat or bones, quickly rinse off meat and wipe with a cloth. Place in a stew-pot with one chopped onion, one sliced tomato, a bay leaf, season with pepper and salt, add a small quantity of hot water, cook, closely covered, several hours. To be tender this meat requires long, slow cooking, when it cooks and browns at the same time. Strain the broth and thicken for gravy and pour around the meat on platter when serving.

"PAPRIKASH"

Two pounds of veal, from leg, cut into small pieces for stewing; 4 good-sized onions, cut rather fine; measure about ½ cup of sweet lard, place onions in pan with some of this lard and fry a light brown. Add meat and cook meat and onions together about one-half hour, adding lard gradually until all is used and the meat is golden brown. Then cover with water and stew, closely covered, about two hours or longer, until meat is ready to serve; then add more water until meat is covered. Season with salt and paprika. Add about three tablespoonfuls of vinegar (not too sour; cook must judge this by tasting); then add ½ pint of sweet cream. Thicken gravy with flour mixed smooth with a little water. Place on platter surrounded with gravy. With this was always served baked or steamed sweet potatoes.

BEEF STEW

Three pounds of the cheaper cut of beef, cut in pieces a couple inches square; brown in a stew-pan, with a sliced onion, a sprig of parsley and a coupe tablespoonfuls of sweet drippings or suet; cook a few minutes, add a little water, and simmer a couple of hours; add sliced turnips and a few medium-sized potatoes. Should there he a larger quantity of broth than required to serve with the meat and vegetables, a cup or more of the broth may form the basis of a palatable soup for lunch the following day.

SAVORY BEEF ROLL

Three and one-half pounds raw beef, or a mixture of beef and veal may be used, run through a food chopper. A cheap cut of meat may be used if, before chopping, all pieces of gristle are trimmed off. Place the chopped meat in a bowl, add 8 tablespoonfuls of fine, dried bread crumbs, 1 tablespoonful of pepper, 1½ tablespoonfuls of salt. Taste the meat before adding all the seasoning specified, as tastes differ. Add 3 raw eggs, 4 tablespoonfuls of sweet milk or cream, 2 tablespoonfuls of butter, a little sweet marjoram or minced parsley. Mix all together and mold into two long, narrow rolls, similar to loaves of bread. Place 1 tablespoonful each of drippings and butter in a large fry-pan on the range. When heated, place beef rolls in, and when seared on both sides add a small quantity of hot water. Place the pan containing meat in a hot oven and bake one hour. Basting the meat frequently improves it. When catering to a small family serve one of the rolls hot for dinner; serve gravy, made by thickening broth in pan with a small quantity of flour. Serve the remaining roll cold, thinly sliced for lunch, the day following.

VEAL CUTLETS

Use either veal chops or veal cutlets, cut in small pieces the size of chops; pound with a small mallet, sprinkle a little finely-minced onion on each cutlet, dip in beaten egg and bread crumbs, well seasoned with salt and pepper. Place a couple tablespoonfuls of a mixture of butter and sweet drippings in a fry-pan; when hot, lay in the breaded cutlets and fry slowly, turning frequently and watching carefully that they do not scorch. These take a longer time to fry than does beefsteak. When a rich brown and well cooked take up the cutlets on a heated platter and serve, garnished with parsley.

MEAT "SNITZEL"

Cut 1½ pounds of thick veal steak into small pieces, dredge with flour, season with salt and pepper, and fry brown in a pan containing bacon fat (fat obtained by frying several slices of fat, smoked bacon). Remove the meat from the pan, add a couple tablespoonfuls of flour to the remaining fat stir until browned, then pour in the strained liquor from a pint can of tomatoes. Add one slice of onion and one carrot, then return the meat to the sauce; cover closely and simmer three-quarters of an hour. When the meat is tender, place on a hot platter, add a pinch of red pepper to the sauce and a little more salt if required, and strain over the meat on the platter. This was a favorite dish of Mary's Uncle, and he said she knew how to prepare it to perfection.

SIRLOIN STEAKS

Procure 2 sirloin steaks, 1½ inches thick, and a small piece of suet. Cut the tenderloin from each steak, and as much more of the steak as required for one meal. Place the finely-cut suet in a hot fry-pan; this should measure 1 tablespoonful when tried out, add one teaspoonful of butter, when the fat is very hot and a blue smoke arises place pieces of steak, lightly dredged with flour, in the pan of hot fat, place only one piece at a time in the fat; sear the meat on one side, then turn and sear on the other side; then place the other pieces of meat in the pan and continue in the same manner, turning the steak frequently. The hot butter and suet sear the steak, thus the juice of the meat is retained, making the meat more palatable; season with salt and pepper, place on a hot platter and serve at once.

MEAT BALLS

Chop meat fine; beef, chicken, lamb or veal; mince a small onion and fry in a tablespoonful of butler; add a tablespoonful of flour, the yolk of one egg, the chopped meat and a little broth, gravy, or milk to moisten, salt and pepper. Stir all together and turn the whole mixture into dish to cool. When cool, shape with well-floured hands into balls the size of a shelled walnut. Dip in beaten white of egg, then into bread crumbs, and fry in deep fat until crisp and brown. Place only three or four meat balls in a frying basket at one time. Too many at a time chills the fat; but if plunged in boiling hot fat, then a crust is formed at once over the outside, which prevents the grease from penetrating. When the meat balls are browned nicely, lay them on brown paper to absorb any grease that may adhere to them. To try whether the fat is the right temperature, drop a small piece of bread in it, and if it browns while you count twenty, the fat is hot enough for any form of croquettes. Garnish with parsley or watercress.

VEAL LOAF

Three pounds raw veal, chopped fine; 1 teaspoonful salt, 1 teaspoonful pepper, 2 tablespoonfuls butter, 2 raw eggs, 2 tablespoonfuls water. Mix all together with 6 tablespoonfuls fine, rolled, dried bread crumbs and mold into a long, narrow loaf. Roll the loaf in two extra tablespoonfuls of bread crumbs. Place in a hot pan, pour 3 tablespoonfuls melted butter over the top, and bake in hot oven two hours or less, basting frequently. Slice thinly when cold. Should the veal loaf be served hot thicken the broth with flour and serve this gravy with it.

SWEETBREADS (BREADED)

Place sweetbreads in cold water, to which ½ teaspoonful salt has been added, for a short time, then drain and put over the fire with hot water. Cook ten minutes. Drain and stand aside in a cool place until wanted. Remove stringy parts, separate into small pieces about the sue of an oyster, dip in beaten white of egg and then in bread crumbs. Put in a pan containing a little hot butter and drippings and fry light brown. Serve hot. Garnish platter with parsley.

FRIED LIVER AND BACON

Have *beef* liver cut in slices about one inch thick; quickly rinse and wipe dry. Remove the thin skin on the edge and cut out all the small, tough fibres. If liver from a *young* beef it can scarcely be told from calves' liver when cooked, and is considerably cheaper. Fry a dozen slices of fat bacon in a pan until crisp and brown. Take from the pan on a warm platter and place in oven. Put the pieces of liver, well dredged with flour, into the pan containing the hot bacon fat, also a little butter, and fry slowly until well done, but not hard and dry. Turn frequently and season with salt and pepper. Take the liver from the pan, add one tablespoonful of flour to the fat remaining in the pan, stir until smooth and brown, then add about one cup of sweet milk or water, stir a few minutes until it thickens and season with salt and pepper. Should the liver be a little overdone, put it in the pan with the gravy, cover and let stand where it will just simmer a few minutes, then turn all on a hot platter and serve the bacon on a separate dish.

BEEFSTEAK SERVED WITH PEAS

Fry quickly a large sirloin steak. Place in the oven, on a warm platter. Add a large tablespoonful of butter to the fry pan, also a can of sifted peas, which have been heated and drained, season with pepper and salt, shake pan to prevent burning and when hot turn on to platter containing steak and serve at once. This makes an appetizing luncheon dish.

CREAMED "DRIED BEEF"

Put a tablespoonful of butter in a frying pan, add ½ cup of chipped beef cut fine and brown it in the butter, then add ¼ cup of water. Let stand and simmer for a short time, then add a cup of sweet milk, thicken to the consistency of thick cream by adding 1 tablespoonful of flour mixed smooth with a small quantity of cold milk, season with salt and pepper. This is an economical way of using small pieces of dried beef not sightly enough to be served on the table. Serve with baked potatoes for lunch, or pour over slices of toasted bread, or over poached eggs for an appetizing breakfast dish.

CREAMED SWEETBREADS

Parboil sweetbreads in water 10 minutes. Remove stringy parts and dry on a napkin. Separate the sweetbreads into small pieces with a *silver knife*, never use *steel*, put in a stewpan with enough cream to cover, add butter, pepper and salt to taste. Flour enough to thicken a little, let all come to a boil. Fill small pattie shells with the mixture and serve hot.

MEAT CROQUETTES

- 2 cups finely chopped meat (beef or veal).
- 1 tablespoonful butter.
- 2 tablespoonfuls flour (or a little more flour).
- 2 tablespoonfuls chopped parsley.
- 1 scant cup of milk.

Put milk on to boil. Mix flour smooth with a little cold milk before adding to boiling milk, add the butter and cook all together until a creamy consistency, then add the chopped meat well seasoned with salt and pepper and the chopped parsley. Mix well and let cool. Shape into croquettes, dip in white of egg and bread crumbs. Let stand until perfectly cold, then fry brown, in deep hot fat.

Chicken, beef, veal and mutton may be prepared in the same manner. When dipping croquettes, 1 tablespoonful of water may be added to the white of egg and 2 tablespoonfuls of water if the whole of the egg is used. Use the whites of eggs for dipping croquettes if possible. Croquettes may be made the day before wanted, and placed in a refrigerator or cool place. Croquettes should be cold before frying.

STEWED RABBIT

After the rabbit has been skinned, and carefully cleaned, wash quickly and let stand over night in cold water to which salt has been added; also a pinch of red pepper. Place on the range in the morning (in a stew-pan with fresh warm water). When it comes to a boil, drain off, add one pint of hot water containing two sliced onions and a little ginger. This prevents the flavor of wild game, objectionable to some. When the meat has cooked tender, drain, dust pieces with flour, and brown quickly in a pan containing a couple tablespoonfuls of hot lard, butter, or drippings.

If you wish the meat of the rabbit white, add a thin slice of lemon to the water when cooking meat.

ROAST LAMB

Select leg or loin, or if a larger roast is wanted, leg and loin together. Carefully rinse the piece of meat. Place in pan, dust lightly with pepper. Have the oven hot and place pan in without putting water in pan. Brown on one side, then turn and brown on the other. Then put about ½ cup of water in roasting pan, and if oven is too hot, leave door open for a few minutes. Allow 25 minutes for each pound of lamb.

"GEFULLTE RINDERBRUST," OR STUFFED BREAST OF BEEF

Take a fillet of beef, rub both sides well with a mixture of finely chopped onion, minced parsley, salt and pepper. Then spread over the fillet a small quantity of raw, chopped, well-seasoned meat, roll together and tie. Place in a stew pan with a small quantity of water, cook closely covered until tender. Serve with gravy.

FRIED PEPPERS WITH PORK CHOPS

Dust four or five pork chops with flour and fry in a pan, not too quickly. When nicely browned, remove to a warm chop plate and stand in warming oven while preparing the following: Slice or cut in small pieces four good-sized, sweet, red peppers and a half teaspoon of finely chopped hot pepper, add to the fat remaining in the pan in which the chops were fried, and cook about ten minutes, until peppers are tender (stirring them frequently). When sufficiently cooked, add one tablespoon of vinegar, pepper and salt to taste, cook one minute longer and serve on the same dish with the chops.

BOILED HAM

When preparing to cook a ham, scrape, wash and trim it carefully. Place ham in a large cook pot or boiler, partly cover with cold water, let come to a boil, then move back on range where the water will merely simmer, just bubble gently around the edge of the boiler. A medium sized ham should be tender in five or six hours. When a fork stuck into the ham comes out readily, the ham is cooked. Take from the boiler and skin carefully, removing all the discolored portions of the smoked end, stick 2 dozen whole cloves into the thick fat, and sprinkle a couple tablespoonfuls of brown sugar and fine bread crumbs over top. Place in a very hot oven a short time, until the fat turns a golden brown. Watch carefully to see that it does not scorch. When cold, slice thin and serve. Aunt Sarah frequently added a pint of cider to the water in which the ham was boiled. She said this improved the flavor of the ham.

SLICED HAM

When about to fry a slice of uncooked ham, do young housewives know how very much it improves the flavor of the ham if it is allowed to stand for ten or fifteen minutes in a platter containing a large teaspoonful of sugar and a little cold water? Turn several times, then wipe quite dry with a clean cloth and fry in a pan containing a little hot drippings and a very little butter (one-half teaspoonful) just enough to prevent its sticking to the pan. Do not fry as quickly as beefsteak. After a slice of ham has been cut from a whole ham, if lard be spread over the end of ham from which the slice has been cut, it will prevent the cut place from becoming mouldy.

ROAST PORK

Place pork roast in a covered roasting pan containing a small cup of hot water, season with pepper and salt and sweet marjoram and sprinkle a little powdered sage over it, and stand in a very hot oven. After the meat has been roasting for a half hour, have less heat in your oven, allow about 25 minutes to every pound of pork, or longer if necessary, but be sure it is *well done*. When served, *underdone* pork is very unwholesome and unappetizing. When meat is sufficiently roasted, pour off all the fat in the pan except a small quantity, to which add ½ cup of boiling water, pepper and salt and serve. Serve baked apples or apple sauce with pork.

PORK CHOPS

Dip pork chops in egg, then into bread crumbs to which has been added salt, pepper, and a very little sage and sweet marjoram. Some prefer chops simply dredged with flour. Fry about 25 minutes or until cooked through and nicely browned, but not scorched. 'Tis said, "The frying of chops in a perfect manner is the test of a good cook."

HOME-MADE SAUSAGE

Nine pounds of fresh pork (lean and fat intermixed as it comes). Cut meat in small pieces, run through a meat cutter. Sprinkle over the finely chopped meat 3 tablespoonfuls salt, 2 tablespoonfuls of black pepper, 4 tablespoonfuls of powdered sage if bought at a chemist's. Aunt Sarah used but three tablespoonfuls of her own home-grown sage, as the flavor was much stronger than dried sage. Some folks add 2 tablespoonfuls of summer savory, but Aunt Sarah did not care for the flavor. Cloves, mace and nutmeg may also be added if one likes highly-spiced food. This is a matter of taste. A good plan is to season the small pieces of meat before chopping, as this distributes the seasoning through the sausage. Fill well cleaned casings, with the finely chopped meat. Or form sausage into small pats, fry brown on both sides and serve with home-made buckwheat cakes.

AUNT SARAH'S METHOD OF KEEPING SAUSAGE

To keep sausage one year, take sausage which has been put in casings (skins in long links) and cook until heated through in a fry pan half filled with hot water. Take sausage from the water, cut in 4-inch length pieces (stick sausage with prongs of a fork, to prevent skins bursting) and fry brown on both sides, as if preparing it for the table. Place, while hot in quart jars, fill jars as compactly as possible, then pour the hot fat remaining in pan over top. Seal air-tight and it will keep well one year if jars are perfectly air-tight.

SOUSE

Two pig's feet, weighing together about 1½ pounds. After thoroughly cleansing with a vegetable brush, place in a stewpan and cover with cold water. Allow water to come to a boil then move stew-pan to place on range where contents will cook slowly for a number of hours, or until the meat is loosened from the bones, then strain liquid, which should measure a scant three cups. (If a lesser quantity of liquid, add hot water until you have the required amount.) Add also 3 tablespoonfuls of sharp cider vinegar, about ¾ teaspoonful of salt and a dust of black pepper.

Pour this mixture over the meat, which should have been separated from bones, allowing a few smaller bones to remain with the meat, which should have been placed in a bowl with several thin slices of lemon, if liked. Stand bowl in a cool place over night or until the "Souse" is of a jelly-like consistency. When cold, remove any surplus grease from the top of "Souse." Turn it from the bowl on to a platter. Serve cold. Garnish with thin slices of lemon and sprigs of parsley. This will furnish about 2¼ pounds of souse.

UTILIZING COLD MEAT "LEFT-OVERS"

Small pieces of cold roast beef, veal or steak may all be utilized by being put through the food chopper. To 1 cup of finely-chopped cold meat add ¼ cup of stale bread, which has soaked for a few minutes in cold water. The water having been squeezed from the bread, it was added to the meat, as was also a small quantity of finely-minced onion or parsley, and either the yolk or while of 1 egg and a seasoning of salt and pepper. Add left-over gravy, to cause the mixture to be soft enough to form into small rolls or cakes, and fry in a pan containing a couple tablespoonfuls of sweet drippings. Mashed potatoes may be substituted for the bread with equally good results. The meat mixture may be formed into small cone shapes, dipped in egg, then rolled in fine bread crumbs and fried in deep fat.

Very appetizing sandwiches may he made from cold pieces of fried ham, run through food chopper. Spread this on thinly-sliced, buttered bread, with a dish of prepared mustard, spread over the prepared ham. Small bits of boiled ham, which cannot be sliced, may also be used in this manner.

The fat was cut from left-over pieces of roast beef (place a couple of tablespoonfuls of fat in a pan on the range until the fat has fried out), then add a little finely-minced onion and the beef cut in pieces the size of a small marble, brown in the fat a few minutes, then add a small quantity of vinegar and water, and thicken to the consistency of cream (with a little flour moistened with cold water, before being added). This Aunt Sarah made frequently, being a frugal housewife, and called "Salmagundi."

FOWL — ROAST CHICKEN OR TURKEY

Singe the fowl, after it has been picked; then with a small vegetable brush quickly scrub it well, with luke-warm water. Do not let it lie in the water. When perfectly clean rinse in cold water, wipe dry, cut out the oil sack, remove craw from neck, draw the fowl, being careful not to break the gall in the process, as that would cause the meat, as well as giblets, to have a bitter taste. Take out the lungs, the spongy red pieces lying in crevices near the bones of the back, and pour cold water through the fowl until you have thoroughly rinsed and chilled it, and no blood remains inside. I think fowls should be rinsed thoroughly inside and outside with cold water (many good cooks to the contrary). Wipe the inside of the fowl perfectly dry with a clean cloth, and it is ready for the "filling." Separate the liver and heart from entrails and cut open the piece containing the gizzard; wash the outer part, and put the giblets on to cook with a little hot water; if wanted to use with the filling. If the fowl is wanted to cook or steam the day following, do not cut in pieces and let stand in water over night, as I have known some quite good cooks to do, as that draws the flavor from the meat and makes it tasteless. If the giblets are not to be cooked and added to dressing, place them inside the fowl, tie feet together, and hang up in a cool place until wanted. When serving a turkey dinner with its accompaniments one finds so many things to be attended to in the morning, especially if the fowl is cooked on a Sunday. It will be found a great help to the cook to have the turkey or chicken stuffed with bread filling the day before it is to be roasted, ready to pop in the oven in the morning.

BREAD FILLING AS AUNT SARAH PREPARED IT

Chop the cold, cooked liver, heart and gizzard into tiny dice; add this to a bowl containing one quart of crumbled stale bread, seasoned with 1 teaspoonful of salt, ¼ teaspoonful pepper, ½ of a small, finely-minced onion, ¼ teaspoonful sweet marjoram and a teaspoonful of chopped parsley. Stir into the crumbs 3 tablespoonfuls of melted butter, moisten all with one egg beaten with 2 tablespoonfuls of milk. Sir all together lightly with a fork. Fill the body of the chicken, put a couple of spoonfuls of this dressing into the space from which the craw was taken, tie the neck with a cord, sew up the fowl with a darning needle and cord, after filling it. (Always keep a pair of scissors hanging from a nail conveniently near the sink in your kitchen, as it saves many steps.) The secret of *good filling* is not to have it *too moist*, and to put the filling into the fowl *very lightly*; on no account press it down when placing it in the fowl, as that will cause the best of filling to be heavy and sodden. Rather put less in, and fill a small cheese cloth bag with what remains, and a short time before the fowl has finished roasting, lay the bag containing the dressing on top of fowl until heated through, then turn out on one side of platter and serve with the fowl. Instead of the chopped giblets, add 2 dozen oysters to the dressing, or a few chestnuts boiled tender, mashed and seasoned with butter, pepper and salt and added to the crumbled bread. This makes a pleasant change. Do not use quite as many crumbs if chestnuts or oysters are added. Place fowl in covered roasting pan, put a couple of pieces of thinly-sliced bacon on the breast of fowl, put two cups of hot water in the pan and set in a very hot oven for the first half hour, then reduce the heat and baste frequently. An ordinary eight-pound turkey takes from two to three hours to roast; a chicken takes about twenty minutes to the pound.

When the fowl has been sufficiently roasted, remove from pan to a hot platter. Pour off some of the fat in the pan and add a small quantity of milk to the broth remaining. Thicken with flour, for gravy, season with salt and pepper and sprinkle one teaspoonful chopped parsley over gravy after being poured into the gravy boat ready to serve. The yolk of one egg added makes a richer gravy to serve with chicken.

FRIED CHICKEN WITH CREAM GRAVY

Cut one small spring chicken in pieces, dip each piece in a batter composed of 1 beaten egg, 1 cup of milk, a pinch of salt, ½ teaspoonful of baking powder, sifted with flour enough to form a batter. Dip the pieces of chicken in this batter, one at a time, and fry slowly in a pan containing a couple tablespoonfuls of hot butter and lard, until a golden brown. Place the fried chicken on a platter.

Make a gravy by adding to the fat remaining in the pan — 1 cup of milk, 1 tablespoonful of corn starch. Allow this to brown and thicken. Then pour the gravy over the chicken and serve garnished with parsley or watercress.

STEWED OR STEAMED CHICKEN

Cut a nicely cleaned chicken into nine pieces. (Do not separate the meat from the breast-bone until it has been cooked.) Put in a cook pot and partly cover with boiling water. Add one small onion and a sprig of parsley, and let simmer about 1½ hours, or until tender. If an old fowl it will take about one hour longer. Add salt and pepper. Strain the broth, if very fat, remove a part from broth. After separating the white meat from the breast-bone, put all the meat on a platter. Add ¼ cup of sweet milk to the strained broth, thicken with a couple tablespoonfuls of flour, mixed smooth with a little cold water. Let come to a boil, and add one teaspoonful of chopped parsley. Pour the chicken gravy over the platter containing the meat, or serve it in a separate bowl. Or you may quickly brown the pieces of stewed chicken which have been sprinkled with flour in a pan containing a little sweet drippings or butter. Should the chicken not be a very fat one, add yolk of one egg to the gravy.

Or, instead of stewing the chicken, place in the upper compartment of a steamer, and steam until tender and serve. The day following that on which stewed or steamed chicken was served, small undesirable left-over pieces of the chicken were added (after being picked from the bones) to the gravy remaining from the day before, heated thoroughly and poured hot over a platter containing small baking powder biscuits broken in half or slices of toasted bread, which is economical, extending the meat flavor.

VEGETABLES—WHITE POTATOES

Potatoes are one of the most valuable of vegetables. White potatoes, after being pared, should be put in a stew-pan over the fire with a little boiling water, but not enough to cover them. The water should be kept boiling continuously. About thirty minutes from the time they commence boiling will be the time required for cooking potatoes of ordinary size. It spoils potatoes to have the water stop boiling even for a short time. Add half a teaspoonful of salt to the potatoes when partly boiled and when cooked sufficiently drain the water from them at *once* and sprinkle a little salt over the dry potatoes. Close the lid of the stew-pan tightly, give it a quick shake, when the potatoes will he found dry and flaky. Mash fine with a potato masher, adding a tablespoonful of butter and a couple tablespoonfuls of milk. Let stand a minute on the hot range to heat the milk, then beat all together with a fork until creamy. Add more salt if necessary. That is quite important, as potatoes require considerable salt. Cover the potatoes with a cloth. Never allow to stand with the lid of the stew-pan over them, as it will draw moisture. Serve white potatoes as soon as possible after being cooked, as they are not appetizing when allowed to stand any length of time.

BAKED POTATOES

All young housewives may not know "that there is more real food value in potatoes baked 'in their jackets' than is found in preparing this well-known tuber in any other way." The secret of a good baked potato lies in having a hot oven, but not too hot. Scrub good sized potatoes, or, for a change, they may be pared before baking, place in a hot oven, and bake about 45 minutes, when they should be a snowy, flaky mass inside the skins, palatable and wholesome. When fully baked they should fed soft to the touch when pressed. Take from oven, pinch one end of potato to break the skin to allow the gas to escape. Always break open a baked potato. Never cut with a knife.

Medium-sized potatoes, pared, cut in half lengthwise, and baked in a hot oven 25 to 40 minutes, until the outside of the potato is a light brown, make a pleasant change from boiled potatoes. When baked the proper length of time and served at once, the inside of potato should be light and flaky. The housewife should occasionally serve rice or macaroni and omit potatoes from the bill of fare, especially in the spring of the year.

Potatoes should always be served as soon as baked, if possible. Potatoes may be baked in less than a half hour in a gas oven.

VARIOUS WAYS OF USING SMALL POTATOES

Early in the season when small, early potatoes are more plentiful and cheaper than large ones, the young housewife will be able to give her family a change, while practicing economy, as there are various ways of using small potatoes to advantage.

First, new potatoes, if about the size of marbles, may be scraped, boiled in salted water, and served with a thin cream dressing, sprinkled liberally with chopped parsley, or the boiled potatoes, while still hot, may be quickly browned in a pan containing a couple tablespoonfuls of hot drippings or butter. They are much better prepared in this manner if the potatoes are put in the hot fat while still warm. Or the small boiled potatoes may be cut in thin slices, browned in a couple tablespoonfuls of butter or drippings and two eggs beaten together stirred over the potatoes a few minutes before they are ready to serve. The small potatoes may also be scraped and dropped in hot, deep fat and fried like fritters.

When possible, the small potatoes should be well cleansed with a vegetable brush and boiled without paring. They may then be easily skinned after they are cooked. Some of the more important ingredients are lost when potatoes are pared, and it is also more economical to boil them before paring. The cold boiled potatoes may be cut up and used for potato salad, or thinly sliced after being skinned and placed in a baking dish alternately with a cream sauce consisting of milk, butter and flour, and seasoned with salt and pepper, having the first and last layer cream sauce. Sprinkle bread crumbs liberally over the top, dot with hits of butter and bake in a moderate oven about 20 minutes until the top is nicely browned. Serve in the dish in which they were baked.

Or peel one-half dozen medium-sized raw potatoes, cut into small, narrow strips about ⅓ inch wide, dry on a napkin and fry in very hot, deep fat about six minutes, then lift from fat, drain, sprinkle salt over and serve hot. These are a nice accompaniment to broiled steak.

Peel and slice, or cut in dice, 6 or 8 cold boiled potatoes, cut into in a stew-pan with 2 tablespoonfuls of butter, salt and pepper to season, heat all together, shaking pan occasionally. Add ½ cup of cream, sprinkle a small teaspoonful of parsley over and serve hot. Instead of slicing or dicing cold boiled potatoes (in the usual manner) to be fried, if they be cut in lengthwise sections like an orange (one potato should make about 8 pieces) and fried quickly in enough hot fat to prevent burning, they can scarcely be distinguished from raw potatoes cut in the same manner and fried in deep fat, and are much easier to prepare. They should be served at once.

Another manner of preparing potatoes is to slice raw potatoes as thinly as possible on a "slaw-cutter," place in a fry-pan with a couple of tablespoonfuls of a mixture of butter and sweet drippings. Watch carefully, as they should be fried quickly over a hot fire, turning frequently. When brown, serve at once.

Raw *sweet* potatoes cut about as thick as half a section of an orange, fried in a couple tablespoonfuls of a mixture of sweet drippings and butter, prove a change, occasionally.

SCALLOPED POTATOES

In a baking dish place layers of pared, thinly sliced, raw white potatoes. Season with a very little salt and pepper and scatter over small bits of butter. A very little finely minced onion or parsley may be added if liked. To 1 quart of the sliced potatoes use a scant half pint of milk, which should almost cover the potatoes. Either sift over the top 1 tablespoon of flour or 2 tablespoons of fine, dried bread crumbs and bits of butter; place in hot oven and bake about ¾ of an hour, until top is browned nicely and potatoes are cooked through. Old potatoes are particularly good prepared in this manner.

CANDIED SWEET POTATOES

Place in an agate pudding dish 6 pared and halved (lengthwise) raw sweet potatoes. Scatter over them three tablespoons of sugar, 2 large tablespoons of butter cut in small bits, and about ½ a cup (good measure) of water. Stand in a hot oven and bake about ¾ of an hour. Baste frequently with the syrup formed in the bottom of the dish. The potatoes when baked should look clear and the syrup should be as thick as molasses. Serve in the dish in which they were baked. Should the oven of the range not be very hot, the dish containing the potatoes may be placed on top the range and cooked about 25 minutes before placing in oven to finish baking.

SWEET POTATO CROQUETTES

To 1 pint of hot mashed potatoes, or cold boiled ones may be used, squeezed through a fruit press; add 1 tablespoon of butter, pinch of salt, 2 eggs, whites beaten separately. When cool, form into small cone-shapes, dip in bread crumbs, then into egg, then into crumbs again, and fry in deep fat. Drain on paper and serve on platter garnished with parsley.

POTATO CHIPS

Aunt Sarah's way of making particularly fine potato chips: She pared six large white potatoes, one at a time. As she wished to slice them to fry, she rinsed the potatoes, rolled them on a clean cloth to dry them. She sliced the potatoes thinly on a "slaw" cutter. She patted the sliced potatoes between old linen napkins, until all moisture was absorbed, then dropped them into hot fat, consisting of two-thirds lard and one-third suet. Place only one layer of potatoes at a time in the fat. The chips quickly turn light brown; then remove with a perforated skimmer to a colander lined with coarse brown paper, to absorb any remaining fat. Should the fat be the right temperature, the chips will be entirely free from grease. Dust salt over the chips while hot. She *never* allowed chips to stand in salt water, as many cooks do. She usually made potato chips when frying doughnuts, and always fried potato chips first; after frying doughnuts in the fat fry several large slices of potato in it, as the potato clarifies it. Six large, thinly sliced potatoes will make about five quarts of potato chips when fried and may be kept several weeks in a dry place. The potato chips may be re-heated by placing in a hot oven a few minutes before serving.

FRIED EGGPLANT

Pare the egg-plant, cut in slices one-half inch thick, sprinkle salt on slices; let stand under heavy weight several hours. Wipe slices dry with a napkin and dip in a mixture of white of one egg, and one tablespoon of water, then dip them in fine rolled bread crumbs and fry a rich brown in deep fat. Drain and serve. Catsup should always be served with eggplant.

BAKED "STUFFED PEPPERS"

Place a fry-pan on stove containing about two tablespoonfuls of butter, add a couple of finely chopped sweet peppers and a finely minced small onion. Let all simmer on stove. Measure the chopped pepper and add an equal amount of finely crumbled bread. Season with salt and pepper and fill (well-washed) peppers from which the stem and seeds have been removed. Stand the peppers in a bake dish containing a small amount of water. Place in a hot oven about twenty-five minutes, or until peppers are tender. Serve hot.

CHILI (AS PREPARED IN NEW MEXICO)

Place hot peppers (well-washed) from which seeds have been removed into a bake dish containing a very little hot water. Stand in a hot oven until tender and skins turn a yellow brown, turning them over occasionally. Remove the outside skin, chop fine, add a small quantity of finely minced onion, pepper and salt and enough vinegar to moisten. If sweet peppers are used add a pinch of cayenne pepper. Serve as a relish in place of pickles or chow-chow. This recipe was given Marry by a friend who had lived in Mexico. The outside skin of the peppers may be more readily removed if upon being removed from the oven the peppers are sprinkled with water, then covered with a cloth and allowed to steam a short time.

BAKED CABBAGE

A half head of cabbage was cut into small pieces and cooked in hot salted water until cabbage was tender. The water was drained from the boiled cabbage, which was placed in an agate pudding dish alternately with cream sauce composed of one cup of milk; one small tablespoonful of flour, 2 tablespoonfuls of butter, seasoned with salt and pepper. Sprinkle a few crumbs and place bits of the butter over top. Bake in oven about 25 minutes and serve hot. This dish is almost equal to cauliflower in flavor, especially if after the cabbage has cooked ten or fifteen minutes the water is drained from it and fresh substituted. And it is said, "Cauliflower is only cabbage with a college education."

CRIMSON CREAMED BEETS

Cut all except two inches from the tops of beets. Scrub thoroughly with a vegetable brush, then pour scalding water over beets. When perfectly cleansed, place in a cook-pot, partly cover with boiling water, stand on range and when beets have cooked tender remove outside skin. Strain and stand aside one cup of water in which beets were boiled, which should be dark wine color. When beets are to be served to the one cup of strained beet juice add one tablespoonful of sugar, one-fourth cup of not *very sharp* vinegar. Add one teaspoon of butter. Thicken this liquid with one and one-fourth tablespoonfuls of a mixture of corn starch and flour. When cooked to the consistency of cream add the quartered beets, season with pepper and salt, stand on back part of range a few minutes, serve hot. To three cups of the quartered beets use one and one-half cups of cream dressing.

BUTTERED BEETS

Wash young beets, cut off tops. Boil one hour or until tender, one tablespoonful of sugar having been added to the water in which beets were boiled. Rub off skins, cut in quarters, strew over them one tablespoon of butter cut in small pieces, stand in oven just long enough for the butter to melt. Or cut the beets in slices one-fourth of an inch thick and while still warm place in a bowl and pour over them half a cup of hot vinegar and water to which had been added one tablespoonful of sugar, a pinch of salt and pepper; serve cold.

PICKLED MANGELWURZEL

A vegetable in taste, similar to very sweet, red beets in shape, greatly resembling carrots. Wash the mangelwurzel and place in a stew-pan with boiling water and cook until tender (allow about an inch of top to remain when preparing to cook). Skin the mangelsurzel, slice and pour over the following, which has been heated in a stew-pan over the fire: One cup of vinegar and water combined, one tablespoonful of sugar, one teaspoonful of salt, a dust of pepper. Stand aside until cold then serve. Or serve hot like buttered beets. Some "Bucks County" farmers raise mangelwurzel simply to feed to their cattle, but Aunt Sarah preferred them when young and tender to beets, and always raised them for her table.

GERMAN STEAMED CABBAGE

Cut one-half head of cabbage fine on a slaw cutter. Place in a stew-pan over fire, with about four tablespoonfuls of water, one tablespoonful of butter, a couple tablespoonfuls of flour, one teaspoonful of sugar and a pinch of salt. Cover and steam twenty minutes. Then add three tablespoons of vinegar. Stir in one beaten egg. Cover and let stand where it will keep hot until ready to serve.

BEAN "SNITZEL"

Place in a pan on the range one tablespoon of diced, smoked bacon, fry a few minutes, watch closely it does not scorch. Add one tablespoonful of sweet lard, when hot, add four thinly sliced, medium-sized onions and four chopped tomatoes and 1½ quarts of string beans, cut in inch lengths. Season with salt and a pinch of red pepper. Simmer all together three hours. After cooking one hour add about one cup of hot water, stirring occasionally to prevent scorching, add a little more water if necessary; when beans are tender and ready to serve there should be a small quantity of liquid, resembling tomato sauce, with the beans.

BOILED SPINACH

Wash one-half peck of spinach thoroughly through a half dozen waters, until free from sand. Place in a stew-pan containing a small quantity of *boiling* water and one teaspoon of butter. Cook until tender, drain, chop fine. Place a large tablespoonful of butter in stew-pan and when hot add chopped spinach, season with salt and pepper; serve in a warmed dish, garnished with either chopped or sliced hard boiled eggs. A German cook, noted for the fine flavor of her cooked spinach and green peas, said her secret consisted in adding a teaspoon of butter to the vegetables while cooking.

FRIED ONIONS AND POTATOES

Another way of utilizing left-over cold boiled potatoes particularly relished by "Pennsylvania Germans," whose liking for the humble onion is proverbial, is to fry onions with potatoes in a fry-pan containing a couple tablespoonfuls of sweet drippings and butter; when heated place a half dozen thinly sliced cold boiled potatoes, half the quantity of thinly sliced raw onions, well seasoned with pepper and salt, cover and steam for ten or fifteen minutes, when uncover and fry until light brown; serve at once. Or the thinly-sliced onions, after skins have been removed, may be sliced thinly across the onion, placed in a fry-pan and partly covered with boiling water; stand on hot range and steam, closely covered, about fifteen minutes, or until onions are tender, then drain off water, should any remain, add a small tablespoonful of butter, salt and pepper to season, fry quickly a light brown; pan should be uncovered. Serve at once with liver or bacon. Onions are considered more wholesome prepared in this manner than if fried.

STEAMED ASPARAGUS (FINE)

Wash asparagus and cut off about an inch of the tough ends, scrape off thin skin. Place pieces of asparagus tips (all in one direction) in the top part of perforated section of a double boiler. Fill lower part of steamer with hot water and steam about three-quarters of an hour or less time, until tender. The fine flavor of the vegetable is retained when steamed. When cooked tender turn out on a hot platter and pour cream sauce over the tips, or the cream sauce may be served separately, or the asparagus may be served on freshly toasted slices of bread, over which the cream sauce should be poured.

"PASTURE" MUSHROOMS

All the members of the Landis family unanimously agreed in declaring the dish "Frau Schmidt" taught Sarah Landis to prepare from the delicious edible Fungi, known as "Pasture" mushrooms (gathered by Professor Schmidt from rich, wind-swept pastures early in the fall of the year until the coming of frost) were good enough to tickle the palate of an epicure.

Sarah Landis was very particular to use *none* unless pronounced *edible mushrooms*, and not poisonous toad-stools, by Professor Schmidt, who was a recognized authority. Said the Professor, "The edible variety may be easily recognized by one having a knowledge of the vegetable. The cap may be readily peeled, and the flesh of the 'Pasture' mushroom, when cut or broken, changes in color to a pale rose pink, and they possess many other distinctive features, easily recognized, when one has made a study of them."

The following is the manner in which the mushrooms were prepared by Fran Schmidt:

STEAMED MUSHROOMS.

One-half pound or about twenty-four small mushrooms were peeled, washed carefully in cold water, placed in a small stew-pan containing two generous tablespoonfuls of butter, covered closely and allowed to simmer or steam for twenty minutes in butter and liquid, drawn from the mushrooms by steaming, then uncover and allow liquid in sauce-pan and mushrooms to cook about ten minutes longer, then sprinkle two teaspoonfuls of flour over the mushrooms, brown a minute, stir into this ½ cup of milk, or enough to make a sauce the consistency of cream, season well with salt and pepper to taste. Have ready prepared six crisply toasted and buttered slices of stale bread. Place four mushrooms and a couple of tablespoonfuls of the mushroom sauce on each slice of bread and serve hot. The combination of toast and mushrooms results in a particularly fine flavor.

STEWED TOMATOES

Scald ripe tomatoes by pouring boiling water over them and allowing them to stand a few minutes. Skin them and cut in small pieces. Place in a stew-pan with 1 tablespoonful of butter, season *well* with pepper and salt, cook about 25 minutes, add ½ teaspoonful of sugar and thicken with 1 teaspoonful of flour mixed smooth with a little water. Let cook a few minutes, then serve. If tomatoes are very tart a small pinch of baking soda, added when cooked, will counteract acidity.

SWEET CORN

Sweet corn on the cob should be cooked as soon as possible after taking it from stalk, as after being removed it soon loses its sweetness. Do not remove the husk until it is to be boiled. Place corn in a kettle of rapidly boiling water, not salted; rather add a pinch of sugar if corn is not as sweet as liked. Cover the kettle to prevent steam escaping. Do not use a *large quantity of water*. Corn is sweeter if steamed. Boil from ten to fifteen minutes. If corn is not cooked in that time, it should be used uncooked for corn fritters, as corn if *not* young and tender may be grated and from it excellent corn fritters may be made.

FRIED TOMATOES WITH CREAM SAUCE

Cut large, solid, ripe tomatoes in half-inch slices; one ordinary tomato makes 3 slices. Dredge thickly with flour. Fry several slices of bacon in an iron pan, take bacon from pan when fried and put in warming oven. Lay the well-floured slices of tomatoes in hot bacon fat and one tablespoon of butter and fry brown on both sides. Serve on hot platter with bacon. Or fry slices of well floured tomato in pan containing just enough butter and drippings to keep them from sticking to the bottom of pan, over a hot fire. Fry quickly, browning on each side. Season with salt and pepper. If the tomatoes are very sour, sprinkle a *very little* sugar over them before frying. When brown, lift the tomatoes carefully from pan and place in a circle around the inside edge of a warm chop plate, add a lump of butter to the pan and a small half cup of sweet milk. Let come to a boil, thicken with a little flour mixed smoothly with a little cold milk, and cook until the consistency of thick cream. Season with salt and pour in centre of chop plate, surrounded with fried slices of tomatoes. Dust pepper over top and serve hot.

This is a delicious way of serving tomatoes. Or slices of the fried tomatoes may be served on slices of crisply toasted bread over which place a couple tablespoons of the cream dressing.

BAKED "STUFFED TOMATOES"

Wash a half dozen ripe red tomatoes. Cut the top from each and remove about the half of the inside of tomato. Sprinkle a very tiny pinch of sugar in each. This small quantity of sugar is not noticed, but counteracts the acidity of the tomato. To one and one-half cups of soft bread crumbs add one small finely minced onion and season highly with salt and pepper, also add one teaspoon of chopped parsley. Mix all together and fill the tomatoes with the mixture. Place a small bit of butter on each tomato. Place in a bake dish containing a half cup of water, a piece of butter, one teaspoonful of sugar, a sprig of parsley and pepper and salt to season. Stand in a hot oven and bake from 25 to 30 minutes. The centres which were removed from tomatoes may be utilized in various ways.

CANNED TOMATOES—FRIED

Place in a bowl a half pint of canned tomatoes, one-fourth teaspoon of sugar and season with salt and pepper. Add about four tablespoonfuls of flour sifted with one-half teaspoon of baking powder and one tablespoon of butter. Use only flour enough to hold the mixture together when fried. Drop spoonfuls some distance apart in a fry-pan containing several tablespoons of hot lard, butter, suet or drippings. Fry on both sides and serve hot. In winter, when the housewife is unable to obtain fresh tomatoes, she will find this dish a good substitute to serve occasionally.

"BUCKS COUNTY" BAKED BEANS

Put one quart of small soup beans to soak over night in cold water to cover. In the morning drain the beans, cover with boiling water, add one tablespoonful of molasses and cook until tender, but not too soft. Drain. Do not use this water. Put the beans in an earthen bake dish. In the centre of the bake dish place one pound of clean, scored smoked bacon, and pour over the beans the water in which the bacon had been simmering for an hour. Add water, if not enough, to almost cover the beans, salt and pepper to taste. Place in oven and bake about three hours, or until beans are tender and a rich brown on top. Add more hot water if beans bake dry, until the last half hour, then allow the water to cook away.

Serve stewed tomatoes, baked apples or apple sauce as an accompaniment to baked beans. This is not a recipe for "Boston Baked Beans." Just a "plain country recipe," but it will be found very satisfactory.

If part of a dish of beans remain after a meal, re-heat the day following in "tomato sauce." Aunt Sarah always baked a pan of corn bread or Johnny cake, to serve hot with baked beans.

When the housewife serves a dish of baked beans at a meal, serve also a quart of stewed tomatoes. The day following a "tomato sauce" may be quickly prepared by adding a well-cooked carrot and an onion to the "left-over" tomatoes. Press all through a coarse sieve, adding a little water if too thick; re-heat beans in this; serve hot. A delicious "cream of tomato soup" may be prepared by substituting milk or cream to which a small pinch of baking soda has been added, omitting the beans.

COOKED HOMINY

Wash one cup of hominy through several waters. (The grains should resemble kernels of corn.) Cover with cold water and stand in a cool place over night. In the morning, drain. Place the hominy in an agate pudding dish holding 2 quarts, cover with boiling water, add more water as the grains swell and water boils away, and 1 teaspoonful of salt. The hominy should be placed on the range to cook early in the morning on the day it is to be served and continue cooking slowly until late afternoon, when all the water should have been absorbed and each grain should be large, white and flaky. The dish should be about three-quarters full.

A half hour before serving the hominy, at a six o'clock dinner, add a generous tablespoonful of butter and about ¾ of a cup of hot milk and stand on back of range until served. This is a remarkably cheap, wholesome and appetizing dish if served properly and is easily prepared.

GRATED "PARSNIP CAKES"

Scrape, then grate enough raw parsnips to fill two cups, put in a bowl and add the yolk of one egg, pinch of salt, 1 tablespoonful of milk, 1 tablespoonful of flour, lastly add the stiffly-beaten white of egg.

Form into small round cakes, dust with flour and fry brown on both sides in a pan containing a tablespoonful of butter and one of drippings. Or these may be crumbed and fried in deep fat. These are much finer flavored than if parsnips had been cooked before being fried.

TO MAKE "SAUER KRAUT"

Cut heads of cabbage in half, after trimming off outside leaves. Cut out centres or hearts, cut cabbage fine on a regular old-fashioned cabbage cutter, which has a square box on top of cutter to hold the pieces of cabbage when being pushed back and forth over the cutter. If not possible to procure this, use small slaw cutter for the purpose.

Partly fill a large pan with the cut cabbage, and mix enough salt, with the hands, through the cut cabbage to be palatable when tasted, no more. This was the rule taught Aunt Sarah by her Grandmother, and always followed by her. Then put the salted cabbage into a wooden cask or small tub to the depth of several inches. Pound the cabbage down well with a long-handled, heavy, wooden mallet, something like a very large wooden potato masher. Then mix another panful of finely cut cabbage, lightly salted, into the tub and pound down well, as before. Continue in this manner until the tub is partly filled with cabbage, pounding down well at the last until the liquid formed by the cabbage and salt rises above the cabbage. Cover the kraut with a layer of large, clean cabbage or grape leaves, then cover top with a clean piece of muslin cloth, place a round, clean board on top and put a well-scrubbed, heavy stone on the board to weight it down. Stand the tub in a warm place several days, to ferment. When fermentation begins, the liquor rises over the top of the board. Remove the scrum which rises to top, in about six days, and stand in a cool part of the cellar after washing stone and cloth with cold water, return to top of kraut and in two weeks the sauer kraut will be ready to use. Should the sauer kraut require extra liquid at any time, add one quart of water in which has been dissolved two teaspoonfuls of salt. Squeeze the sauer kraut quite dry when taking it from the brine to cook. Boil about two quarts of the sauer kraut several hours with a piece of fresh pork and a little water until the pork is thoroughly cooked through, when the sauer kraut should be cooked tender.

Some prefer "frankfurters" cooked with the kraut instead of pork, and others do not care for the German dish without the accompaniment of drop dumplings. Serve mashed potatoes and simple dessert with sauer kraut.

Aunt Sarah taught Mary to save the hearts of the cabbage usually thrown aside when making sauer kraut. The hearts were trimmed all one size, like small triangles. She cooked them in salted water until tender, drained them and served with a cream dressing, and they had much the flavor of a dish of cauliflower.

Frau Schmidt always placed several tart apples among her sauer kraut when making it, and thought it improved the flavor of the kraut; gave it a "winey" flavor, obtained in no other manner. A sour apple, cored and cooked with sauer kraut is considered by some cooks an improvement. The apple, of course, is not eatable. Aunt Sarah *never* placed apples with her sauer kraut.

DUMPLINGS TO SERVE WITH SAUER KRAUT

For these dumplings, 1 egg was broken into a bowl and well beaten. Then a pinch of salt was added and ½ cup of sweet milk. Enough flour was added to make a soft dough, and one tablespoonful of baking powder was sifted with a very little flour into the batter, then a little more flour was added to make the dough the right consistency. Form the dough into small balls, handling as little as possible. Drop on top of the hot cooked "sauer kraut" in cook-pot on range and boil, closely covered, about 20 minutes.

Aunt Sarah taught Mary to cook green vegetables, peas, spinach, etc., in a stew-pan *uncovered*, if she wished them to retain their natural color. Also, that old potatoes may be freshened by being allowed to stand a short time in cold water before being cooked, but they should not stand too long a time in cold water, as it draws the starch from them and causes them to be tasteless, and to lose part of their nourishing qualities.

Also that one teaspoonful of salt will usually season one quart of vegetables, to be put in when the vegetables begin to cook. Cauliflower, cabbage, lettuce and watercress should stand in a pan containing water and a little vinegar for a half hour. This will cause insects to drop to the bottom of the pan.

Changing the water on cabbage and onions when partly cooked will improve their flavor.

PARSLEY DRIED TO PRESERVE ITS GREEN COLOR

Young housewives possessing a bed of parsley in their kitchen gardens, wishing to preserve it for use during the winter, may like to know how Aunt Sarah taught Mary to dry it in a manner to preserve its bright green color.

She washed the parsley in cold water and while still moist placed it on agate pans and dried it *quickly* in a *very hot* oven. Watch carefully as it scorches easily. Place the parsley when dried, in tin cans covered to exclude the dust.

TIME REQUIRED TO COOK VEGETABLES

Bake good-sized potatoes in oven about 45 minutes. Smaller potatoes require less time to bake.

Boil ordinary sized potatoes 25 to 30 minutes.

Steam asparagus from 30 to 40 minutes.

Boil young beets about 60 minutes or longer.

Old beets, two hours, or until tender.

Green corn on cob about 10 or 15 minutes.

Cauliflower, 30 minutes.

Cabbage, 30 to 40 minutes.

Turnips and carrots, 40 minutes.

String beans, 60 minutes to 2 hours.

Lima beans, 45 minutes to 1 hour.

Onions about 1 hour.

Squash about 30 minutes.

Parsnips, 30 to 40 minutes.

Sweet potatoes, good size, 40 minutes.

Spinach, 25 minutes.

Tomatoes, 25 minutes.

Salt should be added to the water when boiling potatoes, carrots, cabbage, parsnips, turnips and onions, even if liquid in which they were boiled is drained from them after being cooked, before being seasoned. Add a small pinch of baking soda to the water in which string beans are boiled, and they will cook tender in less time. Especially should this be done if the beans are not young and tender.

COMMON "CREAM SAUCE"

Young housekeepers will be surprised to learn of the various attractive, appetizing dishes which may be prepared by combining them with a "cream sauce." After cooking vegetables until tender in salted water, they should be drained and served with a cream sauce poured over. The art of making a smooth, creamy sauce of the proper consistency is easily acquired. A good rule for "common cream sauce" is 1 cup of milk, water, or meat broth, thickened with 1 tablespoonful to 1½ tablespoonfuls of flour, or a combination of flour and cornstarch. Mix flour, or cornstarch, with a small quantity of cold milk or water, to a smooth paste, before adding it to liquid; add, usually, one tablespoonful of butter. Place the mixture in a saucepan and cook until the consistency of cream, add ½ teaspoonful of salt just before removing from the fire, and dust pepper over when serving. When mixing gravy to serve with roast beef or veal, omit butter. For a thick sauce use either 2 or 3 tablespoonfuls of flour and the same amount of butter. This thick sauce may be used to mix with meat for croquettes in the proportion of 1 cup of sauce to 2 cups of chopped cold roast lamb, beef, veal or chicken. Should a richer sauce be desired, add 1 or more yolks of eggs to the cream sauce. Some of the numerous dishes which might be served by the young housewife to vary the daily bill of fare by the addition of "cream sauce," are: Small, new potatoes, cauliflower, onions, cabbage asparagus tips, thinly sliced carrots, celery, mushrooms, fish, oysters, chicken, veal and sweetbreads. All of these, when coked, may be served on slices of toasted bread, or served in Pattie-cases, with cream sauce, or served simply with cream sauce.

PREPARATION OF SAVORY GRAVIES

The art of preparing savory gravies and sauces is more important in connection with the serving of the cheaper meats than in connection with the cooking of the more expensive cuts.

There are a few general principles underlying the making of all sauces or gravies, whether the liquid used is water, milk, stock, tomato juice or some combination of these. For ordinary gravy, 2 level tablespoonfuls of flour or 1½ tablespoonfuls of cornstarch, or arrow root, is sufficient to thicken a cup of liquid. This is true excepting in recipes where the flour is browned. In this case, about ½ tablespoonful more should be allowed, for browned flour does not thicken so well as unbrowned. The fat used may be butter or the drippings from the meat, the allowance being 2 tablespoonfuls to a cup of liquid. The easiest way to mix the ingredients is to heat the fat, add the flour and cook until the mixture ceases to bubble, and then to add the liquid. This is a quick method and by using it there is little danger of getting a lumpy gravy. Many persons, however, think it is not a wholesome method, and prefer the old-fashioned one of thickening the gravy by means of flour mixed with a little cold water. (Aunt Sarah was one who thought thus.) The latter method is not "practicable for brown gravies," to quote the *Farmers' Bulletin*.

The *Farmers' Bulletin* further adds:

"Considering the large amount of discussion about the digestibility of fried food and of gravies made by heating flour in fat, a few words on the subject at this point may not be out of order. It is difficult to see how heating the fat before adding the flour can be unwholesome, unless the cook is unskillful enough to heat the fat so high that it begins to scorch. Overheated fat, as has already been pointed out, contains an acrid, irritating substance called 'Acrolein,' which may readily be considered to be unwholesome. It is without doubt the production of this body by overheating which has given fried food its bad name. There are several ways of varying the flavor of gravies and sauces. One should be especially mentioned here. The *flavor of browned flour*—The good flavor of browned flour is often overlooked. If flour is cooked in fat, until it is a dark brown color, a distinctive and very agreeable flavor is obtained.

"This flavor combines very well with that of currant jelly, and a little jelly added to a brown gravy is a great improvement. The flavor of this should not be combined with that of onions or other highly-flavored vegetables."

BUTTER, CHEESE AND SUET—A SUBSTITUTE FOR BUTTER

This formula for preparing a good, sweet, wholesome substitute for butter to be used for baking and frying was given Aunt Sarah by a thrifty German hausfrau, who prepared and used it in her large family many years. Aunt Sarah always kept a supply on hand. It was made as follows:

- 10 pounds of fine solid kidney suet.
- 10 pounds of clean pork fat.
- 10 pounds of butter.

The suet cut in small pieces was put in a large boiler of water, boiled until all was melted, and the fat extracted from the suet. It was then all poured through a fine sieve into a vessel containing hot water (the larger the quantity of hot water the finer the fat will be). Stand aside to become cold and solid. The boiling process prevents the peculiar taste which *fried* lard and suet usually possess. Treat the pork fat in a similar manner. Allow the suet and pork fat to stand until the following morning, when remove the solid fat from the boiler of water, wipe off all moisture and add both pork fat and suet fat to the melted butter, which had been prepared in the following manner: The butter was melted in a porcelain lined boiler and allowed to cook until all salt and other foreign substance had settled and the butter had the appearance of clear oil. At this point the butter should be watched carefully, as when settled it might quickly boil over, when you would be liable to lose your butter, besides suffering serious consequences. Now the liquid butter, suet and pork fat are all put together into a large boiler and allowed to melt together on the back part of the range. This will probably be done in the morning. After the noon meal is finished move the boiler containing fat to front part of range; let come to a boil, skimming it occasionally as it boils up. It needs close watching now, the fat being liable to cook over the top of boiler, when the "fat" will surely be "in the fire." Carefully pour into stone crock, and it may be kept for months in a cool place. The fat which has been first poured off the top, if it has been carefully skimmed, will keep longest. The last taken from the boiler should be put in a stone crock to use first. This may be prepared in lesser quantities, or a smaller quantity of butter might be used to mix with the lard and suet. Although the preparation is to be preferred composed of equal quantities of butter, lard and suet, adding milk to the first water in which the suet is boiled is quite an improvement. After filling the crocks with the fat, take the boiled-out suet and hard scraps and settlings of butter remaining and go through the same process and you will have a small jar of cooking fat for immediate use. A little trouble to do this, I admit, but one is well paid by having good, sweet, inexpensive cooking fat. I should advise a young housekeeper to experiment with one pound each of clarified suet and pork fat after it is rendered, and one pound of butter before attempting the preparation of a larger quantity.

BUTTER — AS IT WAS MADE AT THE FARM, BY "AUNT SARAH"

Aunt Sarah strained fresh, sweet milk into small, brown earthenware crocks kept for this purpose, scrupulously clean. The crocks were kept in the spring-house or cellar in summer (in cold weather the milk should be kept in a warmer place to allow cream to form on the top of the milk). When the cream was thick and sour she skimmed the cream from off the top of milk every day, stirring the cream well together every time she added fresh cream to that on hand. Aunt Sarah churned twice a week; sour cream should not be kept a longer time than one week. The churn was scalded with boiling water, then rinsed with cold water; this prevented the butter adhering to the churn. The cream should be at a temperature of 60 degrees when put in the churn, but this would be almost too cold in Winter. In very hot weather the temperature of the cream should be 56 degrees. Aunt Sarah tested the cream with a small dairy tube thermometer. She churned steadily and usually had butter "come" in about 25 minutes, but should the cream he too cold or too warm it would be necessary to churn a longer time. If the cream is too warm, stand vessel containing cream on ice; if too cold, stand in a warm place near the range. When the sour cream had been churned a certain length of time and granules of butter had formed, she drained off the buttermilk and poured water over the granules of butter. Water should be two degrees colder than the buttermilk. After churning a few minutes the lump of butter was removed from the churn, placed in a bowl, washed thoroughly several times in very cold water, until no buttermilk remained. The butter was worked thoroughly, with a wooden paddle, until all buttermilk had been extracted. One small tablespoonful of salt was added to each pound of butter. She worked the butter well, to incorporate the salt, and molded it into shape. Aunt Sarah did not knead the butter, but smoothed it down, then lifted it up from the large, flat, wooden bowl in which it was molded. When the butter was to be molded into *small shapes*, she scalded the small wooden molds, then dipped them into cold water before using; this prevented the butter adhering to the molds. Before commencing to churn butter, Aunt Sarah was particular to have her hands scrupulously clean. All the utensils used were washed in hot water, then rinsed in cold water, both hands and utensils. She frequently wrapped small pats of freshly-churned butter in small squares of clean cheese-cloth and placed in a stone crock with a cover. Placed in the crock was usually, with the butter, a bunch of sweet clover blossoms, which imparted to the butter a delicious flavor.

"SMIER-KASE" OR COTTAGE CHEESE

Stand a pan containing three quarts of milk in a warm place until it becomes sour and quite thick. Stand the pan containing the thick milk on the back part of the range, where it will heat gradually but not cook. When the "whey" separates from the curd in the centre and forms around the edges it is ready to use. Should the sour milk become *too hot* on the range, or *scald*, the curds, or smier-kase, will not become soft and creamy. When the curd has separated from the "whey," pour the contents of the pan into a cheese-cloth bag and hang in the open air to drip for several hours, when it should be ready to use.

From three quarts of sour milk you should obtain one good pound of smier-kase. To prepare it for the table place one-half the quantity in a bowl and add one teaspoonful of softened butter, a pinch of salt and mix as smoothly as possible. Or the smier-kase may be molded into small rolls, and a small quantity of finely-chopped Pimento added. This will keep fresh several days if kept in a cool cellar or refrigerator.

USES OF "SWEET DRIPPINGS" AND SUET

For deep frying Mary was taught to use lard and kidney suet combined. The latter had been tried out by cutting suet in small pieces. The suet, in an iron pan, was placed in a moderately hot oven until fat was tried out. To prevent suet when rendered having a taste of tallow, place in the upper part of boiler, over one containing hot water, and stand on a hot range until all is tried out, or melted, instead of putting it in oven. Strain into a jar and stand aside in a cool place until wanted. Take one-third of this tried-out suet to two-thirds lard when frying croquettes, oysters, cruellers or fritters. Suet contains food value equal to that of lard and food fried in this fat, combined with lard, is more wholesome than if fried in lard alone—if any food fried in fat *ever is* wholesome. And suet is more economical than lard if rendered at home. Mary was taught by her Aunt to save all the trimmings from steaks, fat left over from roasts, boiled ham, sausage, bacon fat, etc. When different fats have been tried out, to clarify them, add to every pound and a half of combined fat or drippings a half cup of boiling water and a pinch of baking soda. Boil until water evaporates and fat is clear. Strain into a bowl and keep in a cool place. Clean, sweet drippings are preferred by most cooks to lard for many purposes. All young housewives do not know that ham or bacon fat may be substituted for half the shortening called for in many recipes for molasses cakes (where spices are used) with good results. Also that the grease rendered from clean fat of chickens, which greatly resembles butter when tried out and cold, may be combined with an equal quantity of other shortening in making cakes in which spices are used. The difference in the taste of cake made from this fat, if rendered sweet and clean, will not be noticed. Equal parts of ham or bacon fat, pork chops or sausage fat, combined with butter, are excellent for frying cornmeal mush, eggs, sweet potatoes, egg bread and calves' liver. Also sliced tomatoes have a particularly fine flavor if fried in bacon fat. Should fat removed from top of stock pot have a flavor of vegetables, pour boiling water over, strain and stand aside to cool; then remove the clean cake of fat on top of the water and add to bowl of drippings. This is one of the small economies which will, I think, appeal to the frugal young housewife. If possible, procure an iron pot for deep frying. After using, strain the fat remaining, adding sediment in the bottom of cook-pot to the can of soap fat; then return the clean, strained fat to the cook pot. Keep in a cool place, closely covered, and if careful not to scorch the fat. It may be used over and over again, and croquettes, etc., may be prepared in a few moments by simply heating the kettle of fat in which to fry them.

Aunt Sarah frequently filled small glass jars with rendered mutton suet, scented with violet essence, to be used for chapped lips and hands.

EGGS—"EIERKUCHEN" OR OMELETTE

For this excellent omelette or "eierkuchen," as Aunt Sarah called it, she used the following:

- 3 fresh eggs.
- 1 cup sweet milk.

- 3 level tablespoonfuls of flour.

She placed on the range a small fry pan (size of a tea plate), containing one tablespoonful of butter. She then placed 3 tablespoonfuls of flour in a bowl, mixed smoothly with a portion of the cup of milk, then added the three yolks of eggs which had been lightly beaten and the balance of the milk and a pinch of salt. Lastly, she stirred in lightly the stiffly-beaten whites of eggs. Poured all into the warmed fry-pan and placed it in a moderately hot oven until lightly browned on top. The omelette when cooked should be light and puffy, and remain so while being served. Double the omelette together on a hot platter and sprinkle finely chopped parsley over the top. Serve immediately.

HARD BOILED EGGS

Eggs to be hard boiled should be carefully placed in boiling water and cooked 15 minutes from the time the water commences to boil again. If cooked a longer time, the white of egg will look dark and the outer part of yolk will not be a clear yellow, as it should, to look appetizing when served.

SOFT BOILED EGGS

The quicker way to prepare eggs is to drop them in a stew-pan containing boiling water, and let boil 3½ to 4 minutes, when the white part of the egg should be "set" and the yolk soft, but a soft boiled egg is said to be more easily digested if dropped into a stew-pan of rapidly boiling water; remove the stew-pan of boiling water the minute the eggs have been put in from the front part of the range to a place where the water will keep hot, but not allow the eggs to boil. Let the eggs remain in the hot water from 8 to 10 minutes. On breaking the egg open, the yolk will be found soft, and the white of the egg a soft, jelly-like consistency. This latter is the way Aunt Sarah taught Mary.

AN EGG AND TOMATO OMELETTE

Beat the yolks of three eggs until light, then add three tablespoonfuls of water. Beat the whites of the eggs separately. Turn the stiffly-beaten whites of the eggs into the bowl containing the yolks of eggs and water. Stir lightly together and add a pinch of salt. Turn all into a small fry-pan containing a generous tablespoonful of butter and cook on top of stove until the eggs are set, then place the pan containing omelette in a hot oven and finish cooking. When cooked, turn out on a hot platter and spread over the top the following, which was prepared while the omelette was cooking. In a small fry-pan place a tablespoonful of finely-chopped bacon. When fried brown add half a small tomato, finely chopped, ¼ of an onion, chopped fine, and a little chopped green pepper. Cook all together for a short time and season with salt and pepper. After spreading the mixture on the omelette, fold over and serve on a hot platter. This recipe had been given Frau Schmidt years before by a friend and she used no other for making omelette. Always make small omelettes. They are more satisfactory. Use a small pan no larger than a small tea plate, and, if wished, make two small, rather than one large one. Always serve immediately.

MUSHROOM OMELETTE

Place the yolks of three eggs in a bowl and beat until light. Add a teaspoonful of cream and ½ teaspoonful of flour mixed together; ½ cup of chopped mushrooms, salt and pepper and a dust of baking powder. Lastly, the stiffly-beaten whites of the eggs. Turn into a pan containing two tablespoonfuls of melted butter, stand on range a few minutes until eggs are set, then finish cooking in a hot oven. Serve at once.

A few cold, steamed mushrooms (left-overs), if finely chopped, and added to a plain omelette or roast, will improve the flavor.

A CLAM OMELETTE

Two eggs beaten separately, 1 scant cup of milk, 1 tablespoonful of flour, 6 clams run through a food-chopper. Place in a bowl the tablespoon of flour and mix smooth with a little of the milk. Then add the two yolks of eggs and beat well together. Add the milk, salt and pepper, the chopped clams, and lastly the stiffly-beaten whites of eggs, and add a trifle more flour, if necessary. Drop a couple of tablespoons at a time in a large fry-pan containing a couple of table-spoons of butter or drippings. They spread out about the size of a small saucer. Fry as many at a time as the pan will conveniently hold without running together. Turn when browned lightly on one side, and when the other side has cooked fold together and serve at once. Garnish with parsley. These are very easily made for lunch-eon, and are very nice served with fried chicken.

DEVILED EGGS

Boil half a dozen eggs until hard. Remove shells, cut in halves, mash the yolks to a smooth paste with about ½ teaspoon mixed mustard, 1 teaspoon softened butter, pepper and salt to taste. Some like a small quantity of cold boiled minced ham added. When in-gredients are well mixed, press enough of this mixture into the cup-shaped whites of eggs to form a rounding top. Serve on a platter of parsley. To boil eggs uniformly, they should be placed in a wire basket and plunged into boiling water and boiled not longer than 15 to 20 minutes from time water commences to boil, then pour cold water over and shell them.

EGGS IN CREAM SAUCE

Four eggs, boiled hard, cut in halves lengthwise, then across, each egg cut in four pieces. A cream sauce was made using ½ cups sweet milk, 1½ tablespoons flour, 1 generous tablespoon of butter, sea-soned with salt. After letting milk come to a boil and adding flour mixed smoothly with a little cold milk or water, add butter and cook until a thick creamy consistency, then add the quartered eggs to sauce. Stand a few minutes until heated through. Pour the creamed eggs over four or five slices of nicely-toasted bread. Sprinkle a little finely-chopped parsley and a pinch of pepper over top and serve at once. This is a delicious and quickly prepared luncheon dish.

A very wholesome and digestible way to prepare an egg is to put yolk and white of a fresh egg together in a bowl, beat lightly, pour over the egg a pint of rich milk, which has been heated to the boiling point. Add a pinch of salt. Stir constantly while slowly adding the milk. The hot milk should slightly cook the egg. Eat slowly with crackers or toasted bread.

AUNT SARAH'S METHOD OF PRESERVING EGGS WITH LIQUID WATER GLASS

Aunt Sarah for many years preserved eggs in water glass, or soluble glass, also known as "Sodium Silicate," a thick liquid about the consistency of molasses. It is not expensive and may easily be procured at any drug store. She used the water glass in the proportion of 10 quarts of water to one pint of the water glass. The water glass, although in liquid form, is usually sold by the pound, and 1½ pounds equals one pint. The water should always he boiled and allowed to cool before combining with the water glass.

She was particular to use none but perfectly clean, fresh eggs. She placed the eggs, narrow end down, in an earthenware crock which had been well scalded and cooled. When the water glass had been thoroughly mixed through the water she poured the mixture over the eggs in the crock.

A stronger solution might be used to preserve the eggs, but Aunt Sarah declared she used eggs for baking cake which were good at the expiration of a year, which had been preserved in a mixture of 10 quarts of water to a pint of water glass, and she considered this proportion perfectly reliable. So I do not see the need of using a large quantity of the water glass, although many recipes call for a mixture of one pint of water glass to only 8 quarts of water.

Fresh eggs may be added daily until the crock is filled, having the mixture at least one inch above the last layer of eggs. It is best not to wash the eggs before packing, as this removes the natural mucilaginous coating on the outside of the shell. Place clean, fresh eggs carefully into the crock containing the water glass and water, with a long-handled spoon to avoid cracking the shell. Stand the crock containing eggs in a cool place, cover with a cloth tied over top of crock, avoiding frequent change of temperature; they should keep one year. The water glass solution may become cloudy, and resemble a soft-soap mixture, but this is a natural condition and does not affect the eggs.

April is considered the best month for packing eggs. Infertile eggs are to be preferred to others. Carefully remove the eggs from the water glass mixture with a long-handled spoon when wanted to use, as the shells are sometimes not quite as hard as when placed in the crock. The eggs may be used for cooking, baking, in fact, for any purpose except soft-boiled but should you wish to boil them, a tiny puncture should be made in the shell of these eggs before boiling.

Ten quarts of water to one pint of water glass will cover about 12 or 13 dozen eggs.

TO TEST FRESH EGGS.

Place an egg in a tumbler, fill tumbler with cold water. If eggs are fresh they will remain in the bottom of tumbler. If not strictly fresh the egg will float on the top, or near the top of tumbler of water.

SALADS — AUNT SARAH'S SALAD DRESSING

For this she used 1 pint of sour cream, 1½ tablespoonfuls of flour, 1½ tablespoonfuls of mustard (pulverized dry mustard), 3 eggs, ¼ cup butter (or ¼ cup of olive oil may be used instead, if liked), ½ cup good sour vinegar, ½ teaspoonful of black pepper and a pinch of red pepper (cayenne), salt to taste, ½ teaspoonful of sugar. Place in a bowl the 1½ tablespoonfuls of flour with the same quantity of mustard; mix smoothly with a little of the sour cream. Then add the eggs, beaten in one at a time, or use, instead, the yolks of five eggs. When using the whites for angel cake or any white cake Aunt Sarah usually made salad dressing from the remaining yolks of eggs. Add the sour cream and vinegar, salt and pepper. Mix all well together and strain through a fine sieve and cook in a double boiler over hot water until a creamy consistency. Pour in glass jars. This dressing will keep well on ice or in a cool place for two weeks. If too thick, thin with a little vinegar, water or milk when using it. About ¾ of a cup of this dressing was used for mixing with 1 cup of the meat of cold, cooked chicken in making chicken salad. The white meat of chicken was cut in dice and ¾ cup of celery was also cut in small pieces, a couple of hard boiled eggs, cut in dice, were added and the whole was carefully mixed with the salad dressing. Cold boiled veal or pork may be used instead of chicken for salad. Potato salad was sometimes prepared by using a small quantity of this dressing, adding, also, minced onion, parsley and celery. Hot slaw was prepared by heating a couple of tablespoonfuls of the salad dressing and mixing with shredded cabbage. Or use as a dressing for lettuce when not served "Au Natural" with olive oil and vinegar at the table.

Should very *thick*, sour cream be used in making "Aunt Sarah's salad dressing," use a mixture of sour cream and sweet milk, instead of all sour cream.

"DUTCH" CUCUMBER SALAD

Thinly slice one large green cucumber and one medium-sized onion (if liked). Sprinkle over about one teaspoonful of salt. Allow to stand a short time, then place in a piece of cheese-cloth and squeeze out all the moisture possible. Place cucumbers, when drained, in the dish in which they are to be served, add a couple tablespoonfuls of sour vinegar, mix well. Then pour over enough thick sour cream to half cover and a dust of pepper. Cucumbers are considered less unwholesome, prepared in this manner.

CARROT SALAD

Aunt Sarah pared and cut 1½ cups of uncooked carrots in thin strips, not much larger than common match sticks, and cooked in salted water until tender. When drained, pour over them a couple of tablespoonfuls of vinegar. Allowed to stand until cold. When ready to prepare the salad she drained off vinegar remaining. Lined a salad bowl with lettuce leaves or parsley, placed inside this a border of halved or sliced cold hard-boiled eggs; mixed the carrots lightly with salad dressing, placed them in the centre of the bowl and served ice cold. This is a particularly delicious, as well as an appetizing looking, salad. I have never eaten this elsewhere than at Aunt Sarah's home.

"AN OLD RECIPE" FOR CHICKEN SALAD

Two dressed chickens were cooked tender. When cold, meat was removed from bones and cut in dice (not too fine). Cut half the amount of celery you have of meat into small pieces.

Dressing for salad was composed of the following: Three well-beaten yolks of eggs. Pour over these 1 pint of boiling hot cider vinegar, stand on back of range to thicken. Place in a bowl 3 freshly boiled and finely mashed white potatoes, add 1 tablespoonful of dry mustard, 6 teaspoonfuls of olive oil, 1 tablespoonful of salt, 1 tablespoonful of pepper. Mix all well together, then add the thickened vinegar. Beat together until creamy and stand aside until chilled.

Drop the three whites of eggs in hot water, remove when cooked, chop fine and when cold add to the chicken meat and celery.

Pour the dressing over all the ingredients, stir lightly with a fork and stand in a cold place until chilled before serving.

GERMAN POTATO SALAD

Boil one dozen small potatoes without paring. Remove the skin and cut potatoes size of dice, also a small onion, finely minced. Put small pieces of bacon in a pan and fry brown and crisp. Add a large tablespoonful of vinegar and a pinch of salt. Pour the hot bacon fat and vinegar over the diced potatoes, toss them up lightly with a fork and serve hot.

GERMAN TURNIP SALAD

This is the manner in which Aunt Sarah made turnip salad: She pared and sliced thin on a slaw cutter 5 large, solid turnips, put them in a stew-pan which she placed on the range, adding about ¼ cup hot water, 1 teaspoonful of butter and ¼ teaspoonful of sugar (no more). She covered the stew-pan closely and steamed about half an hour until the turnips were tender. Then mixed together 1 teaspoonful of flour with 1 tablespoonful of vinegar and yolk of one egg. This was poured over the stewed turnips, just allowed to come to a boil, then removed from the fire. Add a little salt and serve hot.

GERMAN SALAD DRESSING

For dandelion, watercress, endive or lettuce, a dressing was made thus: The leaves of vegetables used for salad, after being carefully rinsed and looked over, were cut fine, and the following dressing poured over hot and served at once.

A small quantity of bacon was finely minced and fried crisp. To about 2 tablespoonfuls of bacon and fat after being fried, 3 tablespoonfuls of vinegar and 1 of sour cream, were added pepper and salt and a very little flour mixed with cold water, to make it the consistency of cream. The yolk of one raw egg may be added to the dressing if liked. An easier way for the busy housewife to do is to simply add a couple of tablespoonfuls of Aunt Sarah's Salad Dressing, add also a small quantity of water, flour and fried, diced bacon; serve hot at once.

MARY'S POTATO SALAD

A bowl of cold, boiled, diced or thinly-sliced potatoes, three hard boiled eggs, also diced, and about half the quantity of celery chopped in half-inch pieces, and a little minced onion, just enough to give a suspicion of its presence. She mixed all together lightly with a silver fork and mixed through some of the following salad dressing, which is fine for anything requiring a cold salad dressing.

MARY'S SALAD DRESSING

One tablespoonful of flour, 1 tablespoonful of mustard, 2 cups of sweet or sour cream, 1 tablespoonful of sugar, ½ cup of good sharp vinegar, yolks of four eggs, small teaspoonful of salt. Omit sugar when using the dressing for potato or chicken salad. This salad dressing may also be used for lettuce.

"FRUIT" SALAD DRESSING

Three tablespoonfuls of olive oil to 1½ tablespoonfuls of vinegar. Season with salt and pepper. Use this quantity for 1 pint of salad.

GRAPE FRUIT SALAD

Cut the pulp from one grape fruit into small pieces, add an equal amount of chopped apples, a few English walnuts chopped coarsely. Serve on lettuce leaves with fruit salad dressing. This recipe was given Mary by a friend who knew her liking for olive oil.

Grape fruit is delicious, served cut in halves with the addition to each half; of a couple tablespoonfuls of pineapple juice, a tablespoonful of orange juice or tiny pieces of orange pulp, topped with a marachino cherry. A small quantity of sugar should have been added. The sections of grape fruit should each have been cut loose from the white skin inclosing pulp with a small knife or scissors.

A GOOD, INEXPENSIVE SALAD DRESSING

- 1 tablespoonful flour.
- 1 tablespoonful butter.
- 1 tablespoonful mustard.
- ½ tablespoonful sugar.
- 1 teaspoonful salt.
- 1 egg.
- ¾ cup milk.
- ¾ cup vinegar.

Use a double boiler, put in it the first five articles, stir together until smooth; add the well-beaten egg and the milk. Let cook, stirring hard. Then add vinegar, and beat all with an egg-beater until the mixture is smooth and creamy. Let cool before using.

Aunt Sarah frequently used this salad dressing over sliced, cold, hard boiled eggs when other salad materials were not plentiful. Serve on lettuce leaves.

IMITATION LOBSTER SALAD

A bowl was lined with crisp lettuce leaves, over this was spread a layer of cold boiled potatoes, cut in dice, a little finely minced onion, a layer of chopped celery, another layer of diced potatoes, then a layer of sliced tomatoes and one hard boiled egg, thinly sliced. Pour a good salad dressing over and serve ice cold.

"GERMAN" HORSERADISH SAUCE

A sauce to serve with boiled meat was prepared by Aunt Sarah in the following manner: She put half a cup of milk in a stew-pan, let come to a boil, added one large tablespoonful of cracker crumbs, 1 large teaspoonful of butter, 2 large tablespoonfuls of freshly grated horseradish, seasoned with pepper and salt. Also a pinch of salt, sugar and pepper added to grated horseradish, then thinned with vinegar, is an excellent accompaniment to cold meat.

MAYONNAISE DRESSING IN WHICH OLIVE OIL IS USED

Before making this dressing for salads, Mary placed a large soup plate or a shallow bowl in the refrigerator, also a bottle of olive oil and two egg yolks. All should be quite cold. Put the yolks on the cold plate, add ¼ teaspoonful of salt, the same of mustard. Mix well and then, with a fork, stir or blend the olive oil into it drop by drop. After about ½ cup of oil has been blended in, add lemon juice, a drop or two at a time. Then more oil, and when it becomes very thick add more lemon juice. A pint or even more oil may, with care, be blended into two yolks. Care must be taken not to mix oil in too fast, or the egg and oil will separate, making a mixture resembling curdled custard. If this should happen, take another plate, another egg yolk, and begin over again, blending a drop or two at a time in the curdled mixture. Then add more oil and lemon juice as before.

MUSTARD DRESSING TO SERVE WITH SLICED TOMATOES

Two tablespoonfuls mustard, 1 tablespoonful of sugar, ½ cup cream, 1 tablespoon salt, yolks of two eggs and ½ cup of vinegar. Beat all well together, first mixing the mustard until smooth with a small quantity of cream, then add the other ingredients. (Mary used only 1 tablespoonful of mustard, and substituted 1 tablespoonful of flour instead of the second tablespoonful of mustard and thought it improved the dressing.) This mustard dressing may also be served at table, to be eaten with lettuce.

CHICKEN SALAD

The meat of one boiled chicken cut in small pieces, three-fourths as much celery, also cut in small pieces. Three hard boiled eggs cut in dice. Take 2 teaspoonfuls salt, 2 teaspoonfuls pepper, 4 teaspoonfuls mustard, 1 cup of sweet cream and 1 raw egg. Use vinegar to thin the mustard. Beat the raw egg, add to cream, egg and butter (mash yolks of hard boiled eggs and butter together). Mix all the ingredients together and cook until it thickens (all except chicken meat, celery and hard boiled whites of eggs, which should be placed in a large bowl after cutting in small pieces). The salad dressing should he put in another bowl and stood on ice until cold, then mix the salad dressing carefully through the chicken meat, celery, etc., one hour before using. Cover with a plate until ready to serve. Or "Aunt Sarah's Salad Dressing" could be used over the chicken, celery, etc. This is a very old but an excellent recipe used by Aunt Sarah's mother for many years.

PEPPER HASH

Chop fine with a knife, but do not shred with a slaw cutter, 1 pint of finely chopped cabbage, adding 1 teaspoonful of salt, 2 teaspoonfuls of sugar, 1 teaspoonful of whole mustard seed, ½ a chopped red, sweet pepper, a pinch of red cayenne pepper and ½ pint of vinegar. Mix all well together and serve with fried oysters, oyster stew and deviled oysters.

This "pepper hash" is delicious if a couple tablespoonfuls of thick cream be added just before serving.

Should very sour cider vinegar be used in this recipe, the housewife will, of course, dilute it with water.

GERMAN BEAN SALAD

Use small green or yellow string-beans, which snap when broken, called by some "snap beans." String them carefully. (If quite small and tender this should not be necessary.) Rub well with the hands through several waters. This removes the strong bean taste. Have your kettle half filled with boiling water on the range over a brisk fire. Put a tablespoon of butter in the water, add beans by handfuls until all are in and cook until tender. Turn the beans in a colander to drain. When cool add a chopped onion, salt and pour enough good vinegar over to cover, and allow to stand two days, when strain vinegar from beans. Boil vinegar, add water if vinegar is quite sour and pour hot over the beans. Fill quart glass jars with the beans and pour vinegar over, within an inch of top of jar; pour pure olive oil over top of beans, screw on jar covers tightly and stand in a cool place until wanted to use. In the winter, when fresh salads were scarce, Aunt Sarah opened a can of these beans. If they were very sour she poured cold water over, allowed to stand an hour, drained and added a little fresh olive oil. Every one called her "bean salat," as the Pennsylvania Germans call it, delicious. The instructions regarding the preparing and cooking of string beans for salad will answer for beans used as a vegetable, omitting vinegar, of course. There is a great difference in the manner of cooking vegetables. Aunt Sarah always added an onion and a sprig of parsley when cooking beans to serve as a vegetable.

MEAT SALADS

To quote from the *Farmers' Bulletin*: "Whether meat salads are economical or not depends upon the way in which the materials are utilized. If in chicken salad, for example, only the white meat of chicken, especially bought for the purpose, and only the expensive inside stems of expensive celery are used, it can hardly be cheaper than plain chicken. But, if portions of meat left over from a previous serving are mixed with celery grown at home, they certainly make an economical dish, and one very acceptable to most persons. Cold roast pork or tender veal, in fact, any white meat, can be utilized in the same way. Apples cut into cubes may be substituted for part of the celery. Many cooks consider that with the apple the salad takes the dressing better than with the celery alone. Many also prefer to marinate (*i.e.*, mix with a little oil and vinegar) the meat and celery or celery and apples before putting on the final dressing, which may be either mayonnaise or a good boiled dressing."

Celery should not be allowed to stand in water. To keep fresh until used it should be wrapped in a piece of damp cheese-cloth and placed in an ice box or cool cellar.

Lettuce should be broken apart, carefully rinsed, and put loosely in a piece of damp cheese-cloth and placed on ice to crisp before using.

BEVERAGES — COFFEE

Scald coffee pot well before using (never use metal). Place in it five tablespoons ground coffee. (A good coffee is made from a mixture of two-thirds Java to one-third Mocha.) Beat up with the ground coffee one whole egg. Should the housewife deem this extravagant, use only the white of one egg, or peel off the white skin lining inside of egg shells and use. Add three tablespoons cold water and mix well together. Stand on range to heat; when hot add one quart of *freshly-boiled* hot water. Allow coffee to boil to top of coffee pot three times (about eight minutes), pour over one tablespoon cold water to settle. Stand a few minutes where it will keep hot, not boil. Place a generous tablespoon of sweet thick cream in each cup and pour coffee through a strainer over it. Always serve hot.

A larger or smaller amount of coffee may be used, as different brands of coffee vary in strength and individual tastes differ, but five tablespoons of coffee, not too coarsely ground and not pulverized, to one quart of water, will be the correct proportions for good coffee. Use cream and you will have a delicious, rich, brown beverage not possible when milk is used. Better coffee may be made if whole grains of roasted coffee be bought, reheated in oven and freshly ground whenever used, rather finely ground but not pulverized. Coffee, when ground for any length of time, loses strength. If coffee is ground when purchased, always keep it in closely covered cans until used. Or buy green coffee berries and roast them in oven;

when coffee has been roasted, stir one whole raw egg through the coffee berries; when dry, place in covered cans, then no egg will be needed when preparing coffee. As a substitute for cream, use yolk of fresh egg mixed with a couple tablespoonfuls of milk.

COCOA

Mix four tablespoonfuls of cocoa to a smooth paste with one cup of boiling water. Add one more cup boiling water and boil fifteen or twenty minutes. Add four tablespoonfuls of sugar, then add 4 cups of hot boiled milk. A few drops of essence of vanilla improves the flavor. Add a couple tablespoonfuls whipped cream on top of each cup when serving, or, instead of cream, place a marshmallow in each cup before pouring in cocoa. This quantity is for six cups of cocoa.

CHOCOLATE

One square of Baker's unsweetened chocolate shaved thinly or grated, mixed to a smooth paste with 1 cup of boiling water. Boil from fifteen to twenty minutes. Add 1 cup of boiling milk and 2 even tablespoonfuls of sugar. Flavor with a few drops of vanilla, if liked, and add whipped cream to each cup when serving. This is for 2 cups of chocolate.

BOILED WATER

It sometimes becomes necessary to boil drinking water, which usually has a flat, insipid taste. Do young housewives know it is said that after water has been boiled and when quite cool if a bottle be half filled and shaken well the water will become aerated, and have the taste of fresh spring water?

TEA

To make tea always scald the teapot, which should be agate, earthenware or china, never metal. Always use water that has been *freshly* boiled, and use it boiling hot. Never, under any circumstances, boil tea, as tannin is then extracted from the leaves, and the tea will have a bitter taste. Do not allow tea to stand any length of time unless strained from tea leaves. Use one teaspoon of tea for each cup, unless liked stronger, when add one extra teaspoon to each three cups of tea. Some contend that tea is better, if at first a small quantity of boiling water is poured over the leaves, allowing it to steep three minutes — then pour over the remaining quantity of boiling water and let stand about four minutes, when it is ready to serve with cream and sugar, if liked. Should any tea remain after serving do not throw away, but strain at once from tea leaves and when cool place in a glass jar in refrigerator to be used as iced tea.

ICED TEA

For two quarts of delicious iced tea, place in an agate teapot one generous tablespoon of good tea (never buy a cheap, inferior grade of tea). Pour over the tea leaves one quart of freshly boiled, scalding hot water; let stand five minutes, keep hot (not boil), strain from the leaves into a pitcher, then pour over the tea leaves another quart of hot water, allow it to stand a few minutes, then strain as before. Add the juice of one lemon and sugar to taste. When cooled stand on ice and add chipped ice to tumblers when serving.

PUDDINGS

To boil a pudding in a bag, dip the bag, which should be made of thick cotton or linen, in hot water, dredge the inside well with flour before putting batter into the bag. When the pudding has boiled a long enough time, dip the bag quickly in cold water, and the pudding will turn out easily. Allow five large eggs to 1 quart of milk usually to make custard solid enough to keep its shape when turned from the mold. One teaspoonful of extract will flavor one quart. Always stand individual cups in a pan partly filled with hot water. Place pan containing custard cups in a moderate oven and bake slowly forty minutes. Always sift flour over beef suet when chopping it to be used in puddings. Pour boiling water over Pecans (nuts), allow to stand several hours. When cracked, the shell may be easily removed, leaving the nuts whole.

Blanch almonds by pouring boiling water over them. Allow them to stand a short time, when the brown skin may be easily removed. Dry thoroughly by standing in a rather cool oven, then put in glass jars and they are ready to use. Almonds are used particularly by the Germans in various ways. One hausfrau adds chopped almonds to cooked oatmeal for her children's breakfast and they are frequently used as an ingredient; also to decorate the tops of raised cakes. When dried currants and raisins are bought by the frugal housewife they are quickly washed in cold water, carefully picked over, then turned on to a sieve to drain. Raisins are seeded, then spread over pans, placed in a warm oven about 15 minutes, then spread on a plate and allowed to stand in a dry place for several days. When thoroughly dried place in glass jars and stand aside until required. Currants or raisins should always be well floured before adding to cake or pudding. The "German hausfrau" usually serves stewed prunes or raisins with a dish of noodles or macaroni.

RICE PUDDING

One of the simplest and cheapest of desserts depends partly on the quality of the ingredients used, but chiefly on the manner of making for its excellence. If prepared according to directions, you will have a pudding both rich and creamy. Use 1 quart of good sweet milk (do not use either skimmed milk or water), 3 tablespoonfuls of whole uncoated rice (no more), 2½ tablespoonfuls of sugar, pinch of salt, vanilla or almond flavoring.

Wash the rice well, mix all together in a pudding dish, bake from 2½ to 3 hours in an oven with a slow, even heat. When a skin forms on the top of the pudding, carefully stir through the rice. Do this frequently. This gives the pudding a rich, creamy consistency. When grains of rice are tender allow pudding to brown over top and serve either hot or cold. Raisins may be added, if liked, or raisins may be stewed separately and served with the rice, which many think a great improvement to the pudding. Many think rice pudding should always be flavored with grated nutmeg. Aunt Sarah, while using nutmeg flavoring in various other dishes, never used it for her rice pudding.

When mixing a boiled pudding Aunt Sarah frequently substituted a large tablespoon of fine dried bread crumbs instead of the same amount of flour. She said, "'Twas a small economy," and, she thought, "the pudding's improved" by the use of bread crumbs.

FRAU SCHMIDT'S APPLE DUMPLINGS

Prepare a syrup of 1 cup sugar, 2 cups of hot water and 1 tablespoon of butter. Pour all into an agate pudding dish. Add to this syrup 2 heaping cups of pared, sliced sour apples.

Let all come to a boil. For the dumplings, sift together one cup of flour and two even teaspoons of baking powder. Add a pinch of salt. Mix into a soft dough or batter with about ¾ cup of sweet milk or cream. Drop six or eight spoonfuls of this batter into the boiling syrup on top of apples. Cover closely and cook on top of range twenty minutes without uncovering. Serve hot. These dumplings should be light as puff balls. Peaches may be substituted for apples and are delicious.

CARAMEL CUSTARD (AS MARY PREPARED IT)

- 1 pint of milk.
- 2 eggs.
- ½ cup granulated sugar.

Melt ½ cup of sugar in an iron pan on stove and allow it to brown. Add a part of the hot milk, stirring constantly until brown sugar is dissolved. Add balance of the pint of hot milk. Stir all together, then stand aside to cool. When cold, add eggs and bake in oven in custard cups. Stand cups in hot water while baking.

AUNT SARAH'S BREAD PUDDING

Pour 1 quart of boiling milk over 1½ pints of soft bread crumbs. Put the mixture into a buttered pudding dish with 1 teaspoonful salt. Cover closely with a plate and let stand about half an hour. At the end of that time beat into it three eggs, 1 teaspoonful lemon extract, and beat until perfectly smooth. Bake in a moderately hot oven three-quarters of an hour. Serve with the following sauce: 6 tablespoonfuls pulverized sugar, 2 tablespoonfuls butter, 1 tablespoonful lemon juice. Beat all together to a cream; when it is ready to serve. No sugar is needed in this pudding if this sweet sauce is used.

STEAMED BREAD PUDDING

Place 1 cup of fine dried bread crumbs in a bowl. Pour over the crumbs 2 cups of milk and allow to stand a short time. Beat together 2 eggs and scant ½ cup sugar, add 1 tablespoon of butter. Mix all the ingredients together thoroughly; then add ½ cup of chopped raisins, which have been seeded and floured. Pour the batter in the well-buttered top part of a double boiler over hot water. Steam about 2½ to 3 hours. Serve hot with sauce used for cottage pudding, or serve with sugar and cream.

AN ECONOMICAL BREAD AND APPLE PUDDING

Into a well-buttered pudding dish put a layer of sliced sour apples. On the top of these a layer of stale bread crumbs with small bits of butter and sugar sprinkled over them, more sliced apples and bread crumbs, having the crumbs for the top layer. To about three apples use 1 cup of bread crumbs, ½ cup sugar, piece of butter size of walnut and bake in oven until apples are tender. Serve with cream.

CUP CUSTARDS

- 1 quart of sweet milk.
- 5 large eggs.
- 3 tablespoons sugar.
- Grated nutmeg or vanilla flavoring.

Scald milk. Beat whites of eggs separately. Add milk when cooled to the beaten yolks. Add sugar and flavoring. Stir in stiffly beaten whites of eggs, pour into custard cups, stand them in a dripping pan half filled with boiling water. Stand the pan in a moderate oven about twenty minutes, or until custard is "set." This quantity fills about eight small custard cups. The water surrounding the custard cups should not be allowed to boil, but the custard should cook slowly.

Grate nutmeg thickly over top of each custard before placing in the oven. Scalding the milk before using improves the custard.

FRAU SCHMIDT'S GRAHAM PUDDING

Sift into a bowl ¼ cup of pastry flour and 1 teaspoonful of baking powder. Add 1 cup Graham flour, pinch of salt and ½ cup granulated sugar. Mix all thoroughly, then add ½ cup of finely chopped kidney suet. Add 1 cup of seedless raisins mixed with one extra tablespoonful of white flour. Mix into a batter with 1 cup of sweet milk, to which add yolk of one egg. Lastly, add the stiffly beaten white of egg. Flavor with either a little grated nutmeg or essence of vanilla.

Make a strong, unbleached muslin bag 7 by 12 inches. Pour the batter into the bag, which had been previously dipped in cold water, the inside of the bag sifted over with flour, and tie bag at top with a string, allowing room for the pudding to swell. Place the bag in the perforated compartment of a steamer, over boiling water, and boil continuously 1½ hours, or longer, without removing lid of steamer oftener than absolutely necessary.

Serve Graham Pudding hot with sauce used for "cottage pudding," or serve simply with sugar and cream, or a sauce may be served composed of ½ cup of pulverized sugar, creamed with ¼ cup of butter. Add 1 tablespoonful of lemon juice or flavor with vanilla. Stand sauce in a cool place a short time and serve cold on hot pudding.

SPONGE BREAD PUDDING

Place 1¾ cups of soft stale (either white or graham) bread crumbs in a pudding dish. Pour 2 cups of hot milk over the crumbs, cover with a plate and allow it to stand about thirty minutes, then add yolks of 2 eggs, ½ teaspoonful of salt, 1 tablespoonful of sugar and grated yellow rind of orange or lemon for flavoring. Beat the mixture until perfectly smooth, add the stiffly beaten whites of two eggs. Bake in a moderately hot oven. Serve hot with the following sauce:

SAUCE.

Three large tablespoonfuls of pulverized sugar and 1 tablespoonful of butter were beaten together until smooth and creamy, 1 teaspoonful of lemon juice was added. The sauce, when quite cold, was served with the warm pudding.

AUNT SARAH'S COTTAGE PUDDING

Cream together 1 cup of sugar, 2 tablespoonfuls of butter, 1 egg, white beaten separately, and added last, 1 cup of sweet milk, pinch of salt, 2 cups of flour, sifted with 2 heaping teaspoonfuls of Royal baking powder, ½ cup of dried currants, well floured. Add stiffly beaten white of egg. Bake in a small oblong bread pan.

SAUCE.

One cup of milk, ½ cup of water, 1 large teaspoonful of butter, a scant tablespoonful of flour moistened with a small quantity of water, before adding. Sweeten to taste, add ½ teaspoonful of grated nutmeg. Cook all together a few minutes, allow the mixture to partly cool, then stir in the yolk of one egg; stand on stove to heat, but not to cook. Serve hot over freshly baked, warm cottage pudding, cut in squares.

APPLE "STRUDEL"

Aunt Sarah pared and quartered six medium-sized tart apples, placed in the bottom of an agate pudding dish, poured over them one cup of hot water and 2 tablespoonfuls of sugar. She allowed this to stand on the range and cook while she mixed the following dough.

Into a bowl she sifted 1 pint of flour with 2 teaspoons baking powder, one teaspoonful of sugar, a little salt. Cut 1 tablespoonful of butter through the flour. Lightly mixed all together into a soft dough with about ¾ cup sweet milk. Should she have a left-over yolk of egg, that was added to the milk. She rolled dough out lightly on the bread board, cut vents in the crust to allow steam to escape and spread it over the top of the dish containing the hot apples; placed in a hot oven to bake until light brown on top. Serve with sugar and cream.

Aunt Sarah called this "Apple Strudel," but the German recipe for "Apple Strudel," handed down by her Grandmother, was quite different. An ordinary noodle dough was made, placed on a clean cloth on the table and rolled as thin as tissue paper. Small bits of butter were scattered over this, covered with tart apples, thinly sliced, sprinkled with cinnamon, sugar and chopped raisins, rolled up and baked in the oven until brown on top, basting frequently with a thin syrup composed of sugar, butter and water.

"LEMON MERINGUE" PUDDING

- 1 pint of milk.
- ½ cup of sugar
- 1 cup bread crumbs.
- Juice and grated rind of one lemon.
- 2 eggs.
- ⅓ cup of butter.

3 tablespoonfuls of pulverized sugar used for top. Soak the bread crumbs in milk. Beat the butter and sugar together. Add yolks of eggs, soaked bread crumbs and grated lemon rind and about ¾ of the juice of the lemon. Bake in a buttered pudding dish until firm, then cover the pudding with a meringue composed of the stiffly beaten whites of eggs, 3 tablespoonfuls of pulverized sugar and the remaining lemon juice. Place in oven to brown. Stand on ice; serve cold.

SUET PUDDING

- 1 cup suet, chopped fine.
- 1 cup sugar.
- 1 cup sweet milk.
- 2 eggs.
- 1 teaspoonful cinnamon.
- 1 cup raisins.
- 1 cup currants.

- 3 cups flour sifted with 2 teaspoonfuls baking powder.

Steam 2½ hours, then place in oven two or three minutes. This quantity will partly fill three empty 1-pound baking powder cans; allowing room to swell. These puddings are equally as good as when freshly prepared if placed in a steamer a short time before serving until heated through.

SAUCE FOR SUET PUDDING.

One cup of pulverized sugar and 1 large tablespoonful of butter creamed together. One teaspoonful of vanilla. Add one whole egg or the yolks of two eggs, or the whites of two eggs, whichever you happen to have.

STEAMED FRUIT PUDDING

- 1 cup sweet milk.
- 1 cup chopped suet.
- 1 cup molasses.
- 1 cup raisins.
- 1 teaspoonful soda dissolved in a little water.
- 1 teaspoonful salt.

SAUCE FOR PUDDING.

A small quantity of cinnamon, allspice, nutmeg, and a *very little* clove. Flour to make a batter a little thicker than that of ordinary cake. Steam about 3 hours. This pudding is also inexpensive and equally as good as the former recipe.

Beat 1 egg very light, add 1 cup brown sugar, 1 teaspoonful vanilla. Beat all together until creamy. Serve at once.

CORNMEAL PUDDING

Scald 1 quart of sweet milk. While hot stir in 3 tablespoonfuls of cornmeal, 3 tablespoonfuls of flour mixed smooth with a little cold milk. Add 1 tablespoonful of butter. Let cool. Then add to the mixture ½ cup sugar, ¼ cup molasses, 1 well-beaten egg, ½ teaspoonful of ginger, ½ teaspoonful cinnamon, ¼ pint cold milk, a small pinch of soda and ½ cup of floured, seeded raisins. Bake 2 hours in a moderate oven. Serve with sugar and cream.

HUCKLEBERRY PUDDING

Two eggs and 1 small cup of granulated sugar creamed together. Four tablespoonfuls of cold water. Add 1 cup of sifted flour containing 1 teaspoonful of baking powder, and 1 cup of huckleberries, pitted cherries, or raisins and bake. Serve with milk or any sauce liked. This recipe was given Mary by a friend, who called it her emergency pudding, as it may be easily and quickly prepared from canned sour cherries from which liquid has been drained, or any tart fruit, when fresh fruit is not in season.

TAPIOCA CUSTARD

Four tablespoonfuls of pearl tapioca soaked in cold water over night. The next morning drain the tapioca, boil 1 quart of sweet milk, beat the yolks of 4 eggs light, stir them into the tapioca, adding 4 tablespoonfuls of sugar. Beat all together and gradually add the hot milk. Return to the fire and stir until it commences to boil. Take from the range and pour in a glass dish. Flavor with 1 teaspoonful of vanilla. Whip the whites of the eggs to a standing froth and stir into the cooling pudding When cold stand on ice until ready to serve. One-half cup of shredded cocoanut may be added if liked.

DELICIOUS BAKED PEACH PUDDING

For the dough place in a bowl 1 pint of flour sifted with 2 teaspoonfuls of Royal baking powder and a pinch of salt. Cut through this a scant ½ cup of butter. Mix this with sufficient sweet milk to make a soft dough. Roll out dough half an inch thick, cut in strips and in case whole, ripe, pared peaches, leaving top and bottom of the peach exposed. Or solid canned peaches may be used. Put two halves of peach together and place a strip of dough around the peach. Pinch dough well together, place in a bake dish. Prepare a syrup of 2 cups of sugar and 1 cup of water. Let come to a boil, pour around the dumplings and bake a half hour in a moderately hot oven. These are delicious. The recipe was given Mary by a friend who was an excellent cook. From this dough may also be baked excellent biscuits.

CARAMEL CUSTARD

Place 1 pint of milk on the range in a double boiler. Melt half a cup of sugar in an iron pan over the fire until a golden brown. When melted add four tablespoonfuls of boiling water. Allow mixture to cook one minute, then add it to the milk. Remove from the fire and add 1 teaspoonful of vanilla. When cool stir in 4 well-beaten eggs with 2 tablespoonfuls of sugar. Pour the mixture in a small pudding dish. Stand in a pan of boiling water, place in oven to bake until a jelly-like consistency. When cooled serve plain or with whipped cream.

"AUNT SARAH'S" RHUBARB PUDDING

Remove skin from stalks of rhubarb, wash and cut into half-inch pieces a sufficient quantity to half fill a medium-sized agate or earthenware pudding dish. Place in a stew-pan on range, cook slowly with a couple tablespoons of sugar and a very small amount of water. Sift together in a bowl 1 pint of flour, 1½ teaspoons of baking powder and a pinch of salt. With a knife cut through the flour 2 tablespoonfuls of butter, moisten with one beaten egg and sufficient milk added to make a soft dough or batter. Drop table-spoons of this thick batter over top of dish containing hot stewed rhubarb. Place at once in a hot oven, bake quickly until crust is a light brown. Serve on individual dishes, placing over each a couple tablespoonfuls of the following sauce. The combined flavor of rhubarb and vanilla is delicious.

VANILLA SAUCE FOR RHUBARB PUDDING.

Beat 1 egg very light, add 1 cup of light brown sugar and 1 tea-spoon of vanilla flavoring. Beat all together until foamy. Serve at once, cold, on the hot pudding.

RICE CUSTARD

Add 1 cup of cold boiled rice to 2 cups of sweet milk, mix togeth-er slowly. Add ¼ cup sugar, the well-beaten yolks of 2 eggs, let all cook together a few minutes. Remove custard from the fire and pour over the stiffly-beaten whites of two eggs. Beat well with an egg-beater. Place in a glass dish and serve cold.

MARY'S CUP PUDDING (FROM STALE BREAD)

One quart of finely *crumbled stale bread* (not dried crumbs). Fill buttered cups two-thirds full of crumbs and pour over the following custard, composed of one pint of milk and three eggs. Allow to stand a few minutes, then place the cups in a pan partly filled with hot water, place the pan in a moderately hot oven and bake thirty minutes. No sugar is required in this pudding if the following sweet sauce be served with it:

SAUCE FOR PUDDING.

Mix one tablespoonful of cornstarch with a half cup of sugar. Pour over one cup of boiling water, add one generous teaspoonful of butter. Cook all together until clear, take from the fire and add one well-beaten egg and one teaspoonful of vanilla. Serve hot.

"BUCKWHEAT MINUTE" PUDDING

Pour three cups of milk in a stew-pan, place on range and let come to a boil. Then stir slowly into the boiling milk 1¼ cups of buckwheat flour and ¼ teaspoonful of salt. Keep stirring constantly until a thick mush. Serve at once with sugar and cream. I have never eaten this pudding anywhere except in "Bucks County." It is cheap, quickly and easily prepared and well liked by many country folk in Bucks County.

PEACH TAPIOCA

One cup of tapioca soaked in 1 quart of cold water several hours. Place in stew-pan, set on stove and cook until clear. Add sugar to taste and 1 pint can of peaches. Boil two or three minutes, remove from range and pour into the dish in which it is to be served. Stand aside to cool.

AUNT SARAH'S PLAIN BOILED PUDDING

One cup of beef suet chopped fine or run through a food-chopper, ½ cup sour milk, 1 egg, 1 teaspoonful soda, pinch of salt. ½ cup sugar, 1 teaspoonful cinnamon, 1 cup raisins, seeded and floured. Flour enough to make as stiff as ordinary cake batter. Boil or steam in a muslin bag three hours. This is a very inexpensive and good pudding. Dust a small quantity of flour over suet before chopping. Serve with the following sauce:

PUDDING SAUCE.

One large tablespoonful of butter, 1 teacup water, ½ teacup milk, scant tablespoonful of flour, grated nutmeg to flavor. Sweeten to taste, add a pinch of salt. Cook and let cool. Beat up yolk of egg, add to sauce, stand on back of stove to heat, not cook. Serve hot over the pudding.

APPLE TAPIOCA

Pour 1 pint of cold water over ½ cup tapioca. Allow to stand until the following morning, when cook until clean. Slice 6 tart apples. Place in bottom of pudding dish, strew sugar over, then pour over the tapioca; place over this a layer of thinly sliced apples over which dust sugar. Place in oven and bake until the apples are cooked. Serve with sugar and cream. Several thin slices of lemon added before baking impart a fine flavor.

STEAMED WALNUT PUDDING

Place in a bowl ½ cup butter and 1 cup of granulated sugar. Beat to a cream. Add yolks of 2 eggs and ½ cup of syrup molasses or maple syrup, in which had been dissolved 1 teaspoonful baking soda. Then add 1 cup sweet milk, alternately, with about 3½ cups flour, ½ cup of walnut meats, run through food-chopper or crushed with rolling pin, ¾ cup of seeded raisins, ½ teaspoonful ground cinnamon, ½ teaspoonful grated nutmeg, ¼ teaspoonful ground cloves, a pinch of salt and the stiffly beaten whites of the two eggs.

The batter should be placed in two empty one-pound tin coffee cans, about two-thirds full, covered tightly with lid and placed in a pot of boiling water which should be kept boiling constantly for three hours; when steamed the pudding should almost fill the cans.

If the cans were well buttered and flour sifted over, the pudding when steamed may be easily removed to a platter. Slice and serve hot with the following sauce:

Beat one cup of pulverized sugar to a cream with 2 heaping table-spoonfuls of butter. Add white of one egg (unbeaten). Beat all together until creamy. Add ¾ of a teaspoonful of lemon extract and stand sauce in a cold place or on ice one hour before serving on slices of hot pudding. This is a delicious pudding.

"CORNMEAL SPONGE" PUDDING

Crumble cold corn muffins, or corn cake, a quantity sufficient to fill two cups. Soak in 1 quart of sweet milk three or four hours, then add 3 well-beaten eggs, 3 tablespoonfuls of sugar and a pinch of salt. Beat all well together. Place in a pan and bake 1 hour in a moderately hot oven. Serve hot with whipped cream and sugar or with a sauce made by beating to a cream a heaping tablespoonful of butter, 1 cup of granulated sugar, 1 egg and a very little vanilla flavoring.

MARY'S CORN STARCH PUDDING

- 1½ quarts of milk.
- 5 eggs.
- 2 heaping tablespoonfuls of corn starch.
- 1 scant cup of sugar.
- 1 teaspoonful of vanilla.

Pour milk in a double boiler and place on range to cook. Moisten cornstarch with a little cold milk and add to remainder of the milk when boiling hot. Stir thoroughly, then beat yolk of eggs and sugar until light, stir in stiffly beaten whites and when all are mixed stir into the scalding milk. Let come to a boil again and add vanilla or almond flavoring. Pour into individual molds to cool. Serve cold with a spoonful of jelly or preserved strawberry with each serving.

APPLE JOHNNY CAKE (SERVED AS A PUDDING)

This is a good, cheap, wholesome pudding.

- 1 cup corn meal.
- 2 tablespoonfuls of sugar.
- 1 teaspoonful of soda.
- 1 tablespoonful of melted butter.
- ¼ teaspoonful of salt.
- ⅔ cup flour.
- 1 cup sour milk.

Mix batter together as you would for cake, then add 4 pared, thinly sliced, tart apples to the batter. Stir all together. Bake in a quick oven in a bread pan and serve hot with cold cream and sugar. Raisins may be substituted for apples if preferred.

A GOOD AND CHEAP "TAPIOCA PUDDING"

Soak over night in cold water 3 even tablespoonfuls of pearl tapioca. In the morning add tapioca to one quart of milk, 3 tablespoonfuls of sugar, a pinch of salt. Grate nutmeg over top. Bake in a moderate oven about three hours, stirring occasionally.

"GOTTERSPEISE"

Partly fill an earthenware pudding dish with pieces of sponge cake or small cakes called "Lady Fingers;" cut up with them a few macaroons. Place one pint of wine over fire to heat, add to the wine the following mixture, composed of 1 spoonful of cornstarch mixed smooth with a little water, 3 yolks of eggs and 3 spoonfuls of sugar. Mix all together and stir until thickened. Pour the thickened mixture over the cake. When cooled cover with the stiffly-beaten whites of the 3 eggs, spread sliced almonds thickly over top and brown in oven a few minutes. Serve cold.

SPANISH CREAM

Half a box of Knox gelatine, 1 quart of milk, 4 eggs. Put gelatine in milk, let stand 1 hour to dissolve. Set over fire to boil, then add beaten yolks of eggs with 1 cup granulated sugar. Remove from fire while adding this. Stir well. Return to range and let boil. Stand aside to cool. Beat whites of eggs to a froth and beat into custard when cooled. Pour into a glass dish in which it is to be served. Stand in a cold place and serve with cream.

GRAHAM PUDDING

One cup of molasses, 1 egg, 1 cup sweet milk, ½ teaspoonful soda, 1 teaspoonful of salt, 1 tablespoonful brown sugar, 1 cup raisins, 2½ cups Graham flour. Mix all ingredients together. Steam three hours.

"PENNSYLVANIA" PLUM PUDDING (FOR THANKSGIVING DAY)

One cup milk, 2 eggs, 1 cup molasses, ½ teaspoonful nutmeg, ½ teaspoonful salt, 2 teaspoonfuls baking powder, 1 cup bread crumbs, ½ cup corn meal, 1 cup chopped beef suet, ¼ cup finely minced citron, 1 cup seeded raisins, ½ cup currants. Flour to make a stiff batter. Steam fully three hours, turn from the mold, strew chopped almonds over top. Serve pudding hot with sauce for which recipe is given.

Aunt Sarah invariably served this pudding on Thanksgiving Day, and all preferred it to old-fashioned "English Plum Pudding."

SAUCE FOR PUDDING.

Cream together 1 cup of pulverized sugar, scant ½ cup of butter, beat whites of 2 eggs in, one at a time, and one teaspoonful of lemon flavoring; stand on ice a short time before serving. Serve sauce very cold.

"SLICE" BREAD PUDDING

Line the sides of a pudding dish holding two quarts with seven slices of stale bread from which crust had been removed. Beat together 3 eggs, 3 tablespoonfuls of sugar and 3 cups of sweet milk (and add the juice and grated rind of one lemon, or half a grated nutmeg). Pour in the centre of pudding dish. With a spoon dip some of the custard over each slice of bread. Bake about 30 minutes and serve hot with the following sauce:

One cup of water, ½ cup milk, 1 teaspoonful butter, scant tablespoonful of flour mixed smooth with a little water before adding it. Sweeten to taste, add grated nutmeg or vanilla to flavor. Cook all together, then add the yolk of one egg. Place on stove a minute to heat. Add a pinch of salt. Serve hot over the pudding in individual dishes.

CEREALS – OATMEAL PORRIDGE

Oatmeal to be palatable and wholesome should be thoroughly cooked, that is, steamed over a hot fire two hours or longer. Use a double boiler of agateware. Place in the upper half of the boiler about 5 cups of water and stand directly over the hottest part of the range. When the water boils furiously, and is full of little bubbles (not before), stir into the boiling water about 2 cups of oatmeal (if porridge is liked rather thick), and about 1 teaspoonful of salt. (Tastes differ regarding the thickness of porridge.) Let stand directly on the front of the range, stirring only enough to prevent scorching, and cook ten minutes, then stand upper part of double boiler over the lower compartment, partly filled with boiling water; cover closely and let steam from two to three hours. In order to have the oatmeal ready to serve at early breakfast the following morning, put oatmeal on to cook about five o'clock in the evening, while preparing supper, and allow it to stand and steam over boiling water until the fire in the range is dampened off for the night. Allow the oatmeal to stand on range until the following morning, when draw the boiler to front part of range, and when breakfast is ready (after removing top crust formed by standing), turn the oatmeal out on a dish and serve with rich cream and sugar, and you will have a good, wholesome breakfast dish with the flakes distinct, and a nutty flavor. Serve fruit with it, if possible. A good rule for cooking oatmeal is in the proportion of 2½ cups of water to 1 cup of oatmeal.

The cereals which come ready prepared are taking the place of the old-time standby with which mothers fed their growing boys. If you wish your boys to have muscle and brawn, feed them oats. To quote an old physician, "If horses thrive on oats, why not boys who resemble young colts?"

For example, look at the hardy young Scot who thrives and grows hearty and strong on his oatmeal "porritch." Chopped almonds, dates or figs may be added to oatmeal to make it more palatable. Use cup measuring ½ pint for measuring cereals as well as every other recipe calling for one cup in this book.

COOKED RICE

Boil 1 cup of whole, thoroughly cleansed, uncoated rice in 3 quarts of rapidly boiling water (salted) about 25 minutes, or until tender, which can be tested by pressing a couple of grains of rice between the fingers. Do not stir often while boiling. When the rice is tender turn on to a sieve and drain; then put in a dish and place in the oven, to dry off, with oven door open, when the grains should be whole, flaky, white and tempting, not the soggy, unappetizing mass one often sees. Serve rice with cream and sugar. Some prefer brown sugar and others like crushed maple sugar with it. Or rice may be eaten as a vegetable with salt and butter. Rice is inexpensive, nutritious and one of the most easily digested cereals, and if rightly cooked, an appetizing looking food. It is a wonder the economical housewife does not serve it oftener on her table in some of the numerous ways it may be prepared. As an ingredient of soup, as a vegetable, or a pudding, croquettes, etc., the wise housekeeper will cook double the amount of rice needed and stand half aside until the day following, when may be quickly prepared rice croquettes, cheese balls, etc. On the day following that on which rice has been served, any cold boiled rice remaining may be placed in a small bake dish with an equal quantity of milk, a little sugar and flavoring, baked a short time in oven and served with a cup of stewed, seeded raisins which have slowly steamed, covered with cold water, on the back of the range, until soft and plump.

CORN MEAL MUSH

Place on the range a cook-pot containing 9 cups of boiling water (good measure). Sift in slowly 2 cups of yellow granulated corn meal, stirring constantly while adding the meal, until the mixture is smooth and free from lumps. Add 1¼ level teaspoonfuls of salt and ¼ teaspoonful of sugar, and cook a short time, stirring constantly, then stand where the mush will simmer, or cook slowly for four or five hours.

Serve hot, as a porridge, adding ½ teaspoonful of butter to each individual bowl of hot mush and serve with it cold milk or cream. Should a portion of the mush remain after the meal, turn it at once, while still hot, in an oblong pan several inches in depth, stand until quite cold. Cut in half-inch slices, sift flour over each slice and fry a golden brown in a couple tablespoonfuls of sweet drippings and butter. Or dip slices of mush in egg and bread crumbs and fry brown in the same manner. Some there are who like maple syrup or molasses served with fried mush.

This proportion of corn meal and water will make porridge of the proper consistency and it will be just right to be sliced for frying when cold. Long, slow cooking makes corn meal much more wholesome and palatable, and prevents the raw taste of cornmeal noticeable in mush cooked too quickly. The small quantity of sugar added is not noticed, but improves the flavor of the mush.

MACARONI

In early spring, when the family tire of winter foods and it is still too early for vegetables from the home garden, and the high price of early forced vegetables in the city markets prevent the housewife, of limited means from purchasing, then the resourceful, economical housewife serves macaroni and rice in various ways and makes appetizing dishes of the fruits she canned and preserved for Winter use, combined with tapioca and gelatine. Milk and eggs tide her over the most difficult time of the year for young, inexperienced cooks. When the prices of early vegetables soar beyond the reach of her purse, then she should buy sparingly of them and of meat, and occasionally serve, instead, a dish of macaroni and cheese, or rice and cheese, and invest the money thus saved in fruit; dried fruits, if fresh fruits are not obtainable.

Macaroni is such a nutritious food that it should be used frequently by the young housewife as a substitute for meat on the bill of fare. Also occasionally serve a dish of baked beans or a dish composed of eggs, or milk combined with eggs, instead of the more expensive meat dish, all equally useful as muscle-builders, and cheaper than meat. The wise housewife will learn which foods furnish heat for the body and those which produce fat and energy, and those which are muscle-builders, and endeavor to serve well-balanced meals of the foods belonging to the three classes and thus with fruit and vegetables she will make wise provision for her family.

BAKED MACARONI AND CHEESE

Put 2 cups or ½ pound of macaroni (either the long sticks broken in pieces or the "elbow" macaroni, as preferred) in a kettle holding several quarts of rapidly boiling, salted water, and cook about 25 minutes, or until tender. Drain in a colander and allow cold water to run over it for several seconds. This prevents the macaroni sticking together. Place the macaroni in a buttered baking dish and pour over a hot "cream sauce" composed of 1 cup of milk and 1 cup of water, 2 tablespoonfuls of flour, 2 even tablespoonfuls of butter and a pinch of salt. (Too much salt is apt to curdle the milk.) Spread over the top of macaroni about 3 tablespoonfuls of grated cheese, or, if preferred, sprinkle over the top 3 tablespoonfuls of well-seasoned dried bread crumbs and small bits of butter. Stand the bake-dish containing the macaroni in a hot oven ten or fifteen minutes, until lightly browned on top. Serve hot in the dish in which it was baked. Stewed tomatoes are a nice accompaniment to this dish. Double the quantity of macaroni may be cooked at one time and a part of it kept on ice; the following day serve in tomato sauce, thus utilizing any left-over tomatoes.

The macaroni may be cooked while the housewife is using the range, early in the morning. Drain the macaroni in a colander and stand aside in a cool place. It may be quickly prepared for six o'clock dinner by pouring over a hot cream sauce and grated cheese and quickly browning in the oven.

Or the macaroni, when cooked tender in salt water, may be quickly served by pouring over it a hot cream sauce, before the macaroni has become cold. Serve at once.

Housewives should be particular when buying macaroni to get a brand made from good flour.

CAKES — CAKE-MAKING

Sift flour and baking powder together several times before adding to cake batter. Aunt Sarah usually sifted flour and baking powder together four times for cakes. Flour should always be sifted before using. Baking powder should be sifted through the flour dry. Salaratus (or baking soda) should, usually, be dissolved before using in a teaspoonful of hot water, unless stated otherwise. Cream of tartar should be sifted with the flour. Flour should be added gradually and batter stirred as little as possible afterwards, unless directions are given to the contrary. Much beating after flour has been added is apt to make cake tough. Cake will be lighter if baked slowly at first After it has raised increase heat slowly so it will brown nicely on top. The batter, if heated slowly, will rise evenly. This does not mean a cool oven. To prevent cakes sticking to pans, grease pans well with lard, and sift a little flour lightly over pan. Use baking powder with sweet milk. Saleratus is always used with sour milk. Use 1 teaspoonful of saleratus to 1 pint of sour milk. Cream of tartar and saleratus combined may be used with sweet milk instead of baking powder. One heaping teaspoonful of Royal baking powder is equivalent to 1 teaspoonful of cream of tartar and ½ teaspoonful of saleratus combined. Either baking powder or a combination of saleratus and cream of tartar may be used in a cake in which sweet milk is used. Usually take 1½ to 2 scant teaspoonfuls of baking powder to two cups of flour. Saleratus should be used alone with sour milk. Put baking molasses in a stew-pan over fire and allow it to just come to boil; cool before using it. It will not sour as quickly in warm weather, and the cake baked from it will have a better flavor. The cup used in measuring ingredients for cakes holds exactly one-half pint. All cakes are improved by the addition of a pinch of salt. When lard is used instead of butter, beat to a cream and salt well. In mixing cakes, beat butter and sugar together until light and creamy, then add the beaten yolks of eggs, unless stated otherwise as for angel cake, etc., then the flavoring, then mix in the flour and liquid alternately. The baking powder, flour and salt should have been sifted together three or four times before being added. Lastly, fold in lightly the stiffly beaten whites of eggs. Fruit well dredged with flour should be added last, if used. Cool the oven if too hot for baking cakes by placing a pan containing cold water in the top rack of oven. Sponge cake particularly is improved by doing this, as it makes the cake moist. Stir sponge cake as little as possible after adding flour, as too much stirring then will make cake tough. Sift flour several times before using for sponge cake, as tins causes the flour to become lighter. Layer cake, and most small cakes, require a quick oven. The oven door should not be opened for 12 minutes after cake has been placed in oven. Rich cakes, loaf cakes and fruit cakes must bake long and slowly. The richer the cake, the slower the heat required in baking. To test the oven, if the hand can bear the heat of the oven 20 or 25 seconds, the oven then is the right temperature. After placing a loaf cake in oven do not open the oven door for 20 minutes. If oven be not hot enough, the cake will rise, then fall and be heavy. Angel cake, sunshine cake and sponge cake require a moderate oven.

Raisins and dried currants should be washed and dried before using in cake. All fruit should be dredged with flour before being added to cake. Citron may be quickly and easily prepared by cutting on a slaw cutter or it may be grated before being added to cake. When a recipe calls for butter the size of an egg it means two tablespoonfuls. A tablespoonful of butter, melted, means the butter should be measured first, then melted. Aunt Sarah frequently used a mixture of butter and lard in her cakes for economy's sake, and a lesser quantity may be used, as the shortening quality of lard is greater than that of butter. When substituting lard for butter, she always beat the lard to a cream before using it and salt it well. If raisins and currants are placed in oven of range a few minutes to become warmed before being added to cake, then rolled in flour, they will not sink to bottom of cake when baked.

FRAU SCHMIDT'S LEMON CAKE

- 1½ cups sugar.
- ½ cup butter and lard.
- 3 small eggs or 2 large ones.
- ½ cup sweet milk.
- 2 cups flour.
- ½ teaspoonful saleratus.
- 1 teaspoonful cream of tartar.
- Grated yellow rind and juice of half a lemon.

Beat sugar and butter to a cream and add the yolks of eggs. Add the milk, then the flour and cream of tartar and saleratus; and the flavoring. Lastly, the stiffly-beaten whites of eggs.

This makes one loaf cake. The original of this recipe was a very old one which Frau Schmidt had used many years. Every ingredient in the old recipe was doubled, except the eggs, when five were used. Mary thought this cake fine and from the recipe, when she used half the quantity of everything, she baked a fine loaf cake, and from the original recipe was made one good sized loaf and one layer cake. Thinly sliced citron added to this cake is a great improvement.

FINE "KRUM KUCHEN"

One cup sugar, ½ cup butter and lard, mixed; 2 cups flour and 2 teaspoonfuls of baking powder, 2 eggs, ½ cup sweet milk.

Crumb together with the hands the sugar, butter, flour and baking powder sifted together. Take out ½ cup of these crumbs to be scattered over top of cake. To the remainder add the yolks of the eggs, well beaten, and the sweet milk, and lastly the stiffly beaten whites of eggs. Put the mixture in a well-greased pan (a deep custard pie tin will answer), scatter the half cup of crumbs reserved over top of cake and bake about ¾ of an hour in a rather quick oven. When cake is baked, sprinkle over 1 teaspoonful of melted butter and dust top with cinnamon.

AUNT SARAH'S "QUICK DUTCH CAKES"

She creamed together 1 cup of sugar, 1 tablespoonful of lard, 1 tablespoonful of butter and added 1½ cups of luke-warm milk. Add 3 cups flour (good measure), sifted with three scant teaspoonfuls of baking powder. Add a half cup of raisins, seeded and cut in several pieces, if liked, but the cakes are very good without. Spread in two pans and sprinkle sugar and cinnamon on top and press about five small dabs of butter on top of each cake. Put in oven and bake at once. These are a very good substitute for "raised Dutch cakes," and are much more quickly and easily-made and, as no eggs are used, are quite cheap and very good.

A RELIABLE LAYER CAKE

- 1¼ cups granulated sugar.
- 3 eggs.
- ½ cup butter and lard mixed. (Use all butter if preferred.)
- ½ cup sweet milk.
- 2 cups flour sifted with 2 teaspoonfuls Royal baking powder.

Cream together sugar and shortening. Add yolks of eggs, beating well, as each ingredient is added. Then add milk and flour alternately, and lastly the stiffly beaten white of eggs. Stir all together. Bake in two square layer pans, and put together with chocolate or white icing. Or ice the cakes when cold and cut in squares.

BOILED ICING

Boil together 1 cup of granulated sugar and 5 tablespoonfuls boiling water ten or twelve minutes, or until a small quantity dropped from spoon spins a thread. Stir this into the stiffly-beaten white of one egg until thick and creamy. Flavor with lemon, almond or vanilla flavoring and spread on cake. Dip knife in hot water occasionally when spreading icing on cake.

A delicious icing is composed of almonds blanched and pounded to a paste. Add a few drops of essence of bitter almonds. Dust the top of the cake lightly with flour, spread on the almond paste and when nearly dry cover with ordinary icing. Dry almonds before pounding them in mortar, and use a small quantity of rose water. A few drops only should be used of essence of bitter almonds to flavor icing or cake. A pinch of baking powder added to sugar when making boiled icing causes the icing to become more creamy, or add a pinch of cream of tartar when making boiled icing. Or, when a cake iced with "boiled icing" has become cold, spread on top of icing unsweetened, melted chocolate. This is a delicious "cream chocolate icing."

A DELICIOUS "SPICE LAYER CAKE"

- 2 cups light brown sugar.
- 1 cup chopped raisins.
- 2 eggs.
- 1 cup sour milk.
- ½ cup butter.
- 2 cups flour.
- 1 teaspoonful each of soda, cloves, cinnamon, allspice and
- a little grated nutmeg.

Cream sugar and butter together, add yolks of eggs, then the sour milk in which the soda has been dissolved, flour and spices, and lastly stir in the stiffly beaten white of eggs. Bake in two-layer pans.

ICING

Two cups sugar, ¾ cup of milk or cream, 2 tablespoonfuls of butter. Boil until it forms a soft ball when a small quantity is dropped in water, and flavor with vanilla. Beat until cold and spread between layers of cake. Also on top and sides.

AN INEXPENSIVE COCOA CAKE

This is a decidedly good cake and no eggs are required. Cream together 1 cup brown sugar, ¼ cup butter. Add 1 cup of sour milk, 1¾ cups flour, then sift over 1½ tablespoonfuls of cocoa. Add 1 level teaspoonful saleratus, dissolved in a little of the sour milk, and 1 teaspoonful vanilla. Bake in a small loaf. Use the following icing:

¼ cup of grated chocolate, ¾ cup milk, ½ cup sugar, boiled together until thick, and spread on cake.

AUNT SARAH'S WALNUT GINGERBREAD

- ½ cup of New Orleans molasses.
- 1 cup of light brown sugar.
- ½ cup of shortening (composed of butter, lard and sweet
- drippings).
- ½ teaspoonful of ginger, cinnamon and cloves each.
- 2 teaspoonfuls of baking soda (saleratus), sifted with 3½
- cups flour.
- 1 cup boiling water.
- 2 eggs.

Beat to a cream the sugar and shortening in a bowl; add molasses, then pour over all one cup of boiling water. Beat well. Add flour, soda and spices, all sifted together. Beat into this the two unbeaten eggs (one at a time), then add about ¾ of a cup of coarsely chopped *black walnut* meats or the same quantity of well-floured raisins may be substituted for the walnut meats.

The cakes may be baked in muffin pans. In that case fill pans about two-thirds full. The above quantity makes eighteen. They can also be baked in a pan as a loaf cake. This cake is excellent, and will keep fresh several days. These cakes taste similar to those sold in an Atlantic City bake-shop which have gained a reputation for their excellence.

AUNT SARAH'S "GERMAN CRUMB CAKES" BAKED IN CRUSTS

- 3 cups flour.
- 2½ heaping teaspoonfuls baking powder.
- 2 cups sugar (soft A or light brown).
- ½ cup lard and butter mixed.

- 2 eggs.
- 1 cup sweet milk.
- Pinch of salt.
- Flavoring—vanilla or grated orange rind.

Line three small pie tins with pie crust. Sift together into a bowl the flour and baking powder and add light brown or A sugar, and the butter, lard and salt. Rub this all together with the hands until well mixed and crumbly. Take out 1 cupful of these crumbs and stand aside. Add to the rest of the mixture the yolks of eggs, whites being beaten separately and added last. Add slowly 1 cup of sweet milk. Mix it in gradually until the mixture is creamed, then add a small quantity of grated orange peel, lemon or vanilla flavoring. Lastly, stir in the stiffly beaten whites of eggs. Pour the mixture into each one of the three unbaked crusts, then sprinkle the cup of crumbs thickly over the tops. Bake in a moderate oven. These are very good, cheap cakes for breakfast or lunch.

"SOUR CREAM" MOLASSES CAKE

- ½ cup molasses.
- 1 cup sugar.
- ½ cup thick sour cream.
- ½ cup sour milk.
- ½ cup finely chopped peanuts.
- 1 egg.
- 1 teaspoonful soda dissolved in little hot water.
- 2¾ cups flour.
- 1 cup seeded raisins.

Mix together like ordinary cake. Bake in a fruit cake pan in a slow oven about forty minutes. This excellent cake requires no shortening, as cream is used.

ECONOMY CAKE

- 1 egg.
- 1 cup sweet milk.
- 1 cup granulated sugar.
- 2 cups flour.
- ¼ cup butter.
- 2 teaspoonfuls baking powder.

Cream together sugar and yolk of egg, then beat into this mixture the butter and add the milk. Then stir the flour, a small quantity at a time, into the mixture, keeping it smooth and free from lumps. Add the stiffly beaten white of egg. Use any flavoring or spice preferred. Bake in a quick oven.

This is not simply a very cheap cake, but a decidedly good one, and made from inexpensive materials. Follow the recipe exactly or the cake may be too light and too crumbly if too much baking powder is used, or heavy if too much butter is used. By varying the flavor and baking in different forms it is as good as a number of more expensive recipes. It makes three layers of any kind of layer cake, or bake in Gem pans.

GINGER CAKE

- ½ cup brown sugar.
- 1 egg.
- ½ cup lard.
- 2 large cups flour.
- ½ cup New Orleans molasses.
- 1 tablespoonfnl of ginger.
- 1 teaspoonful soda dissolved in half cup lukewarm water.

Beat sugar and lard to a cream, then beat in the yolk of egg, molasses and flour and soda dissolved in water. Lastly, add the stiffly-beaten white of egg. Bake 45 minutes in hot oven.

A VERY ECONOMICAL GERMAN CLOVE CAKE

Place in a stew-pan the following ingredients:

- 1 cup brown sugar.
- 1 cup cold water.
- 2 cups seeded raisins.
- ⅓ cup sweet lard, or a mixture of lard and butter.
- ¼ grated nutmeg.
- 2 teaspoonfuls cinnamon.
- ½ teaspoonful ground cloves.
- Pinch of salt.

Boil all together three minutes. When cold add I teaspoonful of soda dissolved in a little hot water. Add about 1¾ cups flour sifted with ½ teaspoonful of baking powder. Bake in a loaf in a moderately hot oven about thirty minutes. This cake is both good and economical, as no butter, eggs or milk are used in its composition. The recipe for making this excellent, cheap cake was bought by Aunt Sarah at a "Cake and Pie" sale. She was given permission to pass it on.

ICING.

- 1 small cup pulverized sugar.
- 2 tablespoonfuls of cocoa.

Mix smooth with a very little boiling water. Spread over cake.

CAKE ICING FOR VARIOUS CAKES

Cook together 2 cups of granulated sugar, 1¼ cups of water a little less than 12 minutes. Just before it reaches the soft ball stage, beat in quickly 25 marshmallows; when dissolved and a thick, creamy mass, spread between layers and on top of cake.

This is a delicious creamy icing when made according to directions. If sugar and water be cooked one minute too long, the icing becomes sugary instead of creamy. One-half the above quantity will ice the top of a cake nicely.

MARY'S RECIPE FOR "HOT MILK" SPONGE CAKE

For this cake was used:

- 2 cups granulated sugar.
- 4 eggs.
- 2⅛ cups flour.
- 1½ teaspoonfuls of baking powder.
- 1 cup boiling hot milk.

Separate the eggs, place yolks in a bowl, add the sugar and beat until creamy.

Add the stiffly beaten whites of eggs alternately with the sifted flour and baking powder; lastly add the cup of boiling hot milk; should the milk not be rich, add one teaspoon of butter to the hot milk. The cake batter should be thin as griddle cake batter, pour into a tube pan and place at once in a *very moderate* oven; in about 15 minutes increase the heat and in about 25 minutes more the cake, risen to the top of pan, should have commenced to brown on top.

Bake from 15 to 20 minutes more in a moderately hot oven with steady heat; when baked the top of the cake should be a light fawn color and texture of cake light and fine grained.

Mary was told by her Aunt that any sponge cake was improved by the addition of a teaspoon of butter, causing the sponge cake to resemble pound cake in texture.

CHEAP "MOLASSES GINGER BREAD"

- 1 cup New Orleans molasses.
- 1 cup sugar.
- ½ cup shortening (lard and butter mixed).
- 1 cup hot water.
- 1 large teaspoonful soda dissolved in the one cup of hot water.
- 1 teaspoonful of ginger.
- ½ teaspoonful of cinnamon.
- 1 quart of flour.

Stir sugar and shortening together. Add molasses, beat all thoroughly, then add hot water and flour. Stir hard. Bake in two layer pans in quick oven about 30 minutes. Use cake while fresh.

AUNT SARAH'S EXTRA FINE LARGE SPONGE CAKE

- 2 cups granulated sugar.
- 2¼ cups of flour.
- ¾ cup of boiling water.
- 4 large eggs.
- 2 even teaspoonfuls baking powder.
- 1 teaspoonful lemon juice.

Put whites of eggs in a large mixing bowl and beat very stiff. Add sugar (sifted 3 times), then add the well-beaten yolks, flour (sifted 3 times with baking powder), add lemon juice. Lastly, add the hot water. Bake about 50 minutes in a tube pan in a moderately hot oven with a steady heat. Stand a pan of hot water in the upper rack of oven if the oven is quite hot. It improves the cake and causes it to be more moist. This is an excellent sponge cake and easily made, although the ingredients are put together the opposite way cakes are usually mixed, with the exception of angel cake. When this cake was taken from oven, powdered sugar was sifted thickly over the top. Use cup holding ½ pint, as in all other cake recipes.

ANGEL CAKE—AUNT SARAH'S RECIPE

Mary was taught by her Aunt, when preparing a dish calling for yolks of eggs only, to place the white of eggs not used in a glass jar which she stood in a cold place or on ice. When she had saved one even cupful she baked an angel cake over the following recipe:

One heaping cup of pulverized sugar (all the cup will hold), was sifted 8 times. One cup of a mixture of pastry flour and corn starch (equal parts) was also sifted 8 times. The whole was then sifted together 4 times. The one cupful of white of eggs was beaten very stiff. When about half beaten, sprinkle over the partly-beaten eggs one scant teaspoonful of cream of tartar, then finish beating the whites of eggs. Flavor with almond or vanilla. Then carefully sift into the stiffly beaten whites of eggs sugar, flour and corn starch. Fold into the whites of eggs rather than stir. Aunt Sarah always baked this cake in a small, oblong bread pan. This cake should be baked in a *very* moderate oven, one in which the hand might be held without inconvenience while counting one hundred; the oven should be just hot enough for one to know there was fire in the range. Do not open the oven door for 15 minutes, then increase the heat a little; if not too hot, open the oven door a moment to cool and bake slowly for about 55 minutes.

AUNT SARAH'S GOOD AND CHEAP "COUNTRY FRUIT CAKE"

- 1 cup butter and lard, mixed.
- 4 eggs.
- 1 cup New Orleans molasses.

- 1 cup sour milk.
- 1 pound dried currants.
- ¼ pound thinly sliced citron.
- 2 teaspoonfuls baking soda.
- 4 cups flour.
- 2 pounds raisins, seeded.

A little grated nutmeg, ginger, cinnamon and a very small quantity of cloves.

Bake in one large fruit cake pan or in two good sized pans about 1¾ hours. This cake should not be kept as long a time as a more expensive fruit cake, but may be kept several weeks. This was Aunt Sarah's best recipe for an excellent, inexpensive fruit cake.

A "SPONGE CUSTARD" CAKE

- 4 eggs.
- 2 cups granulated sugar.
- 3 cups flour.
- 1 teaspoonful baking soda.
- 1 cup cold water.
- Juice of 1 lemon.
- 2 teaspoonfuls of cream of tartar and pinch of salt.

Beat eggs well, then sift in sugar and half of flour in which cream of tartar has been mixed. Dissolve the soda in a little water and add also the lemon juice and lastly add the balance of flour. Bake in layer cake pans two inches deep.

CUSTARD

Boil 1 pint of sweet milk and add to it, stirring constantly, the following mixture: Two tablespoonfuls corn starch, mixed with a little water before boiling, 1 cup of sugar and 1 well-beaten egg. Allow all to cook a few minutes in a double boiler about 15 minutes. Split the sponge cakes when baked and put custard between when cooled.

GRANDMOTHER'S EXCELLENT "OLD RECIPE" FOR MARBLE CAKE

LIGHT PART.

- 1¾ cups granulated sugar.
- 1 scant cup butter or a mixture of butter and lard.
- Whites of 6 eggs.
- 1 cup milk.
- 3 scant cups flour sifted with 2 teaspoons of baking powder.
- Flavor with essence of lemon.

DARK PART.

- Yolks of 4 eggs.
- ½ cup of a mixture of butter and lard.
- ¾ cup milk (scant measure).
- ½ cup brown sugar.
- 1 tablespoon of molasses.
- 2 tablespoons of cinnamon.
- 1 tablespoon of cloves.

One cup or a little more flour sifted with one teaspoon of baking powder. Place spoonfuls of the dark and light batter alternately in a cake pan until all has been used.

Bake in a moderately hot oven from 45 to 50 minutes.

From this recipe may be made two good sized cakes.

I should advise using one-half the quantity for both dark and light part of cake called for in recipe, which would make one good sized cake. Should this whole recipe be used, the cake baked from it would be of the size of a very large fruit cake.

MARY'S MOLASSES CAKES

She creamed together 1 cup of light brown sugar and 2 table-spoonfuls of butter. Then added 1 cup of New Orleans molasses. The molasses had been allowed to come to a boil, then cooled.

She sifted into the mixture 4 cups of flour alternately with 1 cup of sweet milk in which 2 even teaspoonfuls of soda had been dissolved. She beat all well together, then added yolk of one large egg, and lastly the stiffly beaten white of the egg. Beat the mixture again and bake in 2 square layer cake pans in a hot oven about 25 minutes. This is an excellent cake if directions are closely followed.

CHOCOLATE ICING FOR MOLASSES CAKE.

Boil 1 scant half cup water with 1 cup sugar until it spins a thread, or forms a soft, firm ball in cold water. Pour slowly over the stiffly beaten white of egg, beating while it is being poured. Melt 2 squares or 2 ounces of unsweetened chocolate by standing the bowl containing it in hot water. Add 1 teaspoonful hot water to chocolate. Stir the egg and sugar mixture slowly into the melted chocolate. Beat until stiff enough to spread on cake.

HICKORY NUT CAKE

- 1½ cups sugar.
- ½ cup butter.
- ¾ cup milk.
- Whites of 4 eggs.
- 1 cup hickory nut meats, chopped.
- 2 cups flour sifted with 2 teaspoonfuls baking powder.

Mix together as ordinary cake. Bake in a loaf.

"LIGHT BROWN" SUGAR CAKE

Three cupfuls of light brown sugar, ½ cup of sweet lard and yolk of one egg creamed together until light. Then add 1½ cups sour milk alternately with 4 cups of flour and 1½ teaspoonfuls of cinnamon; 1½ teaspoonfuls of ginger, ½ teaspoonful of cloves and half of a grated nutmeg, 1 tablespoonful of thinly shaved or grated citron is an improvement to cake, but may be omitted. Beat all together, then add 1 teaspoonful of soda dissolved in a small quantity of the sour milk. Lastly, add the stiffly beaten white of one egg and one cup seeded raisins dredged with a little flour. Put the cake batter in a large, well-greased fruit cake pan, lined with paper which had been greased and a trifle of flour sifted over, and bake in an oven with a steady heat about one hour and fifteen minutes. This is a very good, *inexpensive* cake and will keep moist some time if kept in a tin cake box. The fruit might be omitted, but it improves the cake.

"ANGEL FOOD" LAYER CAKE

- 1 cup and 2 tablespoonfuls granulated sugar.
- 1-1\2 cups flour.
- 1 cup and 2 tablespoonfuls scalded milk.
- 3 teaspoonfuls baking powder.
- Pinch of salt.
- Whites of 2 eggs.

Place milk in top part of double boiler and heat to boiling point. Sift dry ingredients together four times and then pour in the hot milk and stir well together. Lastly, add the stiffly beaten whites of eggs. Fold them in lightly, but do not beat. The batter will be quite thin. Do not grease the tins. No flavoring is used. Bake in two square layer tins, put together with any icing preferred. Bake in a moderate oven. This is a good, economical cake to bake when yolks of eggs have been used for other purposes.

MARY'S CHOCOLATE CAKE

One-half cup of brown sugar, ½ cup of sweet milk and ½ cup of grated, unsweetened chocolate. Boil all together until thick as cream; allow it to cool.

Mix ½ cup of butter with ½ cup of brown sugar. Add two beaten eggs, ⅔ of a cup of sweet milk and vanilla flavoring to taste. Beat this into the boiled mixture and add 2 cups of flour sifted with 2 teaspoonfuls of baking powder. Bake in three layers and put together with chocolate icing, or cocoa filling.

COCOA FILLING.

- 1½ cups pulverized sugar.
- 1 tablespoonful butter, melted.
- 2½ tablespoonfuls cocoa.

Place all the ingredients in a bowl and mix to a smooth paste with cold coffee. Flavor with vanilla and spread on cake. Tins cocoa filling should not be boiled.

A CHEAP ORANGE CAKE

- 2 eggs.
- 1½ cupfuls sugar.
- 1 large tablespoonful butter.
- 1 cup milk.
- 2 cups flour sifted with 2 teaspoonfuls baking powder.
- Juice and grated yellow rind of half an orange.

Bake in moderate oven in loaf or layers. If a loaf cake, ice top and sides with the following icing:

1½ cupfuls pulverized sugar, 1 tablespoonful warm water and grated rind and juice of half an orange. Mix all together to a cream and spread over cake.

FRAU SCHMIDT'S MOLASSES CAKE

- 1 pint of New Orleans molasses.
- ¾ cup butter and lard, mixed.
- 4 eggs.
- 1 cup sour milk
- 2 good teaspoonfuls soda.
- 4 cups flour.
- Grated rind of 1 orange.

Bake in a long dripping pan, cut out in square pieces, or it may be baked in a large pan used for fruit cake. It will fill two medium sized cake pans.

APPLE SAUCE CAKE

- ¼ cup butter (generous measure).
- 1 cup light brown sugar.
- 1 cup apple sauce (not sweetened).
- 1 level teaspoonful soda.
- 2 cups flour.
- 1 teaspoonful cinnamon.
- ½ teaspoonful cloves.
- 1 small nutmeg, grated.
- Pinch of salt.
- 1 cup raisins.

Cream together butter, sugar and spices. Add apple sauce and flour. (Dissolve the soda in apple sauce.) Add a cup of seeded raisins or raisins and currants, if preferred. This recipe may be doubled when it makes a very good, cheap fruit cake, as no eggs are required, and it both looks and tastes like a dark fruit cake.

ICING.

One cup pulverized sugar, piece of butter size of a walnut. Moisten with a little water and spread on cake.

"SCHWARZ" CAKE

This delicious black chocolate or "Schwarz" cake, as Aunt Sarah called it, was made from the following recipe:

- 1½ cups of sugar.
- ½ cup butter.
- ½ cup sweet milk.
- 1 even teaspoon of soda (saleratus).
- 3 eggs.
- 1 teaspoonful of vanilla.
- 2 cups flour.
- 1½ teaspoon of Royal baking powder.

Before mixing all the above ingredients place in a stewpan on the range ½ cup of grated chocolate and ½ cup sweet milk; allow them to come to a boil, then stand this mixture aside to cool and add to the cake mixture later.

Cream together sugar and butter, add yolk of eggs; soda dissolved in the milk, then add flour and baking powder sifted together alternately with the stiffly beaten white of eggs. Then beat in last the chocolate and milk mixture which has cooled. Bake in layer cake pans.

Use the following chocolate filling:

- ½ cup sugar.
- ½ cup milk.
- Yolk of one egg.
- ½ teaspoon of corn starch (good measure).
- ¼ cake of Baker's unsweetened chocolate.

Boil all together until quite thick and spread between layers of cake.

APPLE CREAM CAKE

- 2 cups Sugar.
- 2 tablespoonfuls butter.
- 1 cup sweet milk.
- 3 cups flour.
- 3 eggs.
- 2 teaspoonfuls Royal baking powder.

Add the stiffly beaten whites of eggs last and bake in two layers. Flavor with lemon or vanilla.

APPLE CREAM FILLING FOR CAKE.

Beat white of 1 egg very stiff. Add 1 cup of granulated sugar and beat well. Quickly grate one raw apple into the egg and sugar, add the juice of ¼ lemon and beat 20 minutes, when it will be light and foamy. This icing is soft and creamy. Coarsely chopped nut meats may be added if liked. Cake must be eaten with a fork, but is delicious.

A "HALF POUND" CAKE

Cream together ½ pound of sugar and ½ pound of butter. Beat into this the eggs separately, until five eggs have been used. Add flour and 1 small teaspoonful of baking powder. Bake in a moderate oven about 55 minutes; ½ pound of flour is used in this cake. This cake is extra fine.

A DELICIOUS ICING (NOT CHEAP).

Stir to a cream a half cup butter, 1½ cups pulverized sugar, 1 tablespoonful milk and 1 teaspoonful vanilla. It is then ready to use for icing a cake.

COCOANUT LAYER CAKE

- 2 cups sugar.
- ½ cup butter and lard, mixed.
- 3 eggs (yolks only).
- 1 cup milk.
- 3 cups flour, sifted several times with the
- 2 teaspoonfuls cream of tartar and 1 teaspoonful soda (saleratus).

Mix like an ordinary cake.

THE FILLING.

To the stiffly beaten whites of 3 eggs add 1 cup of pulverized sugar. Spread this on each one of the layers of the cake and on top. Strew a half of a grated cocoanut over. To the other half of grated cocoanut add 4 tablespoonfuls of pulverized sugar and strew over top of the cake.

GOLD LAYER CAKE

- Yolks of 6 eggs.
- ½ cup butter.
- 1 large cup granulated sugar.
- ½ cup sweet milk.
- 2½ cups flour.
- 2 heaping teaspoonfuls baking powder.

Cream sugar and butter, add yolks. Beat well, then add milk and flour. Stir all together and bake in square pans in a hot oven.

SUNSHINE SPONGE CAKE

- 1 cup granulated sugar.
- Whites of 7 small fresh eggs and 5 yolks.
- ⅔ cup of flour, or scant cup of flour.
- ⅓ teaspoonful cream of tartar and a pinch of salt.

Beat the yolks of eggs thoroughly, then beat the whites about half; add cream of tartar and beat until very stiff. Stir in sugar sifted lightly through your flour sifter. Then add beaten yolks, stir thoroughly, sift the flour five times. The last time sift into the batter, stirring only enough to incorporate the flour. Bake in a tube pan from 40 to 50 minutes in a very moderate oven. This is a particularly fine cake, but a little difficult to get just right. Place cake in a cool oven; when cake has risen turn on heat. This cake should be baked same as an angel cake.

AN INEXPENSIVE DARK "CHOCOLATE LAYER CAKE"

- 1 cup sugar.
- ½ cup butter.
- 2 eggs.
- ½ cup sweet milk.

- 2 cups flour sifted with
- 2 teaspoonfuls baking powder.
- ½ cup chocolate.

Grate the chocolate, mix with ¼ cup of milk and yolk of 1 egg, sweeten to taste; cook the chocolate; when cooled add to the above mixture. Bake in three layer tins. Put white boiled icing between the layers. The boiled icing recipe will be found on another page.

ANGEL CAKE

- 11 eggs (whites only).
- 1½ cups granulated sugar (sifted 3 times).
- 1 cup flour (sifted 5 times).
- 1 teaspoonful cream of tartar.
- 1 teaspoonful vanilla.

Place white of eggs in a large bowl and beat about half as stiff as you wish them to be when finished beating. Add cream of tartar, sprinkle it over the beaten whites of eggs lightly, and then beat until very stiff. Sift in sugar, then flour very lightly. Fold into the batter, rather than stir, with quick, even strokes with spoon. Put quickly in tube pan, bake in moderate oven from 35 to 50 minutes. Do not open oven door for first 15 minutes after cake has been placed in oven.

If cake browns before it rises to top of pan open oven door two minutes; when cake has risen to top of pan finish baking quickly. The moment cake shrinks back to level of pan remove from oven. This is an old, reliable recipe given Mary by her Aunt, who had baked cake from it for years.

MARY'S CHOCOLATE LOAF (MADE WITH SOUR MILK)

- 2 cups brown sugar.
- ¾ cup lard and butter, mixed.
- 2 eggs.
- ½ cup Baker's chocolate, melted.
- ½ cup sour milk.
- ½ cup warm water.
- 1 teaspoonful vanilla.
- Pinch of salt.
- 1 teaspoonful saleratus.
- 3 cups flour.

Dissolve the saleratus in a little vinegar or warm water. Mix as an ordinary loaf cake.

INEXPENSIVE SUNSHINE CAKE

- 5 eggs.
- 1 cup granulated sugar.
- 1 cup sifted flour.

Beat whites of eggs very stiff and stir in thoroughly, then fold the flour, stirring only just enough to mix it in. If stirred too much, the cake will be tough. Bake in a tube pan. This is a delicious cake if carefully made according to directions. No butter or baking powder is used. Bake in a very moderate oven at first, gradually adding more heat until cake is baked.

MARY'S RECIPE FOR ORANGE CAKE

Grate outside rind of 1 orange into a bowl; 1½ cups sugar and ½ cup butter and lard, mixed. Cream all together. Add yolks of three eggs, 1 cup of sweet milk, 2½ cups flour, sifted with 2¼ teaspoonfuls of baking powder. Lastly, add the stiffly beaten whites of the eggs. Bake in two layers.

FILLING FOR ORANGE CAKE.

Grated rind and juice of half an orange, half the white of one egg, beaten stiff. Add pulverized sugar until stiff enough to spread between cakes and on top. (About two cups of sugar were used.)

ROLL JELLY CAKE

- 1 cup granulated sugar.
- 1¼ cups flour.
- 4 egg yolks.
- Pinch of salt.
- ½ cup boiling water.
- 1 large teaspoonful baking powder.

The yolks of eggs left from making "Pennsylvania Dutch Kisses" may be used for this cake by the addition of an extra yolk of egg. Beat the yolks quite light, then add the sugar and beat until light and frothy. Add the flour sifted with the baking powder and salt. Lastly, add the half cup of boiling water. Bake in a rather quick oven from 25 to 30 minutes in two square layer cake pans. Cover cakes first ten minutes until they have risen. When baked turn cakes out of pans on to a cloth. Take one at a time from the oven, spread as quickly as possible with a tart jelly, either currant or grape, and roll as quickly as possible, as when the cakes become cool they cannot be rolled without breaking. Roll up in a cloth and when cool and ready to serve slice from end of roll. These cakes are very nice when one is successful, but a little difficult to get just right.

AUNT SARAH'S CINNAMON CAKE

- 1 cup sugar.
- 2 cups flour.

- 1 egg.
- 1½ teaspoons baking powder
- Piece of butter the size of egg.
- Pinch of salt.
- 1 cup milk.
- A little grated nutmeg.

Beat the butter to a cream and gradually add the sugar. Then add the unbeaten egg and beat all together thoroughly. Add milk and flour and beat hard for five minutes. Add baking powder, salt and nutmeg. Pour into two small greased pie-tins and before putting in oven sprinkle sugar and cinnamon over top. This is an excellent breakfast cake, easily and quickly made.

"GELB KUCHEN"

Mary's Aunt taught her to make this exceptionally fine cake, yellow as gold, in texture resembling an "angel cake," from the following ingredients: The whites of 6 eggs, yolks of 3 eggs, ¾ cup of fine, granulated sugar, ½ cup of high-grade flour, ½ teaspoonful of cream of tartar (good measure), a few drops of almond extract or ½ teaspoonful of vanilla.

Mix ingredients together in the following manner: Sift sugar and flour separately 3 times. Beat yolks of eggs until light, add sugar to yolks of eggs and beat to a cream. The whites of eggs were placed in a separate bowl and when partly beaten the cream of tartar was sifted over and the whites of eggs were then beaten until dry and frothy. The stiffly beaten whites of eggs were then added alternately with the flour to the yolks and sugar. Carefully fold in, do not beat. Add flavoring, pour batter in a small, narrow bread tin, previously brushed with lard, over which flour had been dusted. The cake when baked may be readily removed from the tin after it has cooled.

Bake cake in a very moderate oven about 60 minutes. After cake has been in oven 15 or 20 minutes increase heat of oven. An extra fine, large cake may be baked from this recipe if double the quantity of ingredients are used.

DEVIL'S FOOD CAKE

- 2 cups brown sugar.
- ½ cup butter and lard, mixed.
- 2 eggs.
- ½ cup boiling water.
- 2 ounces Baker's chocolate.
- 2 cups flour.
- 1 teaspoonful soda.
- ½ cup sour cream or milk.

Cream butter and sugar and add yolks of eggs; then sour milk into which the soda has been dissolved. Add hot water, then the eggs. Bake in layers or loaf. Ice with boiled chocolate icing. If a little of the sour milk is saved until last, the soda dissolved in that, and then added to the cake batter, it will give a brick red appearance. This is an excellent cake.

A CHEAP COCOANUT LAYER CAKE

Cream together 1 cup sugar, ¼ cup butter, 1 egg (white of egg beaten separately), add ¾ cup milk, 2 cups flour sifted with 2 teaspoonfuls baking powder. The stiffly beaten white of egg added last. Bake in two layers. For the filling, to put between layers, beat the white of one egg to a stiff, dry froth; add one tablespoonful of sugar, mix together, spread between layers of cake and on top and over this strew freshly grated cocoanut Grate cocoanut intended for cake the day before using. After it has been grated toss up lightly with a fork and stand in a cool place to dry out before using.

LADY BALTIMORE CAKE

- 1 cup butter.
- 2 cups sugar.
- 3½ cups flour.
- 1 cup sweet milk.
- Whites of 6 eggs.
- 2 level teaspoonfuls baking powder sifted with the flour.
- 1 teaspoonful rosewater.

Mix in the usual way and bake in three layers.

ICING FOR CAKE.

Dissolve 3 cups of sugar in a cup of boiling water. Cook until it spins a thread, about ten or twelve minutes. Take from fire and pour over three stiffly beaten whites of eggs, then add a cup of nut meats (blanched and chopped almonds). One cup of chopped raisins may also be added if liked. Stir until thick and creamy. Allow cake to get cold before icing.

One-half this recipe for icing will be sufficient for an ordinary cake.

AN INEXPENSIVE "WHITE FRUIT CAKE"

- 3 cups sugar.
- 3 eggs.
- 1 lb. seeded raisins.
- 1 cup milk.
- 1 cup butter.
- 1 lb. currants.
- 1 lb. chopped almonds.
- Flavor with almond extract.

- 4 cups flour sifted with 2 teaspoonfuls of Royal baking powder.
- ½ lb. figs.
- ¼ lb. citron.

Beat to a cream sugar, butter and yolks of eggs. Then add milk and flour alternately and fruit and almonds. Lastly, add stiffly beaten whites of eggs. Flour fruit before adding. Chop figs. Cut citron fine or shave it thin. This is a cheaper recipe than the one for a "Christmas fruit cake," but this is a very good cake.

A GOOD AND CHEAP "WHITE CAKE"

- 2 cups sugar.
- ½ cup butter and lard, mixed.
- 1 cup milk.
- Add a few drops of almond flavoring.
- 3 cups flour.
- 2 teaspoonfuls baking powder.
- Whites of five eggs.

Cream together the butter and sugar, add flour sifted with baking powder alternately with the stiffly beaten whites of eggs. The five yolks of eggs left from baking white cake may be used when making salad dressing. Use five yolks instead of three whole eggs, as called for in recipe for salad dressing.

CHOCOLATE ICING (VERY GOOD)

One-quarter cup grated, unsweetened chocolate, ¼ cup milk, half a cup sugar. Boil all together until thick and creamy. This quantity will be sufficient to ice the top of one ordinary cake. Spread icing on cake before icing cools. When this icing is used for layer cake, double the recipe.

TIP-TOP CAKE

- 1 lb. granulated sugar.
- 1 cup butter.
- 1 cup milk.
- 4 eggs.
- 1 lb. chopped raisins. (Citron may be used instead of raisins.)
- ½ a nutmeg, grated.
- 5 scant cups of flour.
- 5 teaspoonfuls baking powder.

Mix together same as ordinary cake and bake in a loaf. This Aunt Sarah considered one of her finest cake recipes. She had used it for years in her family. The friend who gave this recipe to Aunt Sarah said: "A couple of tablespoonfuls of brandy will improve the cake."

ORANGE CAKE

Grate the yellow outside rind of 1 orange into a bowl. Add 1½ cups sugar and ¾ cups butter and beat to a cream. Then add yolks of 3 eggs. Then stir in 1 cup milk, 2½ cups flour with 2 heaping teaspoonfuls baking powder. Lastly, add the stiffly beaten whites of 3 eggs. Bake in three layers.

FILLING.

Use the white of one egg, the grated rind and juice of large orange and enough pulverized sugar to stiffen. Spread between layers.

CHEAP SPONGE CAKE

- 1¼ cups granulated sugar.
- 4 eggs.
- 1½ cups flour.
- 4 tablespoonfuls boiling water.
- 1¼ teaspoonfuls baking powder.
- Pinch of salt; flavor to suit taste.

Cream yolks and sugar thoroughly, then add the stiffly beaten whites of eggs, then flour, then boiling water. Bake in a tube pan about 40 minutes. This is a very easily made cake, which seldom fails and was bought with a set of "Van Dusen cake pans," which Aunt Sarah said: "She'd used for many years and found invaluable."

CARAMEL CAKE AND ICING

1½ cups pulverized sugar, 1 cup of butter, 2 cups flour, ½ cup of corn starch, 2 teaspoons of baking powder sifted through flour and corn starch, 1 cup of milk, the whites of 4 eggs. Mix like ordinary cake. Bake as a loaf cake. Ice top the following: 1 cup of light brown sugar, ¼ cup milk, ½ tablespoonful of butter, ¼ teaspoonful of vanilla. Cook all together until a soft ball is formed when dropped in water. Beat until creamy and spread on top of cake.

A WHITE CAKE

Sift together, three times, the following:

- 1 cup of flour.
- 1 cup of sugar (granulated).
- 3 even teaspoonfuls of baking powder.

Scald one cup of milk and pour hot over the above mixture. Beat well.

Fold into the mixture, carefully, the stiffly beaten whites of 2 eggs. Flavor with a few drops of almond extract. Bake in a *moderate oven*, exactly as you would bake an angel cake.

This is a delicious, light, flaky cake, if directions are closely followed, but a little difficult to get just right.

"DUTCH" CURRANT CAKE (NO YEAST USED)

- 4 eggs.
- 2 cups sugar.
- 1 cup butter.
- 1 cup milk.
- ½ teaspoonful baking soda.
- 1 teaspoonful cream of tartar.
- 1 teaspoonful cinnamon.
- ¼ teaspoonful grated nutmeg.
- 1 cup dried currants.
- 4 to 4½ cups flour.

Make about as stiff as ordinary cake mixture. The butter, sugar and yolks of eggs were creamed together. Cinnamon and nutmeg were added. Milk and flour added alternately, stirring flour in lightly; sift cream of tartar in with the flour. Add the baking soda dissolved in a very little water, then add the well-floured currants, and lastly add the stiffly beaten whites of eggs. Bake in a large cake pan, generally used for fruit cake or bake two medium-sized cakes. Bake slowly in a moderately hot oven. These cakes keep well, as do most German cakes.

AN "OLD RECIPE" FOR COFFEE CAKE

- 5 cups flour.
- 1 cup sugar.

- 1 cup raisins.
- 1 cup of liquid coffee.
- 1 cup lard.
- 1 cup molasses.
- 1 tablespoonful saleratus.
- Spices to taste.

Mix like any ordinary cake. This is a very old recipe of Aunt Sarah's mother. The cup used may have been a little larger than the one holding a half pint, used for measuring ingredients in all other cake recipes.

A CHEAP BROWN SUGAR CAKE

- 1 cup brown sugar.
- I tablespoonful lard.
- 1 cup cold water.
- Pinch of salt.
- 2 cups raisins.
- ½ teaspoonful cloves.
- 1 teaspoonful cinnamon.

Boil all together three minutes, cool, then add 1 teaspoonful of soda and ½ teaspoonful of baking powder sifted with 2 cups of flour.

FRAU SCHMIDT'S "GERMAN CHRISTMAS CAKE"

Cream together in a bowl half a pound of pulverized sugar and half a pound of butter; then add yolks of five eggs, 1 grated lemon rind, 1 pint of milk, 1½ pounds of flour sifted with 4 teaspoonfuls of baking powder, 2 teaspoonfuls of vanilla extract. Bake at once in a moderately hot oven. Mary baked an ordinary-sized cake by using one-half of this recipe. The cake was fine grained, similar to a pound cake, although not quite as rich, and she added a couple tablespoon-fuls of thinly shaved citron to the batter before baking. This is a particularly fine cake.

"AUNT SARAH'S" SHELLBARK LAYER CAKE

- 1½ cups sugar.
- ½ cup butter.
- ¾ cup water.
- 3 eggs.
- 1½ teaspoonfuls baking powder.
- Flour to stiffen.

Save out white of one egg for icing. Bake cake in three layers. Chop 1 cup of hickory nut meats and add to the last layer of cake before putting in pan to bake. Use the cake containing nut meats for the middle layer of cake. Put layers together with white boiled icing.

IMPERIAL CAKE (BAKED FOR MARY'S WEDDING)

- 1 pound sugar.
- 1 pound butter.
- ¾ pound flour.
- 1 pound raisins, seeded.
- 1 pound almonds.
- ½ pound thinly shaved citron.
- 1 lemon.
- 1 nutmeg.
- 12 eggs.

Mix ingredients as for pound cake. A fine cake, but expensive.

A LIGHT FRUIT CAKE (FOR CHRISTMAS)

- 1 pound butter, scant measure.
- 1 pound pulverized sugar.
- 1 pound flour (full pound).
- 10 eggs.
- 1 pound English walnut kernels.
- 1 pound raisins.
- ¾ lb. citron, candied orange and lemon peel.
- 1 cup brandy.
- 1 teaspoonful baking powder.

Bake 2½ to 3 hours. This is an excellent cake.

ENGLISH CAKE (SIMILAR TO A WHITE FRUIT CAKE)

- 5 eggs.
- The weight of 5 eggs in sugar.
- The weight of 4 eggs in flour.
- 1 cup raisins.
- 1 cup currants.
- The weight of 3 eggs in butter.
- ½ teaspoonful baking powder.
- 2 tablespoonfuls of brandy.
- ½ cup finely shaved citron.
- ½ cup English walnut or shellbark meats.
- Small quantity of candied orange and lemon peel.

This recipe was given Mary by an English friend, an excellent cook and cake-baker, who vouches for its excellence.

GRANDMOTHER'S FRUIT CAKE (BAKED FOR MARY'S WEDDING)

- 1 pound butter.
- 1 pound sugar.
- 1 pound flour.
- 2 pounds raisins.
- 2 pounds currants.
- Spices of all kinds.
- ¼ pound thinly sliced citron.
- 8 eggs.
- 1 tablespoonful molasses.
- 1 cup sour milk.
- 1 teaspoonful soda.

Mix together in ordinary manner. Cream butter and sugar, add yolks of eggs, sour milk and soda; add flour alternately with stiffly beaten whites of eggs. Lastly, the well-floured fruit. Bake two hours in a moderate oven. This quantity makes one very large cake, or two medium sized ones, and will keep one year. Line inside of pan with well-greased heavy paper to prevent bottom of cake baking too hard.

Aunt Sarah never cut this cake until one month from time it was baked, as it improves with age and may be kept one year.

AN OLD RECIPE FOR POUND CAKE

Cream together ¾ pound butter and 1 pound sugar and yolks of 10 eggs. Then add 10 whites of eggs well beaten alternately with 1 pound of sifted flour.

Bake in a moderate oven with a steady heat. The bottom of pan should be lined with well-greased paper.

"BUCKS COUNTY" MOLASSES CAKES (BAKED IN PASTRY)

Place in a bowl 1 cup of New Orleans molasses and ¾ of a cup of sweet milk. Add 1 teaspoonful of baking *soda*. (For this cake Aunt Sarah was always particular to use the *Cow*-brand soda), dissolved in a very little hot water. Aunt Sarah always used B.T. Babbitt's saleratus for other purposes.

Stir all ingredients together well, then add gradually three even cups of flour, no more, and beat hard. The cake mixture should not be very thick. Pour into three medium-sized pie-tins lined with pastry and bake in a moderately hot oven. These are good, cheap breakfast cakes, neither eggs nor shortening being used.

BROD TORTE (BREAD TART)

Six yolks of eggs and 1 cup sugar, creamed together. Beat about 15 minutes. Add 1 teaspoonful allspice, 1 teaspoonful cloves, 1 cup Baker's chocolate, which had been grated, melted and cooled; 1 cup stale rye bread crumbs, crushed fine with rolling-pin. Lastly, add the stiffly beaten whites of 6 eggs, a pinch of salt and ½ teaspoonful of baking powder sifted over the batter. Put into a small cake pan and bake half an hour in a moderate oven. When eggs are cheap and plentiful this is an economical cake, as no flour is used. It is a delicious cake and resembles an ordinary chocolate cake.

A DELICIOUS CHOCOLATE CAKE

- ½ cake of Baker's unsweetened chocolate (grated).
- 1 cup granulated sugar.
- ½ cup milk.
- 1 teaspoonful vanilla.
- ½ cup butter.
- 1½ to 2 cups flour.
- 2 eggs.
- 2 teaspoonfuls baking powder.

Boil together chocolate, sugar and milk. Add butter and when cool add yolk of eggs; then the flour, flavoring and stiffly beaten whites of 2 eggs. Beat all thoroughly and bake in a loaf or layers.

CHOCOLATE ICING

Boil together 5 tablespoonfuls grated chocolate, ¾ cup granulated sugar, 2 tablespoonfuls milk, 1 egg.

When the mixture begins to thicken and look creamy, spread on cake. If baked in layers, ice on top and between the two layers.

A WHITE COCOANUT CAKE

Cream together ¾ cup butter and 2 cups sugar. Add whites of 5 eggs, 1 cup milk, 1 teaspoonful cream of tartar, ½ teaspoonful soda sifted with 3 cups flour and 1 grated cocoanut. Bake in a loaf. This is an excellent old recipe of Aunt Sarah's.

A POTATO CAKE (NO YEAST REQUIRED)

Cream together:

- 1 cup of sugar.
- ½ cup lard and butter, mixed.
- Yolk of 2 eggs.
- ½ cup pulverized cocoa.
- ½ cup of creamed mashed potatoes, cold.
- A little ground cinnamon and grated nutmeg.
- A few drops of essence of vanilla.
- ¼ cup of sweet milk.
- ½ cup finely chopped nut meats.

One teaspoonful of baking powder sifted with one cup of flour added to the batter alternately with the stiffly beaten whites of eggs. Bake in two layers, in a moderately hot oven. Ice top and put layers together with white icing. This is a delicious, if rather unusual cake.

A CITRON CAKE

- ½ cup butter.
- 1 cup sugar.
- 4 eggs.
- 2 tablespoonfuls water.
- ¼ pound of thinly shaved citron.
- 1½ cups flour.
- 1¼ teaspoonfuls baking powder.
- Several drops of almond flavoring.

Bake in a loaf in a moderate oven about 45 minutes after mixing ingredients together as for any ordinary cake. This is a very good cake.

AUNT AMANDA'S SPICE "KUCHEN"

- 1 cup butter.
- 2 cups granulated sugar.
- 1 cup of a mixture of sour milk and cream.
- 4 eggs.
- 1 teaspoonful soda.
- ½ teaspoonful cloves.
- 1 teaspoonful cinnamon.
- ½ teaspoonful nutmeg.
- 1 teaspoonful vanilla extract.
- 2 tablespoonfuls cocoa.
- 3 cups flour.

Mix all like any ordinary cake. From one-half this recipe was baked an ordinary sized loaf cake.

A GOOD, CHEAP CHOCOLATE CAKE

One cup of flour, 1 teaspoonful of baking powder and 1 cup of granulated sugar were sifted together. Two eggs were broken into a cup, also 1 large tablespoonful of melted butter. Fill up the cup with sweet milk, beat all ingredients well together. Flavor with vanilla and add 2 extra tablespoonfuls of flour to the mixture. Bake in two layer cake pans.

Place the following mixture between the two layers: ½ cup of grated chocolate, ½ cup sugar and ¼ cup of liquid coffee. Cook together a short time until the consistency of thick cream, then spread between layers.

AN ICE CREAM CAKE

Two cups of pulverized sugar, 1 cup of butter, 1 cup sweet milk, whites of 8 eggs, 1 teaspoonful soda, 2 teaspoonfuls of cream of tartar, 3 cups of flour. From same proportions of everything, only using the 8 yolks instead of whites of eggs, may be made a yellow cake, thus having two good sized layer cakes with alternate layers of white and yellow. Put cakes together with white icing. This was an old recipe of Aunt Sarah's mother, used when cream of tartar and soda took the place of baking powder.

SMALL SPONGE CAKES

For these small cakes take 6 eggs, 1 cup of sugar and ¾ cup of flour and ½ teaspoonful of baking powder, a pinch of salt, flavor with lemon. Beat yolks of eggs separately, then add sugar and beat to a cream, then add the stiffly beaten whites of eggs alternately with the sifted flour and baking powder; add a pinch of salt and flavoring. Bake in small muffin tins in a very moderate oven.

SMALL CAKES AND COOKIES—"AUNT SARAH'S" LITTLE LEMON CAKES

- 2 cups granulated sugar.
- 3 eggs (not separated, but added one at a time to the sugar and shortening which had been creamed together).
- 1 scant cup butter and lard, mixed.
- 2 teaspoonfuls baking powder.
- Pinch of salt
- 1 tablespoonful sweet milk.
- Grated rind of 2 lemons and juice of one.

Stiffen the dough with about 3½ cups flour and use about 1 extra cup of flour to dredge the bake-board when rolling out dough and for sifting over the greased baking sheets so the cakes will come off readily. Roll dough very thin and cut in any desired shape. From this recipe may be made 100 small cakes. The baking sheet (for which I gave measurements in bread recipe) holds 20 of these small round cakes. Do all young housewives know that if dough for small cakes be mixed the day before baking and stood in a cool place, the cakes can be cut out more easily and the dough may be rolled thinner, and as less flour may then be used, the cakes will be richer?

Aunt Sarah always cut these cakes with a small round or heart-shaped cutter and when all were on the baking sheet she either placed a half of an English walnut meat in the centre of each cake or cut out the centre of each small cake with the top of a pepper box lid before baking them.

OATMEAL CRISPS

- 2½ cups rolled oats (oatmeal).

- 1 tablespoonful melted butter.
- ¾ cup sugar.
- 1 teaspoonful baking powder.
- 2 large eggs.
- Pinch of salt.

Beat eggs, add salt and sugar, mix baking powder with oats and stir all together. Drop from a teaspoon on to flat pan or sheet iron, not too close together, as they spread. Flatten very thin with a knife dipped in cold water and bake in a moderate oven a light brown. These cakes are fine and easily made. Did you not know differently, you would imagine these cakes to be macaroons made from nuts, which they greatly resemble.

AUNT SARAH'S GINGER SNAPS

1 cup molasses, 1 cup sugar, 1 cup of a mixture of lard and butter, 1 egg, 1 teaspoonful of ginger, 1 teaspoonful of cinnamon, ½ a grated nutmeg, 1 teaspoonful of soda dissolved in 1 teaspoonful of vinegar. About 3 cups of flour should be added.

Dough should be stiff enough to roll out very thin, and the cakes may be rolled thinner than would be possible otherwise, should the cake-dough stand aside over night, or on ice for several hours, until thoroughly chilled. Cut cakes small with an ordinary cake cutter and bake in a quick oven. These are excellent and will remain crisp some time if kept in a warm, dry place.

GERMAN "LEBKUCHEN"

This is a recipe for good, old-fashioned "German Christmas cakes," from which Aunt Sarah's mother always baked. She used:

- 1 pound dark brown sugar.
- 3 whole eggs and yolks of 3 more.

- ¼ pound citron finely shaved on a "slaw-cutter."
- ½ pound English walnut meats (chopped fine).
- 1 quart flour sifted with 2 teaspoonfuls of baking powder.

Mix well together. Do not roll thin like ginger snaps, but about a half inch thick. Cut out about size of a large coffee cup. Bake in a moderate oven and when cold ice the cakes with the following icing:

ICING FOR GERMAN LEBKUCHEN.

Boil 2 cups of sugar and ½ cup of water seven minutes. Pour over the stiffly beaten whites of three eggs; ice the cakes. Place cakes in a tin box when icing has become cold and these will keep quite a long time. I have eaten high-priced, imported Lebkuchen no better than those made from this recipe.

GRANDMOTHER'S MOLASSES CAKES

One quart of New Orleans molasses, 3 eggs, butter size of an egg. Place all together in a stew-pan on range, allow it to come to boil, stirring constantly, and when cool stir in one tablespoonful of saleratus dissolved in a very little vinegar, and about 3 pounds of flour. Do not have cake dough too stiff. Dough should stand until the following day. Roll out at least ½ inch thick. Cut cakes as large around as an ordinary coffee cup or cut with a knife into small, oblong pieces, a little larger than half a common soda cracker. Bake in a moderate oven. Should too much flour be used, cakes will be hard and dry instead of soft and spongy. This very old and excellent recipe had belonged to the grandmother of Sarah Landis. Cakes similar to the ones baked from this recipe, also those baked from recipe for "honey cakes," were sold in large sheets marked off in oblong sections, seventy years ago, and at that time no "vendue," or public sale, in certain localities throughout Bucks County, was thought complete unless in sound of the auctioneer's voice, on a temporary stand, these cakes were displayed on the day of "the sale," and were eagerly bought by the crowd which attended such gatherings.

ANGEL CAKES (BAKED IN GEM PANS)

The whites of four eggs should be beaten very stiff and when partly beaten sprinkle over ½ teaspoonful of cream of tartan Finish beating egg whites and sift in slowly ½ cup of fine granulated sugar, then sift ½ cup of flour (good measure). Flavor with a few drops of almond flavoring. Bake in small Gem pans, placing a tablespoonful of butter in each. Sift pulverized sugar over tops of cakes. Bake 20 minutes in a *very* moderate oven. The recipe for these dainty little cakes was given Mary by a friend who, knowing her liking for angel cake, said these were similar in taste.

"ALMOND BROD"

Three-fourths cup sugar, 3 eggs, 2½ tablespoonfuls olive oil 2 cups flour, ½ teaspoonfuls baking powder, ½ cup sweet almonds, pinch of salt. A couple of drops of almond extract.

In a bowl place ¾ cup of granulated sugar. Add 3 well-beaten eggs, 2 cups of flour sifted with 1½ teaspoonfuls of baking powder and a pinch of salt. Mix all well together. Add 1 cup whole (blanched) almonds and 2½ tablespoonfuls of good olive oil.

Knead the dough thoroughly. Do not have dough too stiff. Divide the dough into four equal parts, roll each portion of dough on a *well-floured* bake board into long, narrow rolls. Place the four rolls on a baking sheet over which flour had been previously sifted. Place the rolls a short distance apart and bake in a quick oven about twenty minutes or until light brown on top. On removing the baking sheet from the oven cut rolls at once, while the almonds are still warm, into two-inch pieces. From this recipe was made thirty pieces of almond bread. The olive oil, used as shortening, is not tasted when baked. These are a very good little cake, and not bread, as their name would lead one to suppose.

"GROSSMUTTER'S" HONEY CAKES

One quart of boiled honey (if possible procure the honey used by bakers, as it is much cheaper and superior for this purpose than the clear, strained honey sold for table use). Add to the warm honey two generous tablespoonfuls of butter, yolks of four eggs, two ounces of salaratus (baking soda), dissolved in a very small quantity of vinegar, just enough to moisten the salaratus. Add just enough flour to enable one to stir well with a spoon. Work the dough a half hour and allow it to stand until the following day, when cut cakes from the dough which had been rolled out on the bake-board one-half inch thick. The dough should be only just stiff enough to roll out, as should the dough be *too soft* the cakes will become hard and crisp, instead of light and spongy, and if too great a quantity of flour is added the cakes will not be good. As the thickening qualities of flour differ, the exact amount required cannot be given. When about to cut out cakes, the bake-board should be well-floured. Cut the cakes the size of the top of a large coffee-cup, or roll out in one-half inch thick on a well-floured baking sheet and mark in small, oblong sections with a knife, they may then be easily broken apart when baked. These cakes should he baked in a moderately hot oven and not a *hot oven.*

These are the real, old-time honey cakes as made by Aunt Sarah's grandmother on a "Bucks County" farm, and Mary's Aunt informed her she still remembered in her earlier days having bought these cakes at "Bucks County" sales or "vendues," as they were then designated.

LEMON WAFERS OR DROP CAKES

- 2 eggs.
- ½ pound butter.
- ½ pound sugar.
- ½ pound flour.

- Pinch of salt.
- Flavor with lemon essence.

Mix the same as other small cakes. Drop spoonfuls quite a distance apart on the cold pan or tin on which they are to be baked as the dough spreads. These are very thin, delicious wafers when baked.

FRAU SCHMIDT'S SUGAR COOKIES

- 1 cup lard and butter, mixed.
- 2 cups granulated sugar, and
- 2 eggs, all creamed together; then add
- 1 teaspoon soda (mix with a little sour milk).
- Flavor with vanilla.

Beat all well together. Add flour enough that they may be rolled out, no more. Flour bake-board well; cut dough with cake cutter into small round cakes and bake in a rather quick oven. This recipe will make a large number of cakes if dough be rolled thin as a wafer. Frau Schmidt was able to keep these cakes some time — under lock and key. If cake dough be mixed one day and allowed to stand over night, cakes may be rolled out much more easily and cut thinner.

ALMOND MACAROONS (AS PREPARED BY MARY)

Three eggs (whites only), ¾ pound of pulverized sugar, ½ pound of almond paste (which may be bought ready prepared). Beat eggs very stiff, add other ingredients. Drop teaspoonfuls on a baking sheet and bake in a moderate oven 15 or 20 minutes. Macaroons prepared from this recipe are delicious and resemble those sold by confectioners.

"HONIG KUCHEN" (HONEY CAKES)

Two pounds of flour, ½ pound of butter, ⅔ pound of almonds, 2 pounds of honey in liquid form, the grated yellow rind of one lemon, ½ teaspoonful of cloves, ½ teaspoonful of cinnamon, 1 ounce of hartshorn, dissolved in a small quantity of water. Boil together honey and butter, remove from fire, and when mixture has cooled add the hartshorn, coarsely chopped almonds and flour. Allow this mixture to stand several days, roll out ⅓ inch thick. Cut in small round cakes, place a whole almond in centre of each cake. Bake a light brown in a moderate oven.

FRAU SCHMIDT'S MOLASSES SNAPS

Two cups of New Orleans molasses, 1 cup of lard, 1 tablespoonful of ginger, 1 teaspoonful of cinnamon, ¼ teaspoonful of cloves, ½ a grated nutmeg, 1 tablespoonful of saleratus dissolved in a small quantity of hot water. Add enough flour to form a *very* stiff dough. Stand dough aside until the following day, when roll out very thin on a well-floured bake-board. Cut with a small round cake cutter and bake in a hot oven. These are good, cheap small cakes.

HICKORY NUT CAKES

One cup of hickory nut meals, 1 cup of pulverized sugar, 1 egg, a pinch of salt, 2 teaspoons of flour. Mix all ingredients together. Drop small pieces on a sheet-iron and bake.

"LEBKUCHEN" (AS THE PROFESSOR'S WIFE MADE THEM)

Two pounds of sugar, 8 large eggs, ¾ pound of almonds (shelled), ¼ pound of citron, ¼ of a pound each of candied orange and lemon peel, the grated yellow rind of one lemon, 4 teaspoonfuls of cinnamon, 1 teaspoonful allspice, about 2 pounds flour. Separate the eggs. Cream the yolks of eggs and sugar well together. Then add the almonds (which have been blanched by pouring boiling water over them, when the skins may be readily removed), the citron and lemon peel chopped fine. Then add 1 level teaspoonful of different spices. Then add the stiffly beaten whites of eggs, alternately, with the sifted flour. The recipe called for two pounds of flour, but "Frau" Schmidt said; "She was never able to use the whole amount, so she added just enough flour to prevent the mixture spreading when dropped on the baking sheet by tablespoonfuls."

FRUIT JUMBLES

Two cups sugar, 3 eggs (beaten separately), 1 cup butter, 1 cup milk, 3½ cups flour, 3 teaspoonfuls baking powder, ¼ of nutmeg, grated, 1 cup currants. Mix all together and bake in a broad, shallow pan. This is similar to Spanish Bun. When cake is cooled, but not cold, cut in two-inch squares or diamonds before removing from the pan in which the cake was baked.

BROWN "PFEFFERNUSSEN"

For these German cakes Frau Schmidt used the following: 3 pounds of flour, 2 pounds of sugar syrup, ⅛ teaspoonful of black pepper, ¼ pound of lard, ¼ teaspoon of cardamom powder, ¼ pound of butter, ½ teaspoonful of cloves, ½ pound of brown sugar and 2 eggs.

Use as much "Hirschhorn Salz" as can be placed on the point of a knife ("Hirschhorn Salz" translated is carbonate of ammonia and is used for baking purposes). Allow the syrup to heat on the range. Skim off the top. When syrup has cooled mix all ingredients together and stand aside for one week or longer, when form the dough into small balls size of a hickory nut. Place on greased pans and bake half hour in a slow oven.

SMALL OATMEAL CAKES

Cream together 1½ cups of light brown sugar, ½ cup of lard and butter, mixed, and the yolk of one egg. Add ½ cup of hot water and ¾ teaspoonful of saleratus (baking soda) dissolved in a little boiling water; add 2½ cups of oatmeal the stiffly beaten white of egg and 2½ cups of white flour. Mix all together. Dredge the bake board with flour, roll thin. Cut out with a small round cake cutter. Sift a little flour over the well-greased baking sheets, on which place cakes and bake in a moderately hot oven.

FRAU SCHMIDT'S RECIPE FOR "GERMAN" ALMOND SLICES

- ½ pound sugar, ½ pound butter.
- ½ pound of seeded raisins (chopped).
- ½ pound blanched and chopped almonds.
- 1 teaspoonful cinnamon, 1 teaspoonful of allspice.
- Grated rind and juice of 1 lemon.
- 2 cakes German sweet chocolate, grated.
- 3 whole eggs and 2 extra whites of eggs.
- 2 teaspoons baking powder, 3 cups flour.
- 1 tablespoon vanilla, 2 tablespoons of brandy.

Cream butter and sugar, add eggs, one at a time. Then add all the ingredients. Mix with flour. Flour bake board and take a handful of dough and roll with the hands in shape of a sausage roll. This quantity of dough makes eight rolls. Place on greased baking sheets a short distance apart, so they will not touch when being baked. Bake them in a *warm*, not hot, oven. Take from the oven when baked and cut while still warm into small slices across the roll. Slices should be about three-quarters of an inch wide. Cover the three sides with the following icing:

Beat together until smooth and creamy 1 cupful of sweet cream, adding enough confectioners' sugar to make it spread.

You may expedite the work by preparing raisins and almonds the day before.

The Professor's wife always served these almond cakes with coffee when she gave a "kaffee klatch" to her country friends.

"JULY ANN'S" GINGER SNAPS

Two cups of molasses (New Orleans), 1 cup of light brown sugar, 1 egg, 1 tablespoonful of soda, 2 tablespoonfuls of vinegar, 1 tablespoonful of ginger and about 5½ cups of flour.

Place molasses and sugar in a sauce-pan on the range, cook together until sugar is dissolved, no longer.

Mix the soda and vinegar and when foamy add to the sugar and molasses with a portion of the required amount of flour; then add the egg and the flour remaining. Turn dough out on a well-floured bake-beard, roll out into a thin sheet and cut out small cakes with a tin cutter. Bake in a moderately hot oven.

No shortening of any kind was used in these cakes. One hundred cakes were baked from the above ingredients.

COCOANUT COOKIES

Three cups of sugar, 1 cup of butter, 2 eggs, 1 cup of sweet milk, 1 cup of grated cocoanut, 2 teaspoonfuls of baking powder. Mix all together, sift flour with baking powder, add flour to form a dough just stiff enough to roll out, no more. Cut with a small tin cake cutter into round cakes and bake.

CHOCOLATE COOKIES

Two cups of white sugar, 1 cup of grated, unsweetened chocolate, 2 eggs, ½ cup of butter, 2 teaspoonfuls of baking powder. Flavor with vanilla. Mix together sugar, butter and eggs, add melted chocolate and flour to stiffen, just enough flour being used to allow of their being cut with a cake cutter. The baking powder should have been sifted with a small amount of flour before adding.

SMALL "BELSNICKEL" CHRISTMAS CAKES

- 2 cups "A" sugar.
- Pinch of salt.
- 1 cup melted butter.
- 1 teaspoonful baking soda.
- 4 eggs.
- About 3 cups of flour.

Mix in just enough flour so the cake dough may be rolled out quite thin on a floured board, using as little flour as possible. Cut out small cakes and bake lightly in a moderately hot oven.

The butter, when melted, should fill one cup; pour it over the two cups of sugar in a bowl and beat until smooth and creamy; add the eggs, beating one at a time into the mixture. Sift the teaspoonful of baking soda several times through the flour before adding to the cake mixture. Stand this dough in a cold place one hour at least before cutting out cakes. No flavoring is used. Sift granulated sugar thickly over cakes before placing them in oven to bake.

From these ingredients were made over one hundred cakes. One-half this recipe might be used for a small family. The cakes keep well in a dry, cool place.

This old recipe of Aunt Sarah's mother derived its name "Belsnickel" from the fact that the Belsnickels, who invariably visited the houses of "Bucks County" farmers on Christmas Eve, were always treated to some of these delicious little Christmas cakes.

"PENNSYLVANIA DUTCH" KISSES

One cup of pulverized sugar, whites of 3 eggs, 1 heaping cup of nut meats (Mary used hickory nut meats), a pinch of salt. To the very stiffly beaten whites of eggs add sugar, salt and lastly the nut meats. Drop teaspoonfuls of this batter on a greased, floured baking tin. Bake in a moderate oven.

LITTLE CRUMB CAKES

For these small cakes Aunt Sarah creamed together ½ cup of granulated sugar, ¼ cup butter. One quite large egg was used. The egg yolk was added to the creamed sugar and butter and thoroughly beaten, then scant ½ cup of milk was added, and one heaping cup of fine dried bread crumbs sifted with ¾ teaspoonful of baking powder and ¼ cup of finely chopped or rolled *black* walnut meats. Lastly, add the stiffly beaten white of egg. Flavor with grated nutmeg. Bake in small muffin pans in a moderate oven. This makes nine small cakes. No flour is used in these cakes, but, instead of flour, bread crumbs are used.

DELICIOUS VANILLA WAFERS (AS MARY MADE THEM)

- ¼ pound of butter.
- ¼ pound of flour.
- ¼ pound of sugar.
- 2 eggs.

Cream together butter and sugar, add yolks of eggs, beat well, then add stiffly beaten whites of eggs and flour alternately.

Flavor with essence of vanilla, drop from spoon on to *cold* iron pan, not too close together, as the cakes will spread. Bake quickly in a hot oven until outer edge of cakes have browned.

MACAROONS (AS AUNT SARAH MADE THEM)

One-half pound of almonds, blanched and chopped fine, ½ pound of pulverized sugar, whites of 4 eggs. Place sugar and almonds in a pan on the range, until colored a light yellow-brown. Beat whites of eggs very stiff, mix all ingredients together, then drop with a spoon on tins waxed with bees' wax, and bake in a quick oven.

"SPRINGERLES" (GERMAN CHRISTMAS CAKES)

- 4 eggs.
- 1 pound sifted pulverized sugar
- 2 quarts flour, sifted twice.
- 2 small teaspoonfuls baking powder.

Beat whites and yolks of eggs separately, mix with sugar and beat well. Add flour until you have a smooth dough. Roll out pieces of dough, which should be half an inch thick. Press the dough on a floured form or mold, lift the mold, cut out the cakes thus designed and let lie until next day on a floured bread board. The next day grease pans well, sprinkle anise seed over the pans in which the cakes are to be baked; lay in cakes an inch apart and bake in a moderate oven to a straw color. The form used usually makes six impressions or cakes 1½ inches square, leaving the impression of a small figure or flower on surface when dough is pressed on form.

OATMEAL COOKIES

- 1 cup sugar.
- 1 cup butter and lard, mixed (scant measure).

- 1 cup chopped nut meats.
- 1 cup chopped raisins.
- 2 eggs, beaten separately, whites added last.
- 1 teaspoonful baking soda dissolved in 4 tablespoonfuls sour milk.
- 1 teaspoonful vanilla.
- Little grated nutmeg.
- 2 cups oatmeal (uncooked).
- 2 cups white flour.

Drop with tablespoon on well-greased baking sheet over which has been sifted a little flour. Bake in rather quick oven. This recipe makes 65 small cakes.

PEANUT BISCUITS

Sift together 2 cups flour and 3 teaspoonfuls baking powder. Add 1 egg, ½ cup sugar, ½ cup peanuts and pecan nut meats, mixed (run through food-chopper), ½ cup sweet milk, ½ teaspoonful salt. Beat sugar and yolk of egg together add milk, stiffly beaten white of egg, chopped nut meats and flour, alternately. Add salt. Place a large spoonful in each of 12 well-greased Gem pans. Allow to stand in pans about 25 minutes. Bake half an hour.

PLAIN COOKIES

- ½ cup butter.
- 4 tablespoonfuls milk.
- 1 cup sugar.
- ½ teaspoonful grated nutmeg.
- 2 eggs.
- ½ cup chopped walnut meats.
- 3 cups flour.
- 3 teaspoonfuls baking powder.

Cream butter and sugar, add milk slowly, add well-beaten eggs. Beat well, add flour and baking powder, sifted together. Roll thin. Cut with a small cake cutter any desired.

WALNUT ROCKS

Cream together 1½ cups of sugar, ½ cup of butter, a small teaspoonful of salt. Dissolve 1 teaspoonful of soda in 4 tablespoonfuls of warm water, two eggs. Sift 3 cups of flour, add 1 teaspoonful of ginger, 1 teaspoonful of cloves, ½ teaspoonful of grated nutmeg, 1 pound of English walnuts, 1 pound of seeded raisins. Drop by teaspoon on a cold sheet iron and bake in a moderate oven. These are excellent.

CINNAMON WAFERS (AS MADE BY AUNT SARAH)

- 10 eggs
- ¾ pound sugar.
- ¾ pound butter.
- 1 pound flour.

Mix like ordinary cake. Divide this into three parts. Flavor one part with vanilla, 1 with chocolate and the other with cinnamon. These latter will be darker than the first. Place a piece of dough as large as a small marble in a small hot, well-greased waffle or wafer iron. Press two sides of iron together, which flattens out cake, and hold by a long handle over fire, turning it over occasionally until cakes are baked. The cake, when baked, is a delicious, thin, rich wafer, about the size of half a common soda cracker. I have never eaten these Christmas cakes at any place excepting at Aunt Sarah's. The wafer iron she possessed was brought by her Grandmother from Germany. The waffle or wafer irons might be obtained in this country.

ZIMMET WAFFLES (AS MADE BY FRAU SCHMIDT)

- ½ pound butter.
- ½ ounce cinnamon.
- ½ pound sugar.
- 3 eggs.
- Flour.

Work together and form into small balls. Place in hot buttered wafer irons, hold over fire and bake. This is an old German recipe which Frau Schmidt's grandmother used.

"BRAUNE LEBKUCHEN"

- 2 pounds sugar syrup.
- ¼ pound granulated sugar.
- ¼ pound butter.
- ¼ pound coarsely chopped almonds.
- Grate yellow part of one lemon rind.
- ¼ ounce cinnamon.
- ¼ ounce cloves.
- 1 drachm of powdered cardamom.
- 1 ounce of hartshorn, dissolved in a little milk.

Place syrup in stew-pan on range to heat, add butter, almonds, spices, etc.

Remove from range, stir in flour gradually. Use about 10 cups of flour. When cool add the dissolved hartshorn. Allow the cake dough to stand in a warm place eight to ten days before baking.

Then place a portion of the cake dough on a greased baking sheet which has been sprinkled lightly with flour, roll cake dough out on the sheet about ⅓ inch in thickness; place in a *very moderate* oven.

When well dried out and nicely browned on top cut the sheets into small squares, the size of ordinary soda crackers.

This is a very old recipe given Mary by Frau Schmidt.

PEANUT COOKIES

One pint of roasted peanuts, measured, after being shelled. Rub off the brown skin, run through a food-chopper. Cream together 2 tablespoonfuls of butter, 1 cup of sugar. Add 3 eggs, 2 tablespoonfuls of milk, ¼ teaspoonful of salt and the chopped peanuts. Add flour to make a soft dough. Roll out on a floured board, cut with a small cake cutter and bake in a moderate oven. This recipe was given Mary by a friend living in Allentown.

PIES — FLAKY PIE CRUST

Have all the materials cold when making pastry. Handle as little as possible. Place in a bowl 3½ cups flour, ¾ teaspoonful salt and 1 cup good, sweet lard. Cut through with a knife into quite small pieces and mix into a dough with a little less than a half cup of cold water. Use only enough water to make dough hold together. This should be done with a knife or tips of the fingers. The water should be poured on the flour and lard carefully, a small quantity at a time, and never twice at the same place. Be careful that the dough is not too moist. Press the dough with the hands into a lump, but do not knead. Take enough of the dough for one pie on the bake board, roll lightly, always in one direction, line greased pie tins and fill crust. If fruit pies, moisten the edge of the lower crust, cover with top crust, which has been rolled quite thin. A knife scraped across the top crust several times before placing over pie causes the crust to have a rough, flaky, rich-looking surface when baked. Cut small vents in top crust to allow steam to escape. Pinch the edges of fruit pies well together to prevent syrup oozing out. If you wish light, flaky pie crust, bake in a hot oven. If a sheet of paper placed in oven turns a delicate brown, then the oven is right for pies. The best of pastry will be a failure if dried slowly in a cool oven.

When baking a crust for a tart to be filled after crust has been baked, always prick the crust with a fork before putting in oven to bake. This prevents the crust forming little blisters.

Aunt Sarah always used for her pies four even cups of flour, ¼ teaspoonful baking powder and one even cup of sweet, *rich, home-made lard*, a pinch of salt with just enough cold water to form a dough, and said her pies were rich enough for any one. They certainly were rich and flaky, without being greasy, and she said, less shortening was necessary when baking powder was used. To cause her pies to have a golden brown color she brushed tops of pies with a mixture of egg and milk or milk and placed immediately in a hot oven.

Mary noticed her Aunt frequently put small dabs of lard or butter on the dough used for top crust of pies before rolling crust the desired size when she wished them particularly rich.

Aunt Sarah always used pastry flour for cake and pie. A smooth flour which showed the impression of the fingers when held tightly in the hand (the more expensive "bread flour") feels like fine sand or granulated sugar, and is a stronger flour and considered better for bread or raised cakes in which yeast is used, better results being obtained by its use alone or combined with a cheaper flour when baking bread.

AUNT SARAH'S LEMON PIE

This is a good, old-fashioned recipe for lemon pie, baked with two crusts, and not expensive. Grate the yellow outside rind from one lemon, use juice and pulp, but not the white part of rind; mix with 2 small cups of sugar, then add 1 cup of water and 1 cup of milk, and 1 large tablespoonful of corn starch, moistened with a little of the one cup of water. The yolks of 2 eggs were added. Mix all ingredients and add the stiffly beaten whites of eggs. This quantity will fill three small pastry crusts. The mixture will measure nearly one quart. Pour into the three crusts, moisten edges of pies, place top crusts on each pie. Pinch edges of crust together and bake in hot oven.

THE PROFESSOR'S WIFE'S SUPERIOR PASTRY

For superior pastry use 1½ cups flour, 1 cup lard, ½ teaspoonful salt and about ¼ cup of cold water, or three scant tablespoonfuls. Put 1 cup of flour on the bake board, sprinkle salt over, chop ¼ cup of sweet lard through the flour with a knife, until the pieces are about the size of a cherry. Moisten with about ¼ cup of ice cold water. Cut through the flour and lard with a knife, moistening a little of the mixture at a time, until you have a soft dough, easily handled. Roll out lightly the size of a tea plate. Take ⅓ of the lard remaining, put small dabs at different places on the dough (do not spread the lard over), then sprinkle over ⅓ of the remaining half cup of flour and roll the dough into a long, narrow roll, folding the opposite ends in the centre of the roll. Roll out lightly (one way), then add lard and flour; roll and repeat the process until flour and lard have all been used. The pastry may be set aside in a cold place a short time before using. If particularly fine pastry is required, the dough might be rolled out once more, using small dabs of butter instead of lard, same quantity as was used of lard for one layer, then dredged thickly with flour and rolled over and over, and then ends folded together, when it should be ready to use. When wanted to line pie-tins, cut pieces off one end of the roll of dough and roll out lightly. The layers should show plainly when cut, and the pastry should puff nicely in baking, and be very rich, crisp and flaky. When preparing crusts for custards, lemon meringues and pies having only one crust, cut narrow strips of pastry about half an inch wide, place around the upper edge or rim of crust and press the lower edge of the strip against the crust; make small cuts with a knife about ⅓ inch apart, all around the edge of this extra crust, to cause it to look flaky when baked. This makes a rich pie crust.

A very good crust may be made by taking the same proportions as used for superior pastry, placing 1½ to 2 cups flour on the bake board, add salt, cut ½ cup lard through the flour, moistening with water. Roll out crust and line pie-tins or small patty pans for tarts.

This pastry is not quite as fine and smooth as the other, but requires less time and trouble to make.

The Professor's wife taught Mary to make this pastry, but Mary never could learn from her the knack of making a dainty, crimped, rolled-over edge to her pies, which she made easily with a deft twist of her thumb and forefinger.

MARY'S LEMON MERINGUE (MADE WITH MILK)

Line two large pie-tins with pie crust, prick with a fork before placing crusts in oven to bake. When baked stand aside to cool while you prepare the following filling: The juice and grated rind of 1 lemon, 1 pint sweet milk, 1 cup sugar, yolks of three eggs, 3 tablespoonfuls flour, butter size of a walnut. Cream together sugar, flour, yolks of eggs, then add lemon, mix well then add to the scalded milk on the range and cook until thick. Let cool, but do not allow to become quite cold, spread on the two crusts, which have been baked. When quite cold add 3 tablespoonfuls of sugar to the stiffly beaten whites of the three eggs, spread on top of pies, sift 1 tablespoonful pulverized sugar on top of meringue and set in a quick oven until fawn color. Serve cold.

When mixing pie dough, should you have mixed more than needed at one time, line *agate* pie-tins with crust (never stand away in tin). They may be kept several days in a cool place and used later for crumb cakes or custards. Or a crust might be baked and used later for lemon meringues, etc.

APPLE TART

Line pie-tins with rich pie crust, sift over each 1 tablespoonful flour and 2 tablespoonfuls sugar. Place on the crust enough good, tart baking apples, which have been pared, cored, halved and placed (flat surface down) on the crust. Put bits of butter over the top and between the apples, about 1 large tablespoonful altogether, and sprinkle about 2 tablespoonfuls of sugar over, add about 1 tablespoonful of cold water when pies are ready to place in oven. These pies should be baked in a very hot oven. When apples are soft take pies from oven and serve one pie, hot; stand the other one aside until quite cold.

To the stiffly beaten white of one egg add one tablespoonful sugar. Stir together and place a spoonful on the top of each half of apple and place in oven until meringue has browned and serve pie cold. Peach tarts may be made in a similar manner, omitting the meringue and substituting peaches for apples.

RAISIN OR "ROSINA" PIE

"Rosina" pie, as Aunt Sarah called it, was composed of 1 lemon, 1 egg, 1 cup sugar, 1 tablespoonful flour, 1 cup large, blue, seeded raisins. Cover the raisins with one cup of cold water; let soak two hours. Cream egg and sugar together, add juice and grated rind of one quite small lemon, or half a large one. Mix the tablespoonful of flour smooth with a little cold water, add to the mixture, then add raisins and to the water in which they were soaked add enough water to fill the cup and cook until the mixture thickens. When cool fill pie-tins with the mixture, bake with upper and under crust about 20 minutes in hot oven. Aunt Sarah used a *generous* tablespoonful of flour for this pie.

"SNITZ" PIE

Cover a bowlful of well-washed dried apples with cold water and allow to soak over night. The following morning cook until tender and mash through a colander. If quite thick a small quantity of water should be added. Season with sugar to taste. Some apples require more sugar than others. Add cinnamon, if liked. Aunt Sarah never used any spices in these pies. Bake with two crusts or place strips cross-wise over the pie of thinly rolled dough, like lattice work. These are typical "Bucks County" pies.

MARY'S RECIPE FOR PLAIN PUMPKIN PIE

Line a pie-tin, one holding 3 cups of liquid, with rich pastry. For the filling for pie mix together the following: 1 cup of steamed pumpkin, which had been mashed through a colander, 1 egg, beaten separately, 1 tablespoonful of flour, 2½ tablespoonfuls of sugar, ⅛ teaspoonful of salt, ½ teaspoonful of grated nutmeg, same of ginger, 1½ cups of milk (scant measure). The mixture should measure exactly 3 cups, after adding milk. Pour this mixture into the pastry-lined pie-tin and bake in a moderate oven until top of pie is a rich brown.

CHOCOLATE PIE

Melt one square of Baker's unsweetened chocolate, or ¼ cup of powdered cocoa, mix with this ½ cup of granulated sugar and ¼ cup of corn starch. When well mixed add yolks of 3 eggs, a pinch of salt, 2 cups of milk; cook all together in a double boiler until thickened. When cool flavor with vanilla. Fill pastry-lined pie crust with the mixture. Beat the 3 whites of eggs to a froth, mix with a couple tablespoonfuls of pulverized sugar, spread on top of pie, stand in oven until light brown.

"PEBBLE DASH" OR SHOO-FLY PIE

Aunt Sarah made these to perfection and called them "Pebble Dash" pie. They are not really pies, they resemble cakes, but having a crust we will class them with pies. She lined three small sized pie-tins with rich pie crust. For the crumbs she placed in a bowl 3 cups of flour, 1 cup brown sugar and ¾ cup of butter and lard, mixed and rubbed all together with the hands, not smooth, but in small rivels. For the liquid part she used 1 cup baking molasses, 1 cup hot water, 1 teaspoonful baking soda dissolved in a few drops of vinegar and stirred this into the molasses and water. She divided the liquid among the three pans, putting one-third in each crust, over which she sprinkled the crumbs. Bake one-half hour in a moderate oven. These have the appearance of molasses cakes when baked.

VANILLA CRUMB "CRUSTS"

Cook together a short time ½ cup molasses, 1 egg, 1 tablespoonful flour, 1 cup sugar, 2 cups cold water. Moisten the flour with a little cold water before adding to the other ingredients. When cooled add 1 teaspoonful of vanilla. Pour this mixture in the bottom of each of four common sized pie-tins, lined with pastry, and sprinkle over the following crumbs:

THE CRUMBS (FOR VANILLA CRUMB CRUSTS).

Two cups flour, ½ cup butter and lard, mixed, ½ teaspoonful soda and 1 cup sugar, rubbed together with the hands to form crumbs. Scatter these crumbs over the four pies.

These are not thick pies, but simply what the recipe calls them — vanilla "crusts."

"KASHA KUCHEN" OR CHERRY CAKE

Aunt Sarah sometimes filled the bottom crusts of two small pies (either cheese pie or plain custard) with a layer of fresh cherries and poured the custard over the top of the cherries and baked same as a plain custard pie.

Aunt Sarah might be called extravagant by some, but she always made egg desserts when eggs were cheap and plentiful, in the Spring. In Winter she baked pies and puddings in which a fewer number of eggs were used and substituted canned and dried fruits for fresh ones. In summer she used fresh fruit when in season, ice cream and sherbets. She never indulged in high-priced, unseasonable fruits—thought it an extravagance for one to do so, and taught Mary "a wise expenditure in time means wealth."

For banana custard pie she substituted sliced banana for cherries on top of pie.

"RIVEL KUCHEN"

Place in a bowl 1 cup flour, ½ cup sugar (good measure), ½ cup butter and lard, or all butter is better (scant measure). Some like a little grating of nutmeg, especially if part lard is used. Mix or crumb the ingredients well together with the hands to form small lumps, or rivels. Line pie-tins with a rich pastry crust and strew the rivels thickly over and bake in a quick oven. A couple tablespoons of molasses spread over the crumbs is liked by some. This is a favorite pie or cake of many Pennsylvania Germans.

AUNT SARAH'S LEMON MERINGUE

Two cups of water, 1½ cups of sugar, 2 rounding tablespoonfuls of corn starch, 4 eggs, 1 tablespoonful of butter, 2 small lemons. Mix the water, sugar and corn starch dissolved in a little cold water, pour in sauce-pan, place on range and stir mixture until thickened. Beat separately the yolks of 4 eggs and the whites of 2, then add both to the above mixture. Remove from the fire, add the juice of two small lemons and grated rind of one; add butter. Fill two previously baked pastry shells with the cooled mixture. Beat the remaining whites of egg (another white of an egg added improves the appearance of the pie.) Add one tablespoonful of pulverized sugar to each egg used; place the stiffly beaten whites of egg rockily over tops of pies stand in oven until a delicate shade of brown. This is a delicious pie.

A COUNTRY BATTER PIE

Line two medium-sized pie-tins with pastry crust in which pour the following mixture, composed of ½ cup of granulated sugar and one egg, creamed together; then add ½ cup of cold water and the grated yellow rind and juice of one lemon.

For the top of pies: Cream together 1 cup of sugar, ¼ cup of lard and 1 egg, then add ½ cup of sour milk alternately with 1½ cups of flour, sifted with ½ teaspoonful of baking soda and ½ teaspoonful of cream of tartar. Place ½ of this mixture on top of each pie. Bake in oven.

PUMPKIN PIE (AUNT SARAH'S RECIPE)

The best pumpkin for pie is of a deep orange yellow with a rough, warty surface. Remove the soft, spongy pulp and seeds of the pumpkin, pare and cut into small pieces. Steam until tender. Put in a colander to drain, then mash through colander with wooden potato masher. For one deep pie allow one pint of the stewed pumpkin, beat in 2 eggs, one at a time, ½ teaspoonful salt, 1 teaspoonful ginger, ½ teaspoonful grated nutmeg, ½ teaspoonful cinnamon, ⅔ cup sugar, 1 scant pint milk. Beat all together. This mixture should barely fill a quart measure. Pour in a deep pie-tin lined with rich crust, grate nutmeg over the top of pie and bake from 45 to 50 minutes in a moderate oven. Have the oven rather hot when the pie is first put in to bake and then reduce the heat, else the filling in the pie will boil and become watery. If liked, two tablespoonfuls of brandy may be added to the mixture before filling the crust. In that case, use two tablespoonfuls less of milk.

WHITE POTATO CUSTARD (AUNT SARAH'S RECIPE)

Boil one medium-sized potato, mash fine, add 1 large tablespoonful of butter and a generous ½ cup sugar. Beat to a cream. When the mixture has cooled add yolks of 2 eggs, ½ cup sweet milk and grated rind and juice of half a lemon. Lastly, stir in the stiffly beaten whites of the two eggs. Bake in a medium-sized pie-tin with one crust in a moderately hot oven about 25 minutes, until a rich brown on top. This is a delicious pie and would puzzle a "Bucks County lawyer" to tell of what it is composed.

"RHUBARB CUSTARD" PIE

Two cups of rhubarb, uncooked, do not skin it, cut in half-inch pieces. Cream together 1 cup of sugar, 1 tablespoonful of cornstarch, 2 eggs (reserve white of one egg). Add the 2 cups of rhubarb to this mixture and place all in a pie-tin lined with pastry. Place in oven and bake until rhubarb is tender. Remove from oven and when pie has cooled spread over it the stiffly beaten white of the egg, to which had been added one tablespoonful of sugar. Place pie in oven and brown a light fawn color.

"LEMON APPLE" PIE

Grate the yellow rind from a lemon (discard the white part of rind), grate the remainder of the lemon, also pare and grate 1 apple. Add 1½ cups of sugar, then add 2 well-beaten eggs. Pour this mixture into 1 large pie-tin lined with rich pastry; place on a top crust, pinch edges, moistened with water, together; bake in an oven with a steady heat. When pie has baked sift pulverized sugar thickly over top and serve cold. From these materials was baked a fair sized pie.

GREEN CURRANT PIE

Line a pie-tin with rich pastry; place oil this crust 2 tablespoonfuls of flour and 2 tablespoonfuls of sugar; then add 2 cups of well-washed and stemmed green currants, previously mixed with 1 tablespoonful of cornstarch, moistened with a small quantity of cold water. Add 1 cup of sugar (from which had been taken the 2 tablespoonfuls placed on crust;) add 2 tablespoonfuls of water; cover with a top crust, cut small vents in crust, bake in a moderate oven.

When crust loosens from side of pan the pie should be sufficiently baked.

A COUNTRY "MOLASSES" PIE

Place in a mixing bowl ¾ cup flour (generous measure), ½ cup granulated sugar, 1 generous tablespoonful of butter.

Crumble all together with the hands until quite fine. Then to ¼ cup of New Orleans (baking) molasses add ¼ cup of boiling water and ¼ teaspoonful of soda (saleratus). Beat together the molasses, water and soda until the mixture is foamy and rises to top of cup. Then pour into a medium-sized pie-tin, lined with pie crust (the pie-tin should not be small or the mixture, when baking, will rise over top of pan). Sprinkle the prepared crumbs thickly over the molasses mixture and with a spoon distribute the crumbs well through the mixture. Bake in a moderate oven from 25 to 30 minutes and you will have the old-fashioned pie your Grandmother used to bake.

When her baking finished, she had dough remaining for an extra crust. Children always called this "molasses candy pie," as 'twas quite different from the "molasses cake batter" usually baked in crusts.

A MOCK CHERRY PIE

This pie was composed of ¾ cup of chopped cranberries, ¾ cup of seeded and chopped raisins, ¾ cup of sugar, ¾ cup of cold water, 1 tablespoonful of flour, 1 teaspoonful of vanilla all together and bake with two crusts.

AUNT SARAH'S CUSTARD PIE

Line an agate pie-pan (one used especially for custards two inches in depth, holding exactly one quart) with a rich pastry. Break five large eggs in a bowl, heat lightly with an egg-beater and add ½ cup of sugar. Boil 3 cups of sweet milk, pour over the eggs and sugar, add 1 teaspoonful of butter and a pinch of salt, ½ teaspoonful of vanilla. The mixture should fill a one-quart measure. When the custard has cooled, pour either into the deep pie-pan, lined with pastry, holding one quart, or into two ordinary pie-tins holding one pint each. Place the custard pie in a quick oven, that the crust may bake before the custard soaks into the crust; then allow oven to cool and when the custard is "set" (which should be in about 35 minutes) remove from the oven and serve cold. The custard should be the consistency of thick jelly. Scalding the milk produces a richer custard.

PLAIN RHUBARB PIE

Line a pie-tin with rich crust, skin rhubarb and cut into half-inch pieces a sufficient quantity to fill 3 cups. Mix together 1 cup of sugar and ¼ cup of flour. Place a couple tablespoonfuls of this on the bottom crust of pie. Mix sugar and flour remaining with 3 cups of rhubarb and fill the crust. Moisten the edge of crust with water, place on top crust, press two edges of crust together (having cut small vents in top crust to allow steam to escape). Bake in a moderate oven about 30 minutes, when top crust has browned pie should be baked.

MARY'S CREAM PIE

Bake crusts in each of two pie-tins. For filling, 1 pint of milk, 1 generous tablespoonful of corn starch, 2 tablespoonfuls of sugar, 2 yolks of eggs (well beaten), 1 teaspoonful of vanilla. Cook all together until mixture thickens and when cooled put in the two baked crusts.

Mix the stiffly beaten whites of two eggs with two tablespoonfuls of pulverized sugar and spread over cream filling in pies and brown lightly in oven.

Always prick the lower crust of a pie carefully with a fork to allow the air to escape; this will prevent blisters forming in the crusts baked before filling crusts with custards.

APPLE CUSTARD PIE

To 1 cup of hot apple sauce (unsweetened) add a tiny pinch of baking soda, 1 tablespoonful of butter, 1 cup of sugar, grated rind and juice of half a lemon or orange, 2 egg yolks, ½ cup of sweet cream and 1 large teaspoonful of corn starch. Line a pie-tin with pastry, pour in this mixture and bake. When the pie has cooled spread over top a meringue composed of the two stiffly beaten whites of eggs and two tablespoonfuls of pulverized sugar flavored with a little grated orange or lemon peel. Brown top of pie in oven.

LEMON PIE WITH CRUMBS

Place in a bowl 1 cup (good measure) of soft, crumbled stale bread. Pour over this one cup of boiling water, add 1 teaspoonful (good measure) of butter and beat until smooth, then add 1 cup of sugar, the grated rind and juice of 1 lemon and the beaten yolks of 2 eggs. This mixture should measure about 1 pint. Pour into a pie-tin lined with rich pastry and bake. When cold spread over a meringue made of the stiffly beaten whites of the 2 eggs and 3 tablespoonfuls of granulated sugar. Place in the oven until the meringue is a light fawn color and serve cold.

AUNT SARAH'S BUTTER SCOTCH PIE

Boil together 1 cup brown sugar and 2 tablespoons butter until a soft, wax-like consistency. Mix together 2 heaping teaspoons flour, yolk of 1 egg and 1 cup of milk. Beat until smooth; stir this into the sugar and butter mixture and cook until thick. Flavor with lemon or vanilla, pour into baked crust and spread over top the beaten white of 1 egg to which has been added tablespoon sugar and brown in oven.

GREEN TOMATO MINCE MEAT

One peck of green tomatoes, chopped fine; 3 lemons, 2 seeded raisins, 5 pounds of granulated sugar, 1 cup of vinegar, 1 teaspoonful of cloves, 1½ tablespoonfuls of nutmeg, 1 tablespoonful of cinnamon. Cook tomatoes 3¾ hours, then add the other ingredients and cook all together 30 minutes. A small quantity of grated orange peel, finely minced citron, cider, brandy or canned fruit juice may be added to improve the flavor of the mince meat. Fill air-tight jars with the hot mixture and screw on jar-tops. This mince meat may be prepared in season when tomatoes are plentiful; is both good and cheap and is a splendid substitute for old-fashioned mince meat.

ORANGE MERINGUE (A PIE)

Into a bowl grate the yellow outside rind of a large, juicy orange; add the juice and pulp, but not any of the tough part enclosing sections. Add 1 tablespoonful of lemon juice, 1 cup of granulated sugar, which had been beaten to a cream with 2 tablespoonfuls of butter, the yolks of 3 eggs, 2 large tablespoonfuls of corn starch, mixed smoothly with a little cold water, and 1 cup of boiling water. Cook all together until thickened and when cool spread on a rather large pie-tin, lined with a baked crust of superior pastry. Add to the stiffly beaten whites of 3 eggs 3 tablespoonfuls of pulverized sugar. Place meringue over top of pie and place in oven until a light fawn color.

GRANDMOTHER'S RECIPE FOR "MINCE MEAT"

The day preceding that on which mince meat is to be prepared, boil 5 pounds of beef. To the well-cooked, finely-chopped meat add

10 pounds of tart apples, chopped into coarse bits; 2 pounds of finely-chopped suet, 2 pounds of large blue raisins, seeded; 2 pounds of dried, cleaned currants, ½ pound of finely-shaved citron, 2 tablespoonfuls of cinnamon, 1 tablespoonful of cloves, 1 tablespoonful of grated nutmeg, 1 small tablespoonful of salt, 1 pint of baking molasses, 1 pint of brandy or cider which had been boiled down. Mix all well together, add more spices, if liked, also juice of 1 orange or lemon. Place all ingredients in a large preserving kettle, allow the mixture to heat through. Fill glass jars, seal and stand away until used. Add more cider, should it he required, when baking pies.

"TWENTIETH CENTURY" MINCE MEAT

Two pounds lean beef (uncooked), chopped fine, ½ pound beef suet, shredded.

Put the beef and suet in a large stone jar, pour over it ⅔ of a quart of whiskey. Let stand covered with a lid for a week, then add 2 pounds large, seeded raisins, 2 pounds Sultana raisins, 2 pounds currants, ½ pound citron, juice and grated rind of 2 oranges and of 2 lemons, 1 teaspoonful salt, 1 tablespoon ground cinnamon, 2 grated nutmegs, ½ teaspoon ground allspice, 1 pound sugar. Let stand two weeks, then it is ready to use. When you wish to bake pies take out as much of the mince meat as you wish to use and add chopped apples, two parts of mince meat to one part chopped apples, and add more sugar if not as sweet as liked. If too thick, add a little sherry wine and water, mixed. Fill bottom crust with some of the mixture, cover with top crust and bake. There must be just enough liquor in the jar to cover the meat, as that preserves it. This seems like a large quantity of liquor to use, but much of the strength evaporates in baking, so that only an agreeable flavor remains; that is, to those who like liquor in mince meat; some people do not. Others there are who think mince meat not good unless made with something stronger than cider. Mince pies made by this recipe are excellent. This recipe was given Mary by a friend, a noted housekeeper and cook.

A "DUTCH" RECIPE FOR PUMPKIN PIE

Line a medium-sized pie-tin with pastry. Cover the crust thickly with thinly-sliced, uncooked pumpkin, cut in inch lengths. Place on the pumpkin 1 tablespoonful of syrup molasses, 1 tablespoonful of vinegar, 1 tablespoonful flour and sweeten with sugar to taste, dust over the top a little ground cinnamon, cloves and nutmeg; cover pie with a top crust and bake in a moderately hot oven. When baked the pumpkin filling in the pie should resemble diced citron and the pie have somewhat the flavor of green tomato pie. (The vinegar may be omitted and the result be a very good pie.)

MARY'S COCOANUT CUSTARD PIE

Line two medium-sized pie-tins with rich pastry and bake. For the custard filling: 3 egg yolks, 2 cups granulated sugar, 1 quart of milk.

Cook all together, then add 1 tablespoonful of corn starch and one of flour (moistened with a little cold water before adding). Cook all together until the mixture thickens. Flavor with one teaspoonful of vanilla. Allow the mixture to cool.

Grate one good-sized cocoanut, mix half of it with the custard and fill into the two crusts. Spread over the tops of the two pies the stiffly beaten whites of the three eggs to which you have added a small quantity of sugar. Over this sprinkle the remaining half of the grated cocoanut, stand in the oven a few minutes, until top of pie is lightly browned.

GRAPE PIE

Pulp the grapes. Place pulp in a stew-pan and cook a short time. When tender mash pulp through a sieve to remove seeds. Add skins to pulp. Add one scant cup of sugar and rounded teaspoonful of butter. Line a pie plate with rich pastry, sprinkle over one tablespoonful of flour. Pour in the grape mixture and sift another tablespoonful of flour over the top of mixture and cover with a top crust in which vents have been cut, to allow the steam to escape, and bake in a hot oven. Allow two small cups of grapes to one pie.

SOUR CHERRY PIE

One quart of cherries, ½ cup of flour for juicy sour cherries, (scant measure of flour), 1½ cups sugar.

Pit the cherries, saving cherry juice. Mix together sugar and flour and place about ⅓ of this on a pie-tin lined with pastry. Fill with cherries and juice and sprinkle remaining sugar and flour over. Bake with an upper crust, having vents cut in to allow steam to escape.

AUNT SARAH'S STRAWBERRY PIE

Make a rich crust, line a pie-tin and fill with clean, hulled strawberries. Allow one quart to each pie. Sweeten to taste; sprinkle a generous handful of flour over the berries, having plenty of flour around the inside edge of pie. Use ½ cup of flour all together. Cut a teaspoonful of butter into small bits over top of berries, cover with top crust with vents cut in to allow steam to escape, pinch edges of crust together to prevent juice escaping from pie, and bake.

FLORENDINE PIE

To 2 apples, cooked soft and mashed fine (after having been pared and cored) add the yolk of one egg (well beaten) one minute before removing the cooked apple from the range. Then add 1 small cup of sugar, a piece of butter the size of a hickory nut, 1 teaspoonful of flour; flavor with either lemon or vanilla. Line a pie-tin with rich pastry crust. Pour in the mixture and bake in a quick oven. This makes a delicious old-fashioned dessert.

AUNT SARAH'S CHEESE CAKE

Prepare the following for one cheese cake, to be baked in a pie-tin lined with pastry crust:

One heaping cup of rich, creamy "smier kase," or cottage cheese, was placed in a bowl, finely mashed with a spoon until free from lumps. Then mixed smooth with 2 tablespoonfuls of sweet milk, 1 tablespoonful of softened butter was added, a pinch of salt, about ¾ cup of sugar, 1¼ table spoonfuls of flour (measure with an ordinary silver tablespoon). One large egg was beaten into the mixture when it was smooth and creamy, 1 cup of milk was added. After adding all the different ingredients the mixture should measure about 3¾ cups and should be very thin. Pour the mixture into a pastry-lined pie-tin. This is one of the most delicious pies imaginable, if directions given are closely followed. Bake in a moderately hot oven until cheese custard is "set" and nicely browned on top, then allow the oven door to remain open about five minutes before removing the "pie," as I should call it, but Bucks County farmers' wives, when speaking of them, invariably say "cheese cakes." Should the housewife possess "smier kase," *not* rich and creamy, use instead of the one tablespoonful of sweet milk, one tablespoonful of sweet cream.

"FRAU SCHMIDT'S" LEMON PIE

Grated yellow rind and juice of one lemon, 1 cup of sugar, 1 cup of molasses, 1 egg, butter, size of a walnut; 1 tablespoonful of corn starch, ¾ cup of water. Cream together the butter, sugar and egg, add the corn starch moistened with a little cold water, add grated rind and juice of one lemon, molasses, and lastly add water. Cook all ingredients together. When cool fill 2 or 3 small pie-tins lined with rich pastry; cover with top crust and bake.

PICKLES — SPICED CUCUMBERS

- 24 medium-sized cucumbers.
- 6 medium-sized onions.
- 3 red peppers.
- 3 green peppers.

Pare cucumbers, then cut in inch lengths. Slice onions and peppers quite thin. Place all in a large earthenware bowl and sprinkle over about ½ cup of table salt; mix all well together, let stand four or five hours, when place in a colander; cover with a plate and drain off all the salt water possible or squeeze through a cheese-cloth bag.

Boil together for 10 minutes the following; 1 quart of vinegar, 1 tablespoonful of cloves, 1 teaspoonful of turmeric powder (dissolved in a little of the vinegar) and 1 scant cup of sugar. Add the cucumbers, peppers and onions to the hot vinegar. Let come to a boil and allow all to boil two minutes, then place in sterilised jars and seal.

MIXED SAUCE TO SERVE WITH MEATS

- Yolks of 4 eggs.
- ½ cup sugar.
- 1 tablespoonful mixed yellow mustard.
- 1 tablespoonful olive oil.
- 1 teaspoonful salt.
- 1 tablespoonful vinegar with flavor of peppers.

Thin with vinegar and boil until thick. Add 1 teaspoonful of grated horseradish.

To flavor vinegar cover finely-cut green and red peppers with vinegar and allow all to stand about 24 hours, then strain and use the vinegar.

PEPPER RELISH

Chop fine 12 sweet red peppers, 12 sweet green peppers and 8 small onions. Put all in a bowl and cover with boiling water and let stand five minutes. Drain off, cover again with boiling water and let stand ten minutes. Then place in an agate colander or muslin bag and let drain over night. The following morning add 1 quart of good sour vinegar, 1½ cups sugar, 2 even teaspoonfuls salt and boil 20 minutes. While hot fill air-tight jars. This is excellent.

PICKLED RED CABBAGE

Shred red cabbage, not too fine, and sprinkle liberally with salt. Stand in a cool place 24 hours. Then press all moisture from the cabbage, having it as dry as possible; stand the earthen bowl containing the cabbage in the sun for a couple of hours. Take a sufficient quantity of vinegar to cover the cabbage. A little water may be added to the vinegar if too sour. Add 1 cup sugar to a gallon of vinegar and a small quantity of celery seed, pepper, mace, allspice and cinnamon. Boil all about five minutes and pour at once over the cabbage. The hot vinegar will restore the bright red color to the cabbage. Keep in stone jar.

MUSTARD PICKLES

24 cucumbers, 1 quart of small onions, 6 peppers, 2 heads of cauliflower, 4 cups of sugar, or less; celery or celery seed, 3 quarts of good vinegar, ½ pound of ground yellow mustard, 1 tablespoonful turmeric powder, ¾ cup of flour.

The seeds were removed from the cucumbers and cucumbers were cut in inch-length pieces, or use a few medium-sized cucumbers cut in several pieces and some quite small cucumbers. (The quantity of cucumbers when measured should be the same as if the larger ones had been used.) One quart of small whole onions, 6 peppers, red, green and yellow, two of each, cut in small pieces. Place all together in an agate preserving kettle and let stand in salt water over night. In the morning put on the range, the vegetables in agate kettle, let boil a few minutes, then drain well. Take three quarts of good vingar, 4 cups of sugar, if liked quite sweet; 2 teaspoons of either celery seed or celery cut in small pieces. Put the vinegar, sugar and celery in a preserving kettle, stand on stove and let come to a boil; then add the other ingredients. When boiling have ready a half pound of ground mustard, ¾ cup of flour, 1 tablespoon of turmeric powder, all mixed to a smooth paste with a little water. Cook until the mixture thickens. Add all the other ingredients and boil until tender. Stir frequently to prevent scorching. Can while hot in glass air-tight jars.

AUNT SARAH'S CUCUMBER PICKLES

Always use the cucumbers which come late in the season for pickles. Cut small green cucumbers from vine, leaving a half-inch of stem. Scrub with vegetable brush, place in a bowl and pour over a brine almost strong enough to float an egg; ¾ cup of salt to seven cups of cold water is about the right proportion. Allow them to stand over night in this brine. Drain off salt water in the morning. Heat a small quantity of the salt water and pour over small onions which have been "skinned." Use half the quantity of onions you have of cucumbers, or less. Allow the onions to stand in hot salt water on back of range a short time. Heat 1 cup of good sharp cider vinegar, if too sour, add ½ cup of water, also add 1 teaspoonful of sugar, a couple of whole cloves; add cucumbers and onions (drained from salt water, after piercing each cucumber several times with a silver fork). Place a layer at a time in an agate stew-pan containing hot vinegar. Allow them to remain a few minutes until heated through, when fill heated glass jars with cucumbers and onions; pour hot vinegar over until jars are quite full. Place rubbers on jars and screw on tops. These pickles will be found, when jars are opened in six months' time, almost as crisp and fine as when pickles are prepared, when taken fresh from the vines in summer. Allow jars to stand 12 hours, when screw down tops again. Press a knife around the edge of jar tops before standing away to be sure the jars are perfectly air-tight.

"ROT PFEFFERS" FILLED WITH CABBAGE

Cut the tops from the stem end of twelve sweet (not hot) red peppers or "rot pfeffers," as Aunt Sarah called them. Carefully remove seeds, do not break outside shell of peppers. Cut one head of cabbage quite fine on a slaw-cutter; add to the cabbage 1 even tablespoonful of fine salt, 2 tablespoonfuls of whole yellow mustard seed (a very small amount of finely shredded, hot, red pepper may be added if liked quite peppery). Mix all together thoroughly, fill peppers with this mixture, pressing it rather tightly into the shells; place tops on pepper cases, tie down with cord. Place upright in stone jar, in layers; cover with cold vinegar. If vinegar is very strong add a small quantity of water. Tie heavy paper over top of jar and stand away in a cool place until used. These may be kept several months and will still be good at the end of that time.

AN OLD RECIPE FOR SPICED PICKLES

- 500 small cucumbers.
- 2 oz. of allspice.
- 3 gallons vinegar.
- ¼ pound of black pepper.
- 3 quarts salt.
- 1 oz cloves.
- 6 ounces of alum.
- Horseradish to flavor.

Add sugar according to strength of vinegar. Place cucumbers and pieces of horseradish in alternate layers in a stone jar, then put salt over them and cover with boiling water. Allow pickles to stand 24 hours in this brine, then pour off brine and wash pickles in cold water. Boil spices and vinegar together and pour over the pickles. In two weeks they will be ready to use. Pickles made over this recipe are excellent.

AUNT SARAH'S RECIPE FOR CHILI SAUCE

- 18 large red tomatoes.
- 10 medium-sized onions.
- 10 sweet peppers (green or red).
- 1 cup sugar.
- 3 scant tablespoonfuls salt.
- 1½ cups vinegar (cider vinegar).

Tie in a small cheese cloth bag the following:

- 1 large teaspoonful whole allspice.
- 1 large teaspoonful whole cloves.
- About the same quantity of stick cinnamon.

Chop tomatoes, onions and peppers rather finely; add vinegar, sugar and salt and the bag of spices and cook slowly about 2½ hours. Fill air-tight glass jars with the mixture while hot. This is a particularly fine recipe of Aunt Sarah's.

This quantity will fill five pint jars. Canned tomatoes may be used when fresh ones are not available.

TOMATO CATSUP

1½ peck ripe tomatoes, washed and cut in small pieces; also four large onions, sliced. Stew together until tender enough to mash through a fine sieve, reject seeds. This quantity of tomato juice should, when measured, be about four good quarts. Put tomato juice into a kettle on range, add one pint of vinegar, ¼ teaspoon cayenne pepper, 1½ tablespoons sugar, 1½ tablespoons salt; place in a cheese cloth bag 1 ounce of whole black pepper, 1 ounce whole cloves, 1 ounce allspice, 1 ounce yellow mustard seed and add to catsup. Boil down one-half. Bottle and seal while boiling hot. Boil bottles and corks before bottling catsup. Pour melted sealing-wax over corks to make them air-tight, unless self-sealing bottles are used.

PICKLED BEETS

One cup of sharp vinegar, 1 cup of water, 2 tablespoonfuls of sugar, 8 whole cloves and a pinch of black, and one of red pepper. Heat all together and pour over beets which have been sliced after being boiled tender and skins removed, and pack in glass jars which have been sterilized and if jars are air-tight these keep indefinitely.

MARMALADES, PRESERVES AND CANNED FRUITS

Young housewives, if they would be successful in "doing up fruit," should be very particular about sterilizing fruit jars, both tops and rubbers, before using. Heat the fruit to destroy all germs, then seal in air-tight jars while fruit is scalding hot. Allow jars of canned fruit or vegetables to stand until perfectly cold. Then, even should you think the tops perfectly tight, you will probably be able to give them another turn. Carefully run the dull edge of a knife blade around the lower edge of jar cap to cause it to fit tightly. This flattens it close to the rubber, making it air-tight.

To sterilize jars and tops, place in a pan of cold water, allow water to come to a boil and stand in hot water one hour.

For making jelly, use fruit, under-ripe. It will jell more easily, and, not being as sweet as otherwise, will possess a finer flavor. For jelly use an equal amount of sugar to a pint of juice. The old rule holds good—a pound of sugar to a pint of juice. Cook fifteen to twenty minutes. Fruit juice will jell more quickly if the sugar is heated in the oven before being added.

For preserving fruit, use about ¾ of a pound of sugar to 1 pound of fruit and seal in air-tight glass jars.

For canning fruit, use from ⅓ to ½ the quantity of sugar that you have of fruit.

When making jelly, too long cooking turns the mixture into a syrup that will not jell. Cooking fruit with sugar too long a time causes fruit to have a strong, disagreeable flavor.

Apples, pears and peaches were pared, cut in quarters and dried at the farm for Winter use. Sour cherries were pitted, dried and placed in glass jars, alternately with a sprinkling of granulated sugar. Pieces of sassafras root were always placed with dried apples, peaches, etc.

"FRAU" SCHMIDT'S RECIPE FOR APPLE BUTTER

For this excellent apple butter take 5 gallons of cider, 1 bucket of "Schnitz" (sweet apples were always used for the "Schnitz"), 2½ pounds of brown sugar and 1 ounce of allspice. The cider should be boiled down to one-half the original quantity before adding the apples, which had been pared and cored. Cider for apple butter was made from sweet apples usually, but if made from sour apples 4 pounds of sugar should be used. The apple butter should be stirred constantly. When cooked sufficiently, the apple butter should look clear and be thick as marmalade and the cider should not separate from the apple butter. Frau Schmidt always used "Paradise" apples in preference to any other variety of apple for apple butter.

CRANBERRY SAUCE

A delicious cranberry sauce, or jelly, was prepared by "Aunt Sarah" in the following manner: Carefully pick over and wash 1 quart of cranberries, place in a stew-pan with 2 cups of water; cook quickly a few moments over a hot fire until berries burst open, then crush with a potato-masher. Press through a fine sieve or a fruit press, rejecting skin and seeds. Add 1 pound of sugar to the strained pulp in the stew-pan. Return to the fire and cook two or three minutes only. Long, slow cooking destroys the fine flavor of the berry, as does brown sugar. Pour into a bowl, or mold, and place on ice, or stand in a cool place to become cold before serving, as an accompaniment to roast turkey, chicken or deviled oysters.

PRESERVED "YELLOW GROUND CHERRIES"

Remove the gossamer-like covering from small yellow "ground cherries" and place on range in a stew-pan with sugar. (Three-fourths of a pound of sugar to one pound of fruit.) Cook slowly about 20 minutes, until the fruit looks clear and syrup is thick as honey. Seal in pint jars.

These cherries, which grow abundantly in many town and country gardens without being cultivated, make a delicious preserve and a very appetizing pie may be made from them also.

Aunt Sarah said she preferred these preserved cherries to strawberries.

Frau Schmidt preferred the larger "purple" ground cherries, which, when preserved, greatly resembled "Guava" jelly in flavor.

"WUNDERSELDA" MARMALADE

This was composed of 2 quarts of the pulp and juice combined of ripe Kieffer pears, which had been pared and cored, (Measured after being run through a food chopper.) The grated yellow rind and juice of five medium-sized tart oranges, and 6½ cups granulated sugar. Cook all together about forty minutes, until a clear amber colored marmalade. Watch closely and stir frequently, as the mixture scorches easily. This quantity will fill about twenty small jelly tumblers. If the marmalade is to be kept some time, it should be put into air-tight glass jars.

The recipe for this delicious jam was original with the Professor's wife, and Fritz Schmidt, being particularly fond of the confection, gave it the name "Wunderselda," as he said "'twas not 'served often.'"

AUNT SARAH'S SPICED PEARS

Bartlett pears may be used, pared and cut in halves and core and seeds removed, or small sweet Seckel pears may be pared. Left whole, allow stems to remain, weigh, and to 7 pounds of either variety of pear take one pint of good cider vinegar, 3 pounds granulated sugar, a small cheese cloth bag containing several tablespoonfuls of whole cloves and the same amount of stick cinnamon, broken in pieces; all were placed in a preserving kettle and allowed to come to a boil. Then the pears were added and cooked until tender. The fruit will look clear when cooked sufficiently. Remove from the hot syrup with a perforated spoon. Fill pint glass jars with the fruit. Stand jars in a warm oven while boiling syrup until thick as honey. Pour over fruit, in jars, and seal while hot.

PEACH MARMALADE

Thinly pare ripe peaches. Cut in quarters and remove pits. Place peaches in a preserving kettle with ½ cup of water; heat slowly, stirring occasionally. When fruit has become tender mash not too fine and to every three pounds of peaches (weighed before being cooked) allow 1½ pounds of granulated sugar. Cook sugar and fruit together about three-quarters of an hour, stirring frequently, until marmalade looks clear. Place in pint glass, air-tight jars. Aunt Sarah always preferred the "Morris White," a small, fine flavored, white peach, which ripened quite late in the fall, to any other variety from which to make preserves and marmalade.

AUNT SARAH'S GINGER PEARS

- 4 pounds of fruit.
- 2 lemons.
- ¼ pound of ginger root.
- 4 pounds of sugar.
- 1 cup water.

Use a hard, solid pear, not over ripe. Pare and core the fruit and cut into thin slivers. Use juice of lemons and cut the lemon rind into long, thin strips. Place all together in preserving kettle and cook slowly one hour, or until the fruit looks clear. Should the juice of fruit not be thick as honey, remove fruit and cook syrup a short time, then add fruit to the syrup. When heated through, place in pint jars and seal. This quantity will fill four pint jars and is a delicious preserve.

PEAR AND PINEAPPLE MARMALADE

- 2 ripe pineapples,
- 4 quarts Kieffer pears.
- 4 pounds granulated sugar.

Both pears and pineapples should be pared and eyes removed from the latter. All the fruit should be run through food-chopper using all the juice from fruit. Mix sugar with fruit and juice and cook, stirring constantly until thick and clear. (Watch closely, as this scorches easily if allowed to stand a minute without stirring.) Pour into glass pint jars and seal while hot. Any variety of pear may be used, but a rather hard, solid pear is to be preferred. A recipe given Mary which she found delicious.

GRAPE BUTTER

Separate pulp and skins of grapes. Allow pulp to simmer until tender, then mash through a sieve and reject seeds. Add pulp to skins. Take ½ pound of sugar to one pound of fruit. Cook until thick, seal in air-tight jars.

CANNED SOUR CHERRIES FOR PIES

Pit cherries and cover with cold water and let stand over night. Drain in the morning. To 6 heaping cups of pitted cherries take 2 level cups of sugar, ½ cup water. Put all together into stew-pan on range, cook a short time, then add 1 teaspoonful of corn starch mixed with a little cold water and stir well through the cherries; let come to a boil, put in jars and seal. This quantity fills five pint jars. This is the way one country housekeeper taught Mary to can common *sour* cherries for pies and she thought them fine.

CANDIED ORANGE PEEL

Cut orange peel in long, narrow strips, cover with cold water and boil 20 minutes. Pour off water, cover with cold water and boil another 20 minutes, then drain and take equal weight of peel and sugar. Let simmer 1 hour, then dip slices in granulated sugar. Stand aside to cool.

AUNT SARAH'S "CHERRY MARMALADE"

Pitted, red sour cherries were weighed, put through food-chopper, and to each pound of cherries and juice add ¾ pound of granulated sugar. Cook about 25 minutes until syrup is thick and fruit looks clear. Fill marmalade pots, cover with parafine when cool, or use pint glass jars and seal. One is sure of fruit keeping if placed in air-tight jars.

AUNT SARAH'S QUINCE HONEY

Pour 1 quart of water, good measure, in an agate stew-pan on the range with three pounds of granulated sugar. When boiling add 3 large, grated quinces, after paring them. Grate all but the core of quinces. Boil from 20 to 25 minutes, until it looks clear. Pour into tumblers. When cold, cover and stand away until used.

PICKLED PEACHES

Twelve pounds of peaches, 1 quart of vinegar, 3 pounds brown sugar. Rub the fuzz from the peaches. Do not pare them. Stick half a dozen whole cloves in each peach. Add spices to taste, stick-cinnamon, whole doves and mace. Put spices in a small cheese cloth bag and do not remove the bag, containing spices, when putting away the peaches. Scald sugar, vinegar and spices together and pour over the peaches. Cover closely and stand away. Do this twice, one day between. The third time place all together in a preserving kettle. Cook a few minutes, then place fruit in jars, about three-quarters filled. Boil down the syrup until about one-quarter has boiled away, pour over the peaches, hot, and seal in air-tight jars. This is an old and very good recipe used by "Aunt Sarah" many years.

CURRANT JELLY

Always pick currants for jelly before they are "dead ripe," and never directly after a shower of rain. Wash and pick over and stem currants. Place in a preserving kettle five pounds of currants and ½ cup of water; stir until heated through then mash with a potato masher. Turn into a jelly bag, allow drip, and to every pint of currant juice add one pound of granulated sugar; return to preserving kettle. Boil twenty minutes, skim carefully, pour into jelly glasses. When cold cover tops of glasses with melted parafine.

PINEAPPLE HONEY

Pineapple honey was made in a similar manner to quince honey, using one large grated pineapple to one quart of cold water and three pounds of sugar. Boil 20 minutes.

PRESERVED PINEAPPLE

Pare the pineapples, run through a food chopper, weigh fruit, and to every pound of fruit add three-quarters of a pound of sugar. Mix sugar and fruit together and stand in a cool place over night. In the morning cook until fruit is tender and syrup clear; skim top of fruit carefully; fill jars and seal.

GRAPE CONSERVE

Wash and drain ten pounds of ripe grapes, separate the skins from the pulp, stew pulp until soft, mash through a sieve, reject seeds.

Place pulp and skins in a preserving kettle, add a half pound of seeded raisins and juice and pulp of 4 oranges. Measure and add to every quart of this ¾ of a quart of sugar. Cook slowly, until the consistency of jam. A cup of coarsely-chopped walnut meats may be added, if liked, a few minutes before removing jam from the range. Fill pint jars and seal.

MARY'S RECIPE FOR RHUBARB JAM

Skin and cut enough rhubarb in half-inch pieces to weigh three pounds. Add ½ cup cold water and 2 pounds of granulated sugar, and the grated yellow rind and juice of 2 large oranges. Cook all together, stirring occasionally to prevent scorching, a half hour, or until clear. This is a delicious jam.

APPLE SAUCE

When making apple sauce, cut good, tart apples in halves after paring them, cut out the cores, then cook, quickly as possible, in half enough boiling water to cover them. Cover the stew-pan closely. This causes them to cook more quickly, and not change color. Watch carefully that they do not scorch. When apples are tender, turn into sieve. Should the apples be quite juicy and the water drained from the apples measure a half pint, add a half pound of sugar, cook 15 or 20 minutes, until it jells, and you have a glass of clear, amber-colored jelly. Add 1 teaspoonful of butter and sugar to taste to the apple sauce, which has been mashed through the sieve. Apple sauce made thus should be almost the color of the apples before cooking. If the apple sauce is not liked thick, add some of the strained apple juice instead of making jelly; as some apples contain more juice than others.

RHUBARB MARMALADE (AS FRAU SCHMIDT MADE IT)

Cut rhubarb into small pieces, put in stew-pan with just enough water to prevent sticking fast. When cooked tender, mash fine with potato masher, and to three cups of rhubarb, measured before stewing, add 1 cup of granulated sugar, also 1 dozen almonds which had been blanched and cut as fine as possible, and stewed until tender, then added to hot rhubarb and sugar. Cook all together a short time. Serve either hot or cold. A large quantity may be canned for Winter use.

The addition of almonds gave the marmalade a delicious flavor A good marmalade may be made by adding the juice and thinly shaved outside peel of several lemons to rhubarb. Put all together in kettle on range with sugar. Cook over a slow fire until proper consistency. Turn into jars and leave uncovered until day following, when cover and seal air-tight.

GRAPE FRUIT MARMALADE

For this marmalade take 1 large grape fruit, 2 large oranges and 1 lemon. After thoroughly washing the outside of fruit, slice all as thinly as possible, rejecting the seeds. Measure and add three times as much water as you have fruit. Let all stand over night. The next morning boil 15 minutes, stand over night again, in a large bowl or agate preserving kettle. The next morning add 1 pound (scant measure) of sugar to each pint of the mixture and boil until it jells. This is delicious if you do not object to the slightly bitter taste of the grape fruit. Put in tumblers, cover closely with paraffin. This quantity should fill 22 tumblers, if a large grape fruit is used.

ORANGE MARMALADE

Slice whole oranges very thin and cut in short pieces after washing them. Save the seeds. To each pound of sliced oranges add 3 pints of cold water and let stand 24 hours. Then boil all together until the chipped rinds are tender. All the seeds should be put in a muslin bag and boiled with the oranges. Allow all to stand together until next day, then remove the bag of seeds, and to every pound of boiled fruit add a half pound of sugar. Boil continuously, stirring all the time, until the chips are quite clear and the syrup thick as honey on being dropped on a cold dish. The grated rind and juice of 2 lemons will improve the taste of marmalade if added at last boiling. When cooked sufficiently the marmalade should be clear. Pour at once into glass jars and cover closely.

CHERRY RELISH

After sour cherries have been pitted, weigh them and cover with vinegar and let stand 24 hours. Take from the vinegar and drain well, then put into stone crocks in layers, with sugar, allowing 1 pound of sugar to 1 pound of cherries. Stir twice each day for ten days, then fill air-tight jars and put away for Winter use. These are an excellent accompaniment to a roast of meat.

CANNED PEACHES

When canning peaches make a syrup composed of 1 cup of sugar to 2 cups of water.

Place in preserving kettle and when sugar has dissolved cook thinly pared peaches, either sliced or cut in halves, in the hot syrup until clear, watching closely that they do not cook too soft. Place carefully in glass jars, pour hot syrup over and seal in jars.

Aunt Sarah also, occasionally, used a wash-boiler in which to can fruit. She placed in it a rack made of small wooden strips to prevent the jars resting on the bottom of the boiler; filled the jars with un-cooked fruit or vegetables, poured over the jars of fruit hot syrup and over the vegetables poured water, placed the jars, uncovered, in the boiler; water should cover about half the height of jars. Boil until contents of jars are cooked, add boiling syrup to fill fruit jars and screw the tops on tightly.

PEAR CONSERVE

Use 5 pounds of pears, not too soft or over-ripe, cut like dice. Cover with water and boil until tender, then add 5 pounds of sugar. Peel 2 oranges, cut in dice the night before using; let diced orange peel stand, covered with cold water until morning. Then cook until orange peel is tender. Add this to the juice and pulp of the two oranges. Add one pound of seeded raisins and cook all together until thick honey. Put in glass jars and seal.

LEMON HONEY

The juice of 3 lemons, mixed with 3 cups of sugar. Add 3 eggs, beating 1 in at a time. Add 2 cups of water and 2 tablespoonfuls of butter. Cook all together 20 minute, until thick as honey.

CANNED STRING BEANS

Aunt Sarah used no preservative when canning beans. She gathered the beans when quite small and tender, no thicker than an ordinary lead-pencil, washed them thoroughly, cut off ends and packed them into quart glass jars, filled to overflowing with cold water. Placed jar tops on lightly, and stood them in wash boiler in the bottom of which several boards had been placed. Filled wash boiler with luke warm water about two-thirds as high as tops of jars, cooked continuously three to four hours after water commenced to boil. Then carefully lifted jars from wash boiler, added boiling water to fill jars to overflowing, screwed on cover and let stand until perfectly cold, when give jar tops another turn with the hand when they should be air-tight. A good plan is to run the dull edge of a knife around the outer edge of the jar to be sure it fits close to the rubber, and will not admit air. Beans canned in this manner should keep indefinitely.

PRESERVED "GERMAN PRUNES" OR PLUMS

After washing fruit, piece each plum several times with a silver fork, if plums be preserved whole. This is not necessary if pits are removed. Weigh fruit and to each pound of plums take about ¾ pound of granulated sugar. Place alternate layers of plums and sugar in a preserving kettle, stand on the back of range three or four hours, until sugar has dissolved, then draw kettle containing sugar and plums to front of range and boil so minutes. Remove scum which arises on top of boiling syrup. Place plums in glass jars, pour boiling syrup over and seal.

A good rule is about four pounds of sugar to five pounds of plums.

Should plums cook soft in less than 20 minutes, take from syrup with a perforated skimmer, place in jars and cook syrup until as thick as honey; then pour over fruit and seal up jars.

BUCKS COUNTY APPLE BUTTER

A genuine old-fashioned recipe for apple butter, as "Aunt Sarah" made it at the farm. A large kettle holding about five gallons was filled with sweet cider. This cider was boiled down to half the quantity. The apple butter was cooked over a wood fire, out of doors. The cider was usually boiled down the day before making the apple butter, as the whole process was quite a lengthy one. Fill the kettle holding the cider with apples, which should have been pared and cored the night before at what country folks call an "apple bee," the neighbors assisting to expedite the work. The apples should be put on to cook as early in the morning as possible and cooked slowly over not too hot a fire, being stirred constantly with a long-handled "stirrer" with small perforated piece of wood on one end. There is great danger of the apple butter burning if not carefully watched and constantly stirred. An extra pot of boiling cider was kept near, to add to the apple butter as the cider boiled away. If cooked slowly, a whole day or longer will be consumed in cooking. When the apple butter had almost finished cooking, about the last hour, sweeten to taste with sugar (brown sugar was frequently used). Spices destroy the true apple flavor, although Aunt Sarah used sassafras root, dug from the near-by woods, for flavoring her apple butter, and it was unexcelled. The apple butter, when cooked sufficiently, should be a dark rich color, and thick like marmalade, and the cider should not separate from it when a small quantity is tested on a saucer. An old recipe at the farm called for 32 gallons of cider to 8 buckets of cider apples, and to 40 gallons of apple butter 50 pounds of sugar were used. Pour the apple butter in small crocks used for this purpose. Cover the top of crocks with paper, place in dry, cool store-room, and the apple butter will keep several years. In olden times sweet apples were used for apple butter, boiled in sweet cider, then no sugar was necessary. Small brown, earthen pots were used to keep this apple butter in, it being only necessary to tie paper over the top. Dozens of these pots, filled with apple butter, might have been seen in Aunt Sarah's store-room at the farm at one time.

CANNED TOMATOES

When canning red tomatoes select those which ripen early in the season, as those which ripen later are usually not as sweet. Wash the tomatoes, pour scalding water over, allow them to stand a short time, when skins may be easily removed. Cut tomatoes in several pieces, place over fire in porcelain-lined preserving kettle and cook about 25 minutes, or until an orange-colored scum rises to the top. Fill perfectly clean sterilised jars with the hot tomatoes fill quickly before they cool. Place rubber and top on jar, and when jars have become perfectly cold (although they may, apparently, have been perfectly air-tight), the tops should be given another turn before standing away for the Winter; failing to do this has frequently been the cause of inexperienced housewives' ill success when canning tomatoes. Also run the dull edge of a knife blade carefully around the top of jar, pressing down the outer edge and causing it to fit more closely. Aunt Sarah seldom lost a jar of canned tomatoes, and they were as fine flavored as if freshly picked from the vines. She was very particular about using only new tops and rubbers for her jars when canning tomatoes. If the wise housewife takes these pre-cautions, her canned tomatoes should keep indefinitely. Aunt Sarah allowed her jars of tomatoes to stand until the day following that on which the tomatoes were canned, to be positively sure they were cold, before giving the tops a final turn. Stand away in a dark closet.

EUCHERED PEACHES

Twelve pounds of pared peaches (do not remove pits), 6 pounds of sugar and 1 gill of vinegar boiled together a few minutes, drop peaches into this syrup and cook until heated through, when place peaches in air-tight jars, pour hot syrup over and seal.

AUNT SARAH'S METHOD OF CANNING CORN

Three quarts of sweet corn cut from the cob, 1 cup of sugar ¾ cup of salt and 1 pint of cold water. Place these ingredients together in a large bowl; do this early in the morning and allow to stand until noon of the same day; then place all together in a preserving kettle on the range and cook twenty minutes. Fill glass jars which have been sterilized. The work of filling should be done as expeditiously as possible; be particular to have jar-tops screwed on tightly. When jars have become cool give tops another turn, to be positive they are air-tight before putting away for the Winter. When preparing this canned corn for the table, drain all liquid from the corn when taken from the can, pour cold water over and allow to stand a short time on the range until luke-warm. Drain and if not *too* salt, add a small quantity of fresh water, cook a few minutes, season with butter, add a couple tablespoonfuls of sweet milk; serve when hot. This canned corn possesses the flavor of corn freshly cut from the cob. Sarah Landis had used this recipe for years and 'twas seldom she lost a can.

DRIED SWEET CORN

In season when ears of sweet corn are at their best for cooking purposes, boil double the quantity necessary for one meal, cut off kernels and carefully scrape remaining pulp from cob. Spread on agate pans, place in a hot oven a short time (watch closely) and allow it to remain in a cooled oven over night to dry. When perfectly dry place in bags for use later in the season.

When the housewife wishes to prepare dried corn for the table, one cup of the dried corn should be covered with cold water and allowed to stand until the following day, when place in a stew-pan on the range and simmer slowly several hours; add ½ teaspoonful of sugar, 1 tablespoonful of butter, salt and pepper. This corn Aunt Sarah considered sweeter and more wholesome than canned corn and she said "No preservatives were used in keeping it."

When chestnuts were gathered in the fall of the year, at the farm, they were shelled as soon as gathered, then dried and stored away for use in the Winter. Aunt Sarah frequently cooked together an equal amount of chestnuts and dried corn; the combination was excellent. The chestnuts were soaked in cold water over night.

The brown skin of the chestnuts may be readily removed after being covered with boiling water a short time.

PRESERVED CHERRIES

Aunt Sarah's preserved cherries were fine, and this was her way of preparing them: She used 1 pound of granulated sugar to 1 quart of pitted cherries. She placed the pitted cherries on a large platter and sprinkled the sugar over them. She allowed them to stand several hours until the cherries and sugar formed a syrup on platter. She then put cherries, sugar and juice all together in a preserving kettle, set on range, and cooked 10 minutes. She then skimmed out the cherries and boiled the syrup 10 minutes longer, then returned the cherries to syrup. Let come to a boil. She then removed the kettle from the fire, spread all on a platter and let it stand in the hot sun two successive days, then put in glass air-tight jars or in tumblers and covered with paraffin. A combination of cherries and strawberries preserved together is fine, and, strange to say, the flavor of strawberries predominates.

A fine flavored preserve is also made from a combination of cherries and pineapple.

FROZEN DESSERTS—AUNT SARAH'S FROZEN "FRUIT CUSTARD"

One tablespoonful of granulated gelatine soaked in enough milk to cover. Place 2 cups of sugar and ¾ cup of milk in a stew-pan on the range and boil until it spins a thread; that is, when a little of the syrup is a thread-like consistency when dripped from a spoon. Allow it to cool. Add dissolved gelatine and 1 quart of sweet cream. One box of strawberries, or the same amount of any fruit liked, may be added to the mixture; freeze as ordinary ice cream.

This dessert as prepared by Aunt Sarah was delicious as any ice cream and was used by her more frequently than any other recipe for a frozen dessert.

SHERBET

Frau Schmidt gave Mary this simple recipe for making any variety of sherbet:

2 cups of sugar, 1 tablespoonful of flour, mixed with the sugar and boiled with 1 quart of water; when cold, add 1 quart of any variety of fruit.

Freeze in same manner as when making ice cream.

ICE CREAM — A SIMPLE RECIPE GIVEN MARY

When preparing this ice cream Mary used the following: Three cups of cream and 1 cup of milk, 1 egg and 1 cup of pulverized sugar (were beaten together until light and creamy). This, with 1 teaspoonful of vanilla flavoring, was added to the milk and cream. The cream should be scalded in warm weather. The egg and sugar should then be added to the scalded milk and cream, stirring them well together. When the mixture has cooled, strain it into the can of the freezer. Three measures of cracked ice to one of salt should be used. The ice and salt, well-mixed, were packed around the freezer. The crank was turned very slowly the first ten minutes, until the mixture had thickened, when it was turned more rapidly until the mixture was frozen.

FRAU SCHMIDT'S ICE CREAM

This recipe for ice cream is simple and the ice cream is good. A boiled custard was prepared, consisting of 1 quart of milk, 4 eggs, between 3 and 4 cups of granulated sugar. When the custard coated the spoon she considered it cooked sufficiently. Removed from the fire. When cold she beat into the custard 1 quart of rich cream and 1 teaspoonful of vanilla, turned the mixture into the freezer, packed outside tub with ice and salt. It was frozen in the ordinary manner.

MAPLE PARFAIT

For this rich, frozen dessert Mary beat 4 eggs lightly, poured slowly over them 1 cup of hot maple syrup, cooked in a double boiler, stirring until very thick. She strained it, and when cold added 1 pint of cream. She beat all together, poured into a mold, packed the mold in ice and salt, and allowed it to stand 3 hours. This is a very rich frozen dessert, too rich to be served alone. It should be served with lemon sherbet or frozen custard with a lemon flavoring, as it is better served with a dessert less rich and sweet.

ICE CREAM MADE BY BEATING WITH PADDLE

This recipe for a delicious and easily prepared ice cream was given Mary by a friend living in Philadelphia and is not original. She found the ice cream excellent and after having tried the recipe used no other. A custard was made of 1 quart of scalded milk, 6 eggs, 3 cups of sugar. The eggs were beaten light, then sugar was added, then the hot milk was poured over and all beaten together. She put all in a double boiler and stirred about ten minutes, until thick and creamy. A small pinch of soda was added to prevent curdling. When the custard was perfectly cold she stirred in three cups of sweet, cold cream, flavored with either vanilla or almond flavoring, and beat all together five minutes, then turned the mixture into the freezer, packed well with pounded ice and coarse salt. She covered the freezer with the ice and salt and threw a heavy piece of old carpet or burlap over the freezer to exclude the air. She let it stand one hour, then carefully opened the can containing the cream, not allowing any salt to get in the can. With a long, thin-handled knife she scraped down the frozen custard from the sides of the freezer, and with a thin wooden paddle beat it hard and fast for about five minutes. This made the cream fine and smooth. Any fruit may now be added, and should be mixed in before the cream is covered. The cream should be beaten as quickly as possible and covered as soon as the fruit has been added. Aunt Sarah usually made peach ice cream when peaches were in season. Fine ripe peaches were pared and pitted, then finely mashed, 2 small cups of sugar being added to a pint of mashed peaches. She allowed the peach mixture to stand one hour before adding to the beaten cream. When the mashed peaches had been added to the cream, she fastened the lid and drained off part of the water in outer vessel, packed more ice and salt about the can in the freezer, placed a weight on top to hold it down, covered closely with a piece of old carpet to exclude the air, left it stand three or four hours. The beating was all the labor required. The dasher or crank was not turned at all when making the ice cream, and when frozen it was delicious.

Mary was told by her Aunt of a friend in a small town, with a reputation for serving delicious ice cream, who always made ice cream by beating with a paddle, instead of making it by turning a crank in a freezer.

AUNT SARAH'S RECIPE FOR FROZEN CUSTARD

One quart of rich, sweet milk, 2 tablespoons of corn starch, 4 eggs, 1 cup of sugar, small tablespoon of vanilla. Cook the milk in a double boiler, moisten corn starch with a little milk. Stir it into the hot milk until it begins to thicken. Beat sugar and eggs together until creamy, add to the hot milk, cook a minute, remove from fire, add the vanilla, and when cool freeze. Crush the ice into small pieces, for the finer the ice the quicker the custard will freeze, then mix the ice with a fourth of the quantity of coarse rock salt, about 10 pounds ice and 2 pounds salt will be required to pack sides and cover top of a four-quart freezer. Place can in tub, mix and fill in ice and salt around the can, turn the crank very slowly until the mixture is thoroughly chilled. Keep hole in top of tub open. When mixture is cold, turn steadily until it turns rather hard. When custard is frozen, take out inside paddle, close the freezer, run off the salt water, repack and allow to stand several hours. At the end of that time it is ready to serve.

PINEAPPLE CREAM

This is a delicious dessert, taught Mary by Aunt Sarah. She used 1 quart sweet cream, 1½ cups sugar, beaten together. It was frozen in an ice cream freezer. She then pared and cut the eyes from one ripe pineapple and flaked the pineapple into small pieces with a silver fork, sprinkled sugar over and let it stand until sugar dissolved. She then stirred this into the frozen cream and added also the beaten white of one egg. Packed ice and salt around freezer and allowed it to stand several hours before using. Mary's Aunt always cooked pineapple or used canned pineapple with a rich syrup when adding fruit before the cream was frozen.

MARY'S RECIPE FOR PEACH CREAM

Mary made ice cream when peaches were plentiful; she used 1 quart of sweet cream, sweetened to taste (about 2 cups sugar) and 2 quarts of ripe peaches mashed and sweetened before adding to cream. Freeze in ordinary manner. If peaches were not fine flavored, she added a little almond flavoring.

LEMON SHERBET

This is the way Frau Schmidt taught Mary to make this dessert. She used for the purpose 1 quart of water, 5 lemons, 2 tablespoons gelatine, 2 large cups sugar. She soaked the gelatine in about 1 cup of water. She squeezed out the juice of lemons, rejecting seeds and pulp. She allowed a cup of water out of the quart to soak the gelatine. This mixture was put in an ice cream freezer and frozen.

FRAU SCHMIDT'S FROZEN CUSTARD

- 1½ quarts milk.
- 2 cups sugar.

- 5 eggs.
- 2½ tablespoonfuls of flour.

Scald the milk in a double boiler. Moisten flour (she preferred *flour* to corn starch for this purpose) with a small quantity of cold milk, and stir into the scalded milk. Beat together egg yolks and sugar until light and creamy, then add the stiffly beaten whites of eggs and stir all into the boiling milk. Cool thoroughly, flavor with vanilla and freeze as you would ice cream. When partly frozen crushed strawberries or peaches may be added in season. A little more sugar should then he added to the fruit, making a dessert almost equal to ice cream. In Winter one cup of dried currants may be added, also one tablespoonful of sherry wine, if liked.

CARAMEL ICE CREAM

Scald one pint of sweet milk in a double boiler. Stir into it one cup of sugar and one rounded tablespoonful of flour, which had been mixed smoothly with a small quantity of the milk before scalding. Add two eggs which had been beaten together until light and creamy. At the same time the milk was being scalded, a fry-pan containing one cup of granulated sugar was placed on the range; this should be watched carefully, on account of its liability to scorch. When sugar has melted it will be brown in color and liquid, like molasses, and should then be thoroughly mixed with the foundation custard. Cook the whole mixture ten minutes and stand aside to cool; when perfectly cold add a pinch of salt, one quart of sweet cream, and freeze in the ordinary manner.

CHERRY SHERBET

Aunt Sarah taught Mary to prepare this cheap and easily made dessert of the various berries and fruits as they ripened. Currants, strawberries, raspberries and cherries were used. They were all delicious and quickly prepared. The ice for freezing was obtained from a near-by creamery. The cherries used for this were not the common, sour pie cherries, so plentiful usually on many "Bucks County Farms," but a fine, large, red cherry, not very sour. When about to prepare cherry sherbet, Mary placed over the fire a stew-pan containing 1 quart of boiling water and 1 pound of granulated sugar. Boiled this together 12 minutes. She added 1 tablespoonful of granulated gelatine which had been dissolved in a very little cold water. When the syrup had cooled, she added the juice of half a lemon and 1 quart of pitted cherries, mixed all together. Poured it in the ice cream freezer, packed around well with coarse salt and pounded ice. She used 1 part salt to 3 parts ice. She turned the crank slowly at first, allowed it to stand a few minutes, then increased the speed. When the mixture was firm she removed the dasher. She allowed the water to remain with the ice and salt, as the ice-cold water helped to freeze it. She filled in ice and salt around the can in the freezer and on top of the can; covered the top of the freezer with a piece of old carpet and allowed it to stand a couple of hours, when it was ready to serve. Almost any fruit or fruit juice, either fresh or canned, may be made into a delicious dessert by this rule.

One quart of boiling water and 1 pound of sugar boiled together to form a syrup, then add 1 quart of juice or fruit and juice to measure exactly one quart. Mix together according to directions and freeze.

GRAPE SHERBET

Grape sherbet was made in this manner: The grapes were washed, picked from the stems and placed in a stew-pan over the fire. When hot remove from the fire and mash with a potato-masher and strain through a jelly bag, as if preparing to make jelly. Boil together 1 pound of granulated sugar and 1 quart of water, about 12 minutes. While hot add 1 pint of grape juice and 1 teaspoonful of granulated gelatine, which had been dissolved in a very little cold water, to the hot syrup. When the mixture was partly frozen add the stiffly beaten white of 1 egg and 1 tablespoonful of pulverized sugar, beaten together. All were stirred together, covered and stood away until cold. Then placed in a freezer, iced as for ice cream, and frozen in the same manner as for cherry sherbet. The juice of all berries or fruits may be extracted in the same manner as that of grapes.

WINES AND SYRUPS — UNFERMENTED GRAPE JUICE

To 6 pounds of stemmed Concord grapes add 1 quart of water, allow them to simmer on range until grapes have become soft. Strain through a piece of cheese-cloth, being careful to press only the juice through, not the pulp of the grapes. Return the grape juice to the preserving kettle and add ¾ of a pound of sugar. Allow the juice to just commence to boil, as cooking too long a time spoils the flavor of the juice. Bottle at once, while juice is hot. Bottles must be sterilized and air-tight if you expect grape juice to keep. Cover corks with sealing wax.

VINEGAR MADE FROM STRAWBERRIES

"Aunt Sarah" Landis possessed the very finest flavored vinegar for cooking purposes, and this is the way it was made. She having a very plentiful crop of fine strawberries one season, put 6 quarts of very ripe, mashed strawberries in a five-gallon crock, filled the crock with water, covered the top with cheese-cloth and allowed it to stand in a warm place about one week, when it was strained, poured into jugs and placed in the cellar, where it remained six months, perhaps longer, when it became very sharp and sour, and had very much the appearance of white wine with a particularly fine flavor. This was not used as a beverage, but as a substitute for cider in cooking.

BOILED CIDER FOR MINCE PIES

In Autumn, when cider was cheap and plentiful on the farm, 3 quarts of cider was boiled down to one, or, in this proportion, for use in mince meat during the Winter. A quantity prepared in this manner, poured while hot in air-tight jars, will keep indefinitely.

LEMON SYRUP

Boil two cups of granulated sugar and one cup of water together for a few minutes until the sugar is dissolved, then add the juice of six well-scrubbed, medium-sized lemons; let come to a boil and add the grated yellow rind of three of the lemons. Be careful not to use any of the white skin of the lemons, which is bitter. Put in air-tight glass jars. This quantity fills one pint jar. A couple tablespoonfuls added to a tumbler partly filled with water and chipped ice makes a delicious and quickly prepared drink on a hot day.

EGG NOGG

Add to the stiffly beaten white of one egg the slightly beaten yolk of egg. Pour into glass tumbler, fill with cold sweet milk, sweeten with sugar to taste and a little grated nutmeg on top or a tablespoonful of good brandy. This is excellent for a person needing nourishment, and may be easily taken by those not able to take a raw egg in any other form. The egg nogg will be more easily digested if sipped slowly while eating a cracker or slice of crisply toasted bread.

ROSE WINE

Gather one quart of rose leaves, place in a bowl, pour over one quart of boiling water, let stand nine days, then strain, and to each quart of strained liquid add one pound of granulated sugar. Allow to stand until next day, when sugar will be dissolved. Pour into bottles, cork tightly, stand away for six months before using. Aunt Sarah had some which had been keeping two years and it was fine.

DANDELION WINE

Four good quarts of dandelion blossoms, four pounds of sugar, six oranges, five lemons. Wash dandelion blossoms and place them in an earthenware crock. Pour five quarts of boiling water over them and let stand 36 hours. Then strain through a muslin bag, squeezing out all moisture from dandelions. Put the strained juice in a deep stone crock or jug and add to it the grated rind and juice of the six oranges and five lemons. Tie a piece of cheese-cloth over the top of jug and stand it in a warm kitchen about one week, until it begins to ferment. Then stand away from stove in an outer kitchen or cooler place, not in the cellar, for three months. At the end of three months put in bottles. This is a clear, amber, almost colorless liquid. A pleasant drink of medicinal value. Aunt Sarah always used this recipe for making dandelion wine, but Mary preferred a recipe in which yeast was used, as the wine could be used a short time after making.

DANDELION WINE (MADE WITH YEAST)

Four quarts of dandelion blossoms. Pour over them four quarts of boiling water; let stand 24 hours, strain and add grated rind and juice of two oranges and two lemons, four pounds of granulated sugar and two tablespoonfuls of home-made yeast. Let stand one week, then strain and fill bottles.

GRAPE FRUIT PUNCH

Two cups of grape juice, 4 cups of water, 1½ cups of sugar, juice of 3 lemons and 3 oranges, sliced oranges, bananas and pineapples. Serve the punch in sherbet glasses, garnished with Marachino cherries.

A SUBSTITUTE FOR MAPLE SYRUP

A very excellent substitute for maple syrup to serve on hot grid-dle cakes is prepared from 2 pounds of either brown or white sugar and 1¾ cups of water, in the following manner: Place the stew-pan containing sugar and water on the back part of range, until sugar dissolves, then boil from 10 to 15 minutes, until the mixture thickens to the consistency of honey. Remove from the range and add a few drops of vanilla or "mapleine" flavoring. A tiny pinch of cream of tartar, added when syrup commences to boil, prevents syrup granu-lating; too large a quantity of cream of tartar added to the syrup would cause it to have a sour taste.

SALTED ALMONDS OR PEANUTS

Blanch 2 pounds of shelled almonds or peanuts (the peanuts, of course, have been well roasted) by pouring 1 quart of boiling water over them. Allow them to stand a short time. Drain and pour cold water over them, when the skin may be easily removed. Place in a cool oven until dry and crisp. Put a small quantity of butter into a pan. When hot, throw in the nuts and stir for a few minutes, sprin-kle a little salt over. Many young cooks do not know that salted peanuts are almost equally as good as salted almonds and cheaper. Peanuts should always be freshly roasted and crisp.

PEANUT BUTTER

When peanuts have been blanched, are cold, dry and crisp, run them through a food chopper. Do not use the *very finest* cutter, as that makes a soft mass. Or they may be crushed with a rolling pin. Season with salt, spread on thinly-sliced, buttered bread. They make excellent sandwiches. Or run peanuts through food chopper which has an extra fine cutter especially for this purpose. The peanuts are then a thick, creamy mass. Thin this with a small quantity of olive oil, or melted butter, if preferred. Season with salt and you have "peanut butter," which, spread on slices of buttered bread, makes a delicious sandwich, and may frequently take the place of meat sandwiches. Nuts, when added to salads, bread or cake, add to their food value.

A CLUB SANDWICH

On a thinly-cut slice of toasted bread lay a crisp lettuce leaf and a thin slice of broiled bacon. On that a slice of cold, boiled chicken and a slice of ripe tomato. Place a spoonful of mayonnaise on the tomato, on this a slice of toasted bread. Always use stale bread for toast and if placed in a hot oven a minute before toasting it may be more quickly prepared.

CANDIES-WALNUT MOLASSES TAFFY

Place 2 cups of New Orleans molasses and ¾ cup of brown sugar in a stew-pan on the range and cook; when partly finished cooking (this may be determined by a teaspoonful of the mixture forming a soft ball when dropped in water), add 1 tablespoonful of flour, moistened with a small quantity of water, and cook until a teaspoonful of the mixture becomes brittle when dropped in cold water; at this stage add 1 scant teaspoonful of baking soda (salaratus). Stir, then add 1 cup of coarsely chopped black walnut meats; stir all together thoroughly, and pour into buttered pans to become cool.

COCOANUT CREAMS

Grate 1 medium-sized cocoanut, place in a bowl, add 2 pounds of confectioners' sugar, mix with the cocoanut; then add the stiffly beaten white of 1 egg and 1 teaspoonful of vanilla; knead this as you would bread for 10 or 15 minutes. If the cocoanut is a large or a dry one, about ½ pound more sugar will be required. Shape the mixture into small balls, press halves of English walnut meats into each ball, or have them plain, if preferred. Stand aside in a cool place a half hour. Melt a half cake of Baker's unsweetened chocolate, add a half teaspoonful of paraffin, roll the small balls in this chocolate mixture until thoroughly coated. Place on waxed paper to dry. From the ingredients in this recipe was made 3 pounds of candy.

FUDGE (AS MADE BY MARY)

Two cups of granulated sugar, 1 cup of sweet milk, ¼ cup of butter, ¼ cake or 2 squares of Baker's unsweetened chocolate. Cook all together until when tried in water it forms a soft ball. Remove from fire, flavor with vanilla, beat until creamy, pour in buttered pan and when cooled cut in squares.

A DELICIOUS "CHOCOLATE CREAM" CANDY

Place in an agate stew-pan 2 cups of granulated sugar, 1 cup of sweet milk, butter size of an egg. Cook all together until it forms a soft ball when a small quantity is dropped into cold water. Then beat until creamy. Add a half a cup of any kind of chopped nut meats. Spread on an agate pie-tin and stand aside to cool.

For the top layer take 1 cup of sugar, ½ cup milk and butter size of an egg, 2 small squares of a cake of Baker's unsweetened chocolate. Cook together until it forms a soft ball in water. Beat until creamy. Add half a teaspoonful of vanilla, spread over top of first layer of candy and stand away until it hardens and is quite cold.

MARY'S RECIPE FOR MOLASSES TAFFY

Four tablespoonfuls New Orleans molasses, 9 tablespoonfuls sugar, 3 tablespoonfuls water, 2 teaspoonfuls butter, 1 teaspoonful vanilla. Boil all together until it becomes brittle when a small quantity is dropped in water. Pour the mixture into buttered pans and when cool enough to handle, pull with the hands until a light creamy yellow shade. Pull into long, thin strips, cut into small pieces with scissors. This taffy is fine if boiled a long enough time to become crisp and brittle, and you will be surprised at the quantity this small amount of sugar and molasses will make.

RECIPE FOR MAKING HARD SOAP WITHOUT BOILING

To make hard soap without boiling, empty a can of "Lewis Perfumed Lye" (or any other good, reliable brand of lye) into a stone jar with 1 tablespoonful powdered borax. Add 2½ pints of cold water to the lye. Stir until dissolved. Be very careful not to allow any of the lye to touch hands or face. Wear old gloves when emptying can and stirring lye. Stand the dissolved lye in a cool place. The tin cans containing the fat to be used for soap (which have accumulated, been tried out, strained, and put in empty tin cans at different times) should be placed in the oven of range for a few minutes. When warm they may be turned out readily into a large stew-pan. Put over fire and when all has dissolved and melted, strain through cheese-cloth bag into an agate dish pan. When weighed you should 5½ pounds of clear fat. A recipe telling exact quantity of fat and lye usually comes with can of lye. When temperature of fat is 120 degrees by your thermometer (luke-warm), the lye should have been allowed to stand about 1 hour from the time it was dissolved. It should then be the right temperature to mix with strained, luke-warm fat or grease not over 80 degrees by thermometer. Now slowly pour the dissolved lye over the fat (a half cup of ammonia added improves soap), stir together until lye and grease are thoroughly incorporated, and the mixture drops from the stirrer like honey. The soap may be scented by adding a few drops of oil of cloves, if liked. Stir the mixture with a small wooden paddle or stick. Stir slowly from 5 to 10 minutes, not longer, or the lye and fat may separate. Pour all into a large agate dish pan lined with a piece of clean muslin. Throw an old piece of carpet over the top and stand near the range until evening, when, if made early in the morning, a solid cake of soap, weighing 8½ pounds, may be turned out on a bakeboard (previously covered with brown paper) and cut into 20 pieces of good hard soap. Lay the pieces of soap in a basket, cover to protect from dust, and stand in a warm room to dry thoroughly before using. Soap made according to these directions should be solid and almost as white as ivory if the fat used has not been scorched.

This soap is excellent for scrubbing and laundry purposes. The greater length of time the soap is kept, the better it will become. The grease used may be clarified by adding water and cooking a short time. Stand away and when cool remove fat from top, wiping off any moisture that may appear. Soap-making is a *small economy*. Of course, the young housewife will not use for soap *any fat* which could be utilized for frying, etc., but she will be surprised to find, when she once gets the saving habit, how quickly she will have the quantity of fat needed for a dollar's worth of soap by the small outlay of the price of a can of lye, not counting her work. The young, inexperienced housewife should be careful not to use too small a stew-pan in which to heat the fat, and should not, under any circumstance, leave the kitchen while the fat is on the range, as grave results might follow carelessness in this respect.

TO IMITATE CHESTNUT WOOD

Before painting the floor it was scrubbed thoroughly with the following: One-half cup of "household ammonia" added to four quarts of water. The floor, after being well scrubbed with this, was wiped up with pure, clean water and allowed to get perfectly dry before painting. For the ground color, or first coat of paint on the floor, after the cracks in floor had been filled with putty or filler, mix together five pounds of white lead, one pint of turpentine and about a fourth of a pound of yellow ochre, add 1 tablespoon of Japan dryer. This should make one quart of paint a light tan or straw color, with which paint the floor and allow it to dry twenty-four hours, when another coat of the same paint was given the floor and allowed to dry another twenty-four hours, then a graining color, light oak, was used. This was composed of one pint of turpentine, one teaspoon of graining color and two tablespoons of linseed oil, and 1 tablespoon of Japan dryer, all mixed together. This was about the color of coffee or chocolate. When the wood had been painted with this graining color, before drying, a fine graining comb was passed lightly over to imitate the grain of wood. This was allowed to dry twenty-four hours, when a coat of floor varnish was given. The room was allowed to dry thoroughly before using. The imitation of natural chestnut was excellent.

MEASURES AND WEIGHTS

When a recipe calls for one cup of anything, it means one even cup, holding one-half pint, or two gills.

One cup is equal to four wine glasses.

One wine glass is equal to four tablespoons of liquid, or one-quarter cup.

Two dessertspoonfuls equal one tablespoonful.

Six tablespoonfuls of liquid equal one gill.

Two tablespoonfuls dry measure equal one gill.

Two gills equal one cup.

Two cups, or four gills, equal one pint.

Four cups of flour weigh one pound and four cups of flour equal one quart.

One even cup of flour is four ounces.

Two cups (good measure) of granulated sugar weigh one pound and measure one pint.

Two cups butter equal one pound.

A pint of liquid equals one pound.

A cup of milk or water is 8 ounces.

Two tablespoonfuls liquid equal one ounce.

One salt spoonful is ¼ teaspoonful.

Four tablespoonfuls equal one wine glass.

Piece of butter size of an egg equals two ounces, or two tablespoons.

A tablespoonful of butter melted means the butter should be first measured then melted.

One even tablespoonful of unmelted butter equals one ounce.

One tablespoonful sugar, good measure, equals one ounce.

Ordinary silver tablespoon was used for measuring, not a large mixing spoon.

COOKING SCHEDULE TO USE WITH THE OVEN THERMOMETER OF A GAS STOVE

To Cook—	Temperature—	Cook for—
Bread, white	280°	40 minutes
Biscuit, small	300°	30 minutes
Biscuit, large	300°	30 minutes
Beef, roast rare	300°	15 minutes per pound
Beef, roast well done	320°	15 minutes per pound
Cake, Fruit	260°	2 hours
Cake, Sponge	300°	30 minutes
Cake, Loaf	300°	40 minutes
Cake, Layer	300°	15 minutes
Cookies	300°	5 minutes
Chickens	340°	2 hours
Custards	260° to 300°	20 minutes
Duck	340°	3 hours
Fish	260° to 300°	1 hour
Ginger Bread	260° to 300°	20 minutes
Halibut	260° to 300°	45 minutes
Lamb	300°	3 hours
Mutton, rare	260° to 300°	10 minutes per pound
Mutton, well done	300°	15 minutes per pound
Pie crust	300°	30 minutes
Pork	260° to 300°	2½ hours
Potatoes	300°	1 hour

Pudding, Bread	260° to 300°	1 hour
Pudding, Plum	260° to 300°	1 hour
Pudding, Rice	260° to 300°	30 minutes
Pudding, Tapioca	260° to 300°	30 minutes
Rolls	260° to 300°	20 minutes
Turkeys	280°	3 hours
Veal	280°	2½ hours

When a teacher of "Domestic Science," the Professor's wife was accustomed to using a pyrometer, or oven thermometer, to determine the proper temperature for baking. She explained its advantages over the old-fashioned way of testing the oven to Mary and gave her a copy of the "Cooking Schedule," to put in her recipe book, which Mary found of great assistance, and said she would certainly have a range with an oven thermometer should she have a home of her own, and persuaded Aunt Sarah to have one placed in the oven door of her range.

THE END.